AF822543

Manhattan Review®

Test Prep & Admissions Consulting

Turbocharge Your DSAT: DSAT Math Guide

Published on February 11th, 2025

Version 50

- ☐ *Designed as per the Digital SAT*
- ☐ *Text-cum-graphic explanations of concepts*
- ☐ *300 DSAT-like practice questions*
- ☐ *Classified into four sub-sections:*
  - *Algebra*
  - *Advanced Math*
  - *Problem Solving & Data Analysis*
  - *Geometry & Trigonometry*
- ☐ *Many questions with Alternate Approaches*

www.manhattanreview.com

Copyright and Terms of Use

Copyright and Trademark

Terms of Use

No Warranties

Limitation on Liability

International Standard Book Number (ISBN): 978-1-62926-116-4

Published on February 11th, 2025.

Manhattan Review, 2 Park Avenue, Suite 2010, New York, NY 10016.
Phone: +1 (212) 316-2000. E-Mail: info@manhattanreview.com. Web: www.manhattanreview.com

About the Turbocharge your SAT Series

The Turbocharge Your SAT Series was created to provide students with comprehensive and highly effective SAT preparation for maximum SAT performance. Thousands of students around the world have received substantial score improvements by using Manhattan Review's SAT prep books. Now in its updated 2nd edition for the new SAT in 2016, the full series of 12 guides is designed to provide SAT students with rigorous, thorough, and accessible SAT instruction for top SAT scores. Manhattan Review's SAT prep books precisely target each testing area and deconstruct the different test sections in a manner that is both student-centered and results-driven, teaching test-takers everything they need to know in order to significantly boost their scores. Covering all of the necessary material in mathematics and verbal skills from the most basic through the most advanced levels, the Turbocharge Your SAT Series is the top study resource for all stages of SAT preparation. Students who work through the complete series develop all of the skills, knowledge, and strategies needed for their best possible SAT scores.

- □ **DSAT Math Essentials (ISBN: 978-1-62926-115-7)**
- ■ **DSAT Math Guide (ISBN: 978-1-62926-116-4)**
- □ **DSAT Reading & Writing Guide (ISBN: 978-1-62926-117-1)**
- □ **DSAT Math Question Bank (ISBN: 978-1-62926-118-8)**

About the Company

Manhattan Review's origin can be traced directly back to an Ivy League MBA classroom in 1999. While teaching advanced quantitative subjects to MBAs at Columbia Business School in New York City, Professor Dr. Joern Meissner developed a reputation for explaining complicated concepts in an understandable way. Prof. Meissner's students challenged him to assist their friends, who were frustrated with conventional test preparation options. In response, Prof. Meissner created original lectures that focused on presenting standardized test content in a simplified and intelligible manner, a method vastly different from the voluminous memorization and so-called tricks commonly offered by others. The new methodology immediately proved highly popular with students, inspiring the birth of Manhattan Review.

Since its founding, Manhattan Review has grown into a multi-national educational services firm, focusing on preparation for the major undergraduate and graduate admissions tests, college admissions consulting, and application advisory services, with thousands of highly satisfied students all over the world. Our SAT instruction is continuously expanded and updated by the Manhattan Review team, an enthusiastic group of master SAT professionals and senior academics. Our team ensures that Manhattan Review offers the most time-efficient and cost-effective preparation available for the SAT. Please visit www.ManhattanReview.com for further details.

About the Founder

Professor Dr. Joern Meissner has more than 25 years of teaching experience at the graduate and undergraduate levels. He is the founder of Manhattan Review, a worldwide leader in test prep services, and he created the original lectures for its first test preparation classes. Prof. Meissner is a graduate of Columbia Business School in New York City, where he received a PhD in Management Science. He has since served on the faculties of prestigious business schools in the United Kingdom and Germany. He is a recognized authority in the areas of supply chain management, logistics, and pricing strategy. Prof. Meissner thoroughly enjoys his research, but he believes that grasping an idea is only half of the fun. Conveying knowledge to others is even more fulfilling. This philosophy was crucial to the establishment of Manhattan Review, and remains its most cherished principle.

International Phone Numbers and Official Manhattan Review Websites

Manhattan Headquarters	+1-212-316-2000	www.manhattanreview.com
USA & Canada	+1-800-246-4600	www.manhattanreview.com
Argentina	+1-212-316-2000	www.review.com.ar
Australia	+61-3-9001-6618	www.manhattanreview.com
Austria	+43-720-115-549	www.review.at
Belgium	+32-2-808-5163	www.manhattanreview.be
Brazil	+1-212-316-2000	www.manhattanreview.com.br
Chile	+1-212-316-2000	www.manhattanreview.cl
China	+86-20-2910-1913	www.manhattanreview.cn
Czech Republic	+1-212-316-2000	www.review.cz
France	+33-1-8488-4204	www.review.fr
Germany	+49-89-3803-8856	www.review.de
Greece	+1-212-316-2000	www.review.com.gr
Hong Kong	+852-5808-2704	www.review.hk
Hungary	+1-212-316-2000	www.review.co.hu
India	+1-212-316-2000	www.review.in
Indonesia	+1-212-316-2000	www.manhattanreview.id
Ireland	+1-212-316-2000	www.gmat.ie
Italy	+39-06-9338-7617	www.manhattanreview.it
Japan	+81-3-4589-5125	www.manhattanreview.jp
Malaysia	+1-212-316-2000	www.review.my
Netherlands	+31-20-808-4399	www.manhattanreview.nl
New Zealand	+1-212-316-2000	www.review.co.nz
Philippines	+1-212-316-2000	www.review.ph
Poland	+1-212-316-2000	www.review.pl
Portugal	+1-212-316-2000	www.review.pt
Qatar	+1-212-316-2000	www.review.qa
Russia	+1-212-316-2000	www.manhattanreview.ru
Singapore	+65-3158-2571	www.gmat.sg
South Africa	+1-212-316-2000	www.manhattanreview.co.za
South Korea	+1-212-316-2000	www.manhattanreview.kr
Sweden	+1-212-316-2000	www.gmat.se
Spain	+34-911-876-504	www.review.es
Switzerland	+41-435-080-991	www.review.ch
Taiwan	+1-212-316-2000	www.gmat.tw
Thailand	+66-6-0003-5529	www.manhattanreview.com
Turkey	+1-212-316-2000	www.review.com.tr
United Arab Emirates	+1-212-316-2000	www.manhattanreview.ae
United Kingdom	+44-20-7060-9800	www.manhattanreview.co.uk
Rest of World	+1-212-316-2000	www.manhattanreview.com

Contents

Chapter 1

Welcome

Dear Students,

Here at Manhattan Review, we constantly strive to provide you with the best educational content for standardized test preparation. We make a tremendous effort to keep making things better and better for you. This is especially important for an examination such as the SAT. As you know, from Spring'24 on, the SAT went through a major change. The Digital SAT is even more challenging.

Manhattan Review's DSAT-Math Guide has a collection of 300 questions classified into four subsections: **Algebra, Advanced Math, Problem Solving & Data Analysis, and Geometry & Trigonometry**.

In a nutshell, Manhattan Review's "DSAT-Math" Guide is holistic and comprehensive for the practice; it was created so because we listen to what students need. Should you have any queries, please feel free to write to us at *info@manhattanreview.com*.

Happy Learning!

Professor Dr. Joern Meissner
& The Manhattan Review Team

Chapter 2

Introduction to The Digital SAT

In 2023-24, the College Board changed the previous SAT to a digital format. The digital version of the SAT is reduced to 2 hours, 14 minutes, as opposed to the previously 3 hours, 15 minutes. This change is welcome because it validates the relevance and significance of standardized testing.

When Can a Student Take the Digital SAT?

The College board publishes test dates for the digital SAT, as before.

What changes did the College Board make to the digital SAT?

In addition to reducing testing time to 2 hours, 14 minutes from the previous 3 hours, 15 minutes, the digital SAT will allow more time per question.

The following are the significant changes compared to the previous SAT content:

- Shorter reading passages with a single (discrete) question per text
- Texts from a broader range of college-level exams
- Calculator access for the entire math section
- The digital SAT scores are provided more quickly, allowing students and universities to make quicker college decisions.

Advantages of digital SAT

Safe: The paper-and-pencil version of the test could be canceled if there are problems outside or inside the school. The digital SAT is safer and better protected because each student will have a unique test form and will use their own or a test center-issued device.

Faster Score: Students get their scores in days as opposed to weeks.

What are the differences between the digital SAT and the previous SAT?

The digital SAT is a simpler version of the previous SAT. The new structure will place more emphasis on college-ready abilities and less on tenacity and speed. The previous SAT Reading section, with its long passages and frequently vague history passages, demanded patience to solve. The digital SAT incorporates texts from the humanities and sciences that dive into more modern topics in finance, economics, and the environment, as well as a greater variety of short passages.

Also, there is relatively greater emphasis on using both textual and graphic content to answer questions.

Since calculators are used for the entire math section, it seems like the test has been changed in a big way. In addition to the focus on algebra and word problems on the SAT, there may be a shift to more straightforward questions, like those on the ACT, and more geometry and trigonometry questions.

As the College Board changed test content to make it better for digital distribution, the digital SAT still tests the same skills and knowledge learned in high school that are most important for college and job readiness.

Digital SAT Vs. Previous SAT: What Has Not Changed?

Despite the fact that the digital SAT comes up with a number of changes that is beneficial to both students and colleges, many of the SAT's most essential components stay unaltered.

- The digital SAT still tests students on the skills and knowledge they need to be ready for college and careers.
- The SAT is scored on the same 1600-point scale, so teachers and students are able to see how much they have improved over time.
- The digital SAT cannot be taken at home but rather in a school or testing center in a proctored setting.

Does a student Need a Device to Take the Digital SAT?

The digital SAT is taken on a laptop or tablet. Students can use their personal or test center-issued devices. If a student does not have a laptop, they may request one from the College Board, and it is provided to them on their SAT Day. If a student has any device or connectivity issues on the test day, the College Board has designed the test application so that it survives internet outages. Students are able to continue with the test without interruption if the internet goes down; their responses are stored, and they will not lose any testing time.

How Challenging Is It to Take the SAT Online?

The good news is that most students who took the digital SAT found the new version of the test easier to understand and use. When the College Board says that the digital version of the test is more "intuitive," it implies that it is "easier to navigate" and "easier to answer without becoming exhausted."

The digital SAT interface has a provision for a clock, which makes it easier for students to keep track of time.

In addition to highlighters, strikethrough, and bookmarks for review, there is an integrated calculator and a provision for a standard math formula sheet. At a single location, test-takers find all the resources necessary to answer questions.

What Tools Are Available for Students Taking the Digital SAT?

The test application includes many testing tools. Here are two examples:

- You can mark questions for review using the bookmark feature.
- Test-takers have the option to display or hide the digital clock placed at the top of their testing screen.

- During the math section, students may use the integrated graphing calculator (or one can bring their own basic calculator).
- A reference sheet with standard mathematical formulas for each question is provided.

When and How to Study for the Digital SAT?

Start preparing for the digital SAT as quickly as possible. Focus on making noticeable gains in every area: reading, writing, and math. Don't be hesitant to monitor your progress using the previous SAT tests. There is no need to take full-length tests, but you may always attempt to earn a perfect score on practice tests. Before starting to read, emphasize the need to develop comprehension skills and knowledge of previous global issues and concerns. This will gradually boost your vocabulary and reading speed, preparing you for future SAT reading success.

How to Study for the Digital SAT?

Reading, writing, grammar, and math skills are still the most important parts of the SAT for measuring college readiness.

This is how to begin preparing for each section of the digital SAT:

- **Reading and Writing (RW)**: Start with a variety of humanities- and science-related texts. You should read the New York Times, the Economist, the New Scientist, and the Washington Post. You should start reading the editorials. Learn about contemporary environmental and geopolitical concerns. Focus on being a knowledgeable reader in general. Instead of being a passive recipient of information, develop views. Application of grammar and punctuation rules is still tested on the SAT.
- **Math**: Start practicing questions on all the math topics on the previous SAT. In addition, pay close attention to the geometry and trigonometry questions on the official SAT material and Manhattan Review guides. Learn the formulae and their application to the previous SAT topics, as well as the rationale behind their operation.

How Would These Changes Impact College Admissions?

Since the SAT's basic content, level of difficulty, and scoring system did not change much, we expect schools to evaluate the digital version in the same way they evaluate the previous version. In addition, colleges make an attempt not to fault students for situations beyond their control, so taking one version of the SAT will not negatively affect your college applications relative to those who took the other version.

Even with these changes, colleges have a very different view of the SAT and ACT than they had a few years ago. Several schools are paying less attention to standardized test scores and more attention to other parts of an application, such as a student's grade point average (GPA), the courses they took in high school, and extracurricular activities. The COVID-19 epidemic made this worse by making it difficult or impossible for many students to take standardized tests for several months. Almost every school made testing optional in 2020 and 2021, and hundreds of colleges have decided to stay test-optional permanently. If a school is test-optional, you have the choice of submitting SAT or ACT scores with your application. If you do not, the other application aspects will carry more weight.

The College Board is aware of these trends, and many of the changes to the SAT were made to make it easier for students who might have been turned off by the length and difficulty of the test. Still, a College Board survey showed that 83% of students wanted to be able to send test scores with their college applications, and almost all schools still let them, so the SAT is here to stay.

Key Takeaways

The following are the six most significant takeaways:

- The SAT is administered digitally.
- The duration of the SAT is reduced (to about 2 hours instead of about 3 hours).
- Students have more time to answer each question.
- Calculator is permitted for the entire math section of the test.
- Reading passages is of shorter length. Each passage has a single question.
- Students get their digital SAT scores within days rather than weeks.

Digital SAT Math content

Content	**Topics**	**Number of questions**
Algebra	Fundamental concepts used in Algebra; arranging formulae, linear equations, inequalities, etc.	13–15
Advanced Math	Advanced concepts in Algebra, incl. absolute value, quadratic, exponential, polynomial, rational, radical, and other nonlinear equations	13–15
Problem Solving & Data Analysis	Understanding qualitative & quantitative data, analyzing relationship, incl. graphical; Ratio & Proportion, Percents, and units of measurements	5–7
Geometry & Trigonometry	Geometry (angles, triangles, area, and volume), Trigonometry (Trigonometric ratios)	5–7
	Total time (70 minutes)	**44**

Chapter 3

Algebra Concepts

3.1 Identities

Let us first look at a few basic terms.

Polynomial: A polynomial is an expression of variables that can only have non-negative integer coefficients.

For example, $x^3 + 2xy^2 + 6z$ is a polynomial.

Order and Degree: The Order of a polynomial refers to the number of variables involved in the polynomial.

For example, $x^2 + 2xy$ is a polynomial of order '2' since there are two variables, x and y.

The Degree of a polynomial refers to the highest value of the exponent of any variable in the polynomial.

For example, $x^2 + 2xy$ has degree '2' since the highest exponent is 2.

Identities: An identity is a statement equating two expressions which are equal for all values of the variables involved.

For example, $5x = 2x + 3x$ is an identity, as $5x$ and $2x + 3x$ are equal for all values for x.

Some important identities are as follows:

- $(a \pm b)^2 = a^2 \pm 2ab + b^2$
- $\left(x \pm \frac{1}{x}\right)^2 = x^2 + \frac{1}{x^2} \pm 2$
- $a^2 - b^2 = (a + b)(a - b)$
- $(a \pm b)^3 = a^3 \pm 3a^2b + 3ab^2 \pm b^3 = a^3 \pm b^3 \pm 3ab(a \pm b)$

Let us take an example:

If $x + \frac{1}{x} = 2$, what is the value of $x^4 + \frac{1}{x^4}$?

We have: $\left(x + \frac{1}{x}\right)^2 = 2^2 => x^2 + \frac{1}{x^2} + 2 = 4$

$=> x^2 + \frac{1}{x^2} = 4 - 2 = 2$

$=> \left(x^2 + \frac{1}{x^2}\right)^2 = 2^2 => x^4 + \frac{1}{x^4} + 2 = 4$

$=> x^4 + \frac{1}{x^4} = 4 - 2 = 2.$

Let us take another example:

What is the value of $\left\{\frac{(0.21 + 0.17)^2 + (0.21 - 0.17)^2}{\left(0.21^2 + 0.17^2\right)}\right\}$?

Let $a = 0.21$ and $b = 0.17$

Thus, we have:

$$\frac{(0.21 + 0.17)^2 + (0.21 - 0.17)^2}{\left(0.21^2 + 0.17^2\right)}$$

$$= \frac{(a + b)^2 + (a - b)^2}{(a^2 + b^2)}$$

$$= \frac{a^2 + 2ab + b^2 + a^2 - 2ab + b^2}{a^2 + b^2}$$

$$= \frac{2\,(a^2 + b^2)}{a^2 + b^2}$$

$$= 2$$

3.2 Linear Equations

Linear equations are equations in which the highest exponent/index of the variables is 1, i.e. the degree is 1.

For example, $2x + 5 = 11$ is a linear equation with a single variable, x, whose index is one.

Similarly, $3x + 5y = 9$ is a linear equation with two variables, x and y, as the indices of both are 1.

Solving linear equations with a single variable:

While solving any linear equation with a single term, we need to group all terms containing the variable on the left and all other constant terms on the right.

While shifting terms from left to right as addition or subtraction, the sign of the terms are negated. Similarly, while shifting terms from left to right as a multiplication or division, the processes are interchanged.

Let us take an example:

$$\frac{2x+1}{3} - 4 = \frac{3}{2} - x$$

We group all terms with x to the left and the other constant terms to the right. Thus, the term $-x$, when taken to the left becomes $+x$, similarly, the term -4, when taken to the right, becomes $+4$:

$$\frac{2x+1}{3} + x = \frac{3}{2} + 4$$

Simplifying the left and right sides:

$$\frac{(2x+1)+3x}{3} = \frac{3+2\times 4}{2}$$

$$=> \frac{5x+1}{3} = \frac{11}{2}$$

The '3' on the left and the '2' on the right are both 'divisions', and hence, on changing sides, they become 'multiplication':

$=> 2\,(5x + 1) = 11 \times 3$

$=> 10x + 2 = 33$

We take the constant '2' to the right:

$=> 10x = 33 - 2 = 31$

The '10' as a product on the left is taken to the right as a division:

$$=> x = \frac{31}{10}$$

Number of possible solutions:

For any linear equation of the form $Ax + B = Px + Q$, there can be three possible scenarios:

Case I: $A = P$ and $B = Q$

=> Any value of x would satisfy the above equation.
Thus, there are infinite possible solutions.

Case II: $A = P$ but $B \neq Q$

=> The x terms cancel out, leaving behind two unequal constants.
Thus, there is no possible solution.

Case III: $A \neq P$
=> There is a unique solution for x.

Solving Linear Equations with two variables:

There are two possible ways of solving such equations:

- Method of substitution: From the first equation, we express one variable in terms of the other and substitute in the second equation.
- Solving simultaneously: Using both equations simultaneously we eliminate one of the two variables.

Let us take an example:

$2x + 3y = 15 \ldots \text{(i)}$

$5x - 2y = 28 \ldots \text{(ii)}$

- Solving by the method of substitution:

  From (i): $x = \frac{15 - 3y}{2}$

  Substituting in (ii):

  $$\frac{5(15 - 3y)}{2} - 2y = 28$$

  The above can be solved as discussed above for a single variable:

  $=> 75 - 15y - 4y = 56 => -19y = -19$

  $=> y = 1$

  Substituting $y = 1$ in the expression for x, we have:

  $$x = \frac{15 - 3 \times 1}{2} = 6$$

- Solving simultaneously:

  We need to eliminate one of the variables by multiplying the equations with a suitable constant and then adding/subtracting them:

  Equation (i) $\times$ 2 + equation (ii) $\times$ 3:

  $4x + 6y = 30$

  $15x - 6y = 84$

  $=> 19x = 114 => x = 6$

  Substituting $x = 6$ in (i):

  $2 \times 6 + 3y = 15$

  $=> y = \frac{15 - 12}{3} = 1$

Graphically, a pair of linear equations with two variables represents two straight lines, the solution of which represents their point of intersection.

Let us take two general equations:

$ax + by = c \ldots$(i)

$Ax + By = C \ldots$(ii)

Here, $a,\ b,\ c,\ A,\ B,\ C$ are the constants and x, y are the variables.

There are three important results possible:

- $\frac{a}{A} = \frac{b}{B} = \frac{c}{C}$: This implies that the equations are identical equations, i.e. they are overlapping lines and hence, they have infinite solutions.

- $\frac{a}{A} = \frac{b}{B} \neq \frac{c}{C}$: This implies that the equations are equations of parallel lines and since parallel lines never intersect, there exists no possible solution.

- $\frac{a}{A} \neq \frac{b}{B}$: This refers to the case of intersecting lines, i.e. a unique solution can be obtained from the two equations.

  - $\frac{a}{A} \neq \frac{b}{B} = \frac{c}{C}$: Since the ratio of coefficients of y is the same as the ratio of the constant terms, the value of x in the equation is zero.

  - $\frac{b}{B} \neq \frac{a}{A} = \frac{c}{C}$: Since the ratio of coefficients of x is the same as the ratio of the constant terms, the value of y in the equation is zero.

Solving for x and y from a single equation given x and y are integers:

Let us take an example:

Find all the possible values of x and y if $6x + 32y = 102$, where x, y are positive integers:

- First we need to reduce the coefficients to their lowest terms.

  Thus, we have: $3x + 16y = 51$

- We find any possible integer solution of the above equation (it is not important to keep both x and y positive at this stage).

  We can see that $x = 17$ and $y = 0$ obviously satisfies.

- It is clear that we need to increase the value of y to make it positive. Thus, we add the coefficient of x to the value of y and simultaneously, we subtract the coefficient of y from the value of x to generate the next possible integer solution. Thus, we get:

x	y
17	0
$17 - 16 = \mathbf{1}$	$0 + 3 = \mathbf{3}$
$1 - 16 = -15$	$3 + 3 = 6$

We see that on repeating the same process a second time, the value of x becomes negative and hence, inadmissible. Thus, there is only one possible solution: $x = 1,\ y = 3$.

Determining the value of an expression involving the variables without solving:

Let us take an example:

$2x + 5y = 19 \ldots$(i)

$x + 2y = 8 \ldots$(ii)

We need to determine the values of:

(1) $3x + 7y$

(2) $x + 3y$

(3) $4x + 9y$

For (1): Simply adding (i) and (ii): $3x + 7y = 27$

For (2): Simply subtracting (ii) from (i): $x + 3y = 11$

For (3): (i) + 2 × (ii): $4x + 9y = 35$

Additional solved problems:

(1) In the GCAT test, for every correct answer, a student is awarded '+3' marks, and for every wrong answer, the student is awarded '−1' marks. There are 40 questions, and the student must answer all of them.

I. If the student gets 40 marks, how many questions did he get correct?
II. How many different scores can the student get, if he randomly marks the answers?

Explanation:

Let the student got x correct and y wrong answers.

I. Thus, we have:
$x + y = 40$ (There are 40 questions and he answered all questions)

$3x - y = 40$ (Each correct answer is of 3 marks and each wrong answer is of -1 mark)

Adding the above two equations:

$4x = 80 => x = 20$

II. If the student answers ALL questions randomly, x changes from 0 to 40 and y, correspondingly changes from 40 to 0.

Thus, there are 41 possible values of the score: 120, 116, ... − 36, −40.
(Note that the score forms a series with the difference between their consecutive terms equals to 4)

(2) A and B have a number of marbles. The total number of marbles is 24. If A gives B 6 marbles, the ratio of the number of marbles with A and B is reversed. How many marbles does A have?

Explanation:

Let the number of marbles with A be x.

Thus, the number of marbles with B = $(24 - x)$.

After A gives 6 marbles to B, we have:

Number of marbles with A = $(x - 6)$.

Number of marbles with B = $(24 - x + 6) = (30 - x)$.

Since the ratio of the number of marbles with A and B is reversed, we have:

$$\frac{x-6}{30-x} = \frac{24-x}{x}$$

=> $x^2 - 6x = 720 - 54x + x^2$

=> $48x = 720$

=> $x = 15$

Alternate Approach:

As we can see, the above approach is a bit complicated.

However, we can use a much simpler logic.

After A gives B 6 marbles, the total number of marbles with A and B remain unchanged.

Since the ratio of the number of marbles has reversed, we can simply say that the number of marbles with A and B have been interchanged.

Thus, we have:

$x - 6 = 24 - x$

=> $x = 15$

(3) Abe has 20 coins with him. A few of the coins are dimes (a dime is equal to 10 cents), and the rest are quarters (a quarter is equal to 25 cents). If the total amount with him is \$4.25, how many dimes and quarters does he have separately? (Note: 100 cents make a dollar.)

Explanation:

Let us solve this problem in three different ways.

Approach 1: Using two variables

Let the number of dimes be x and the number of quarters be y.

Thus, we have:

$x + y = 20 \ldots$(i)

Total worth of dimes is $10x$ cents and the total worth of quarters is $25y$ cents.

Since the total amount with him is \$4.25, i.e. 425 cents, we have:

$10x + 25y = 425$

=> $2x + 5y = 85 \ldots$ (ii)

(ii) – (i) × 2:

$3y = 45$

=> $y = 15$

=> $x = 5$

Approach 2: Using one variable

Let the number of dimes be x.

Since there are 20 coins, the number of quarters = $(20 - x)$

Since the total amount is 425 cents, we have:

$10x + 25(20 - x) = 425$

=> $10x - 25x = 425 - 500$

=> $-15x = -75$

=> $x = 5$

=> Number of quarters = $20 - 5 = 15$

Approach 3: Using no variables

Let us assume that all his coins are dimes.

Thus, he has 20 dimes, which account for $20 \times 10 = 200$ cents.

However, he actually has 425 cents.

Thus, there is a shortage of $425 - 200 = 225$ cents.

Thus, we can conclude that some of his dimes are actually quarters.

In that case, each such dime would increase in value from 10 cents to 25 cents, i.e. an increase of $25 - 10 = 15$ cents.

The number of times such a conversion from dime to quarter required to fulfill the shortage of 225 cents

$= \frac{225}{15} = 15$ times

Thus, of the 20 dimes, 15 need to be converted to quarters and the rest should remain as dimes.

Thus, number of quarters is 15 and the number of dimes is 5.

(4) Joe purchases seven identical gift packs from Shop A and four identical gift packs from Shop B. Each gift pack has a number of pens. The total number of pens in the 11 gift packs is 26. If he wants an additional 18 pens, what is the maximum number of gift packs, similar to the ones he purchased initially, he needs to buy?

Explanation:

Let the number of pens present in gift-packs from Shop A = x

Let the number of pens present in gift-packs from Shop B = y

Thus, total number of pens = $(7x + 4y)$

Thus, we have:

$7x + 4y = 26$

Since x and y must be positive integers, $4y$ is even. Thus, $7x$ must also be even since the sum of $7x$ and $4y$ is 26, an even number.

Since $7x$ is even, x must be even.

Thus, the only value of $x = 2$ (any even value of x greater than 3 is not possible since $7x$ would exceed 26 and y would become negative).

Thus, we have:

$y = \frac{26 - 7 \times 2}{4} = 3$

Thus, the gift-packs from Shop A have 2 pens each and those from Shop B have 3 pens each.

Thus, to maximize the number of gift-packs, he should purchase the ones from Shop A.

Thus, to buy 18 pens, the maximum number of gift-packs required

$= \frac{18}{2} = 9$

3.3 Inequalities

Inequalities: It deals with cases where variables or numbers are less than or greater than other variables or numbers. The following symbols are used:

- $>$: More than;
- $<$: Less than;
- $\leq$: Less than or equal to;
- $\geq$: More than or equal to

Let us look at some of the important rules:

(1) In any given inequality, we can add or subtract any quantity from both sides keeping the inequality the same.

For example: If we have: $x > y$

Adding z to both sides: $x + z > y + z$

Subtracting z from both sides: $x - z > y - z$

(2) In any given inequality, we can multiply or divide any other 'positive' quantity on both sides keeping the inequality the same.

For example: If we have: $x > y$

Multiplying a positive quantity z on both sides: $x \times z > y \times z$

Dividing by a positive quantity z on both sides: $\frac{x}{z} > \frac{y}{z}$

(3) In any given inequality, we can multiply or divide any other 'negative' quantity on both sides, thus, reversing the inequality.

For example: If we have: $x > y$

Multiplying a negative quantity $(-z)$ on both sides: $-x \times z < -y \times z$

Dividing by a negative quantity $(-z)$ on both sides: $-\frac{x}{z} < -\frac{y}{z}$

(4) Any two inequalities having the same inequality can be added.

For example: If we have: $x > y$ and $p > q$

Adding the inequalities, we have: $x + p > y + q$

(5) In order to subtract one inequality from another, the former has to be negated, thus reversing the inequality and then the inequalities can be added.

For example: If we have $x > y$ and $p > q$

Negating the former inequality, we have: $-x < -y => -y > -x$

Thus, on adding the inequalities $p > q$ and $-y > -x$, we have:

$p - y > q - x$

(6) Any term with an even exponent is always non-negative, its minimum value being '0'. Thus, we have:

$x^2 \geq 0$, $y^4 \geq 0$, $z^6 \geq 0$, etc.

Thus, if we have: $x^5y^2z > 0$, we have:

=> $(x^4y^2)\,xz > 0$

Since x^4y^2 is always non-negative, we have:

=> $xz > 0$

=> $x > 0$ and $y > 0$

OR

$x < 0$ and $y < 0$

(7) Inequality of ratio of two quantities and the inequality of the product of those quantities are similar.

For example: If we have: $\frac{x}{y} > 0$

=> $\frac{x}{y} \times y^2 > 0$ (since $y^2 > 0$)

=> $xy > 0$

Note: However, there is a slightly different result for "greater than or equal to" or "less than or equal to" type of inequalities, as shown:

If we have: $\frac{x}{y} \geq 0$

=> $\frac{x}{y} \times y^2 \geq 0$

=> $xy \geq 0$

On careful observation, it can be seen that $y = 0$ is a possible solution for $xy \geq 0$, however, not so in the case of $\frac{x}{y} \geq 0$.

3.4 Absolute value

Modulus or absolute value of a number is a function that returns the magnitude of the number ignoring the sign.

Thus, we have: $|2| = |-2| = 2$

Hence, we have: $|x| = 2 => x = \pm 2$

Another way of interpreting the modulus of a number is distance of the number from a point on the number line.

For example, $|x - 3|$ denotes the distance of x from the point '3' on the number line.

Thus, $|x - 3| = 4$ implies the distance of the point x from the point '3' on the number line is 4 units. Since the distance can be measured on either side of '3', we attain the points $3 + 4 = 7$ and $3 - 4 = -1$ on the number line, as shown below.

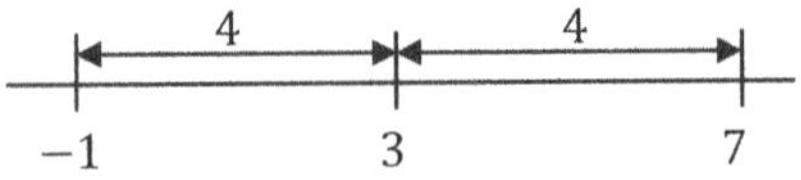

Thus, using the above concept, we have:

$|x - a| = b => \pm(x - a) = b => x - a = \pm b$

$=> x = a \pm b$

Modulus based inequality: We have the following two rules:

- $|x - a| > b$

  $=> x - a > b$ OR $x - a < -b$

  $=> x > a + b$ OR $x < a - b$

  Alternately, the given inequality implies that the distance of x from a is greater than b units. If the distance were equal to b units, we would have got the points $(a + b)$ and $(a - b)$. Since the distance should be greater than b units, x should overshoot the above two points, i.e. x is greater than the greatest point or x is less than the smallest point.

  Thus, we have: $x > a + b$ OR $x < a - b$

- $|x - a| < b$

  $=> -b < x - a < b$

  $=> a - b < x < a + b$

  Alternately, the given inequality implies that the distance of x from a is lesser than b units. If the distance were equal to b units, we would have got the points $(a + b)$ and $(a - b)$. Since the distance should be lesser than b units, x should remain within the above two points.

Thus, we have: $a - b < x < a + b$

Another way to solve this is to remove the modulus and put $\pm$ with the x term:

- $|x - a| > b \Rightarrow \pm(x - a) > b \Rightarrow x - a > b$ or $x - a < -b \Rightarrow x > a + b$ or $x < a - b$
- $|x - a| < b \Rightarrow \pm(x - a) < b \Rightarrow x - a < b$ or $x - a > -b$
  $\Rightarrow x < a + b$ or $x > a - b \Rightarrow a - b < x < a + b$

Additional solved problems:

(1) What are the integer values of x satisfying $\left(\frac{2x-3}{x+1}\right) \leq 0$?

Explanation:

We have: $\left(\frac{2x-3}{x+1}\right) \leq 0$

Thus, we have:

Case I:

$2x - 3 \leq 0$ AND $x + 1 > 0$

=> $x \leq \frac{3}{2}$ AND $x > -1$

=> $-1 < x \leq \frac{3}{2}$

=> The integer values of x are 0 and 1.

Case II:

$2x - 3 \geq 0$ AND $x + 1 < 0$

=> $x \geq \frac{3}{2}$ AND $x < -1$, which is not a possible scenario

Thus, the only possible integer values of x are 0 and 1.

(2) What are the integer values of x satisfying $|x^2 - 6| < 3$?

Explanation:

We have: $|x^2 - 6| < 3$

=> $-3 < x^2 - 6 < 3$

=> $3 < x^2 < 9$

Case I: $x^2 < 9$

=> $-3 < x < 3$

=> The possible integer values of x are $-2,\ -1,\ 0,\ 1,\ 2$

Case II: $x^2 > 3$

=> $x > \sqrt{3}$ OR $x < -\sqrt{3}$

=> The possible integer values of x are $\cdots -4,\ -3,\ -2,\ 2,\ 3,\ 4,\ \ldots$

Thus, the common set of values of x are -2 and 2.

(3) If $-5 < x < 3$, $-9 < y < -5$, $-13 < p < 8$, $-4 < q < 6$ and $1 \leq r \leq 8$, where x, y, p, q, and r are integers, what are the range of values of

I. $x + 2y$
II. $2x - y$
III. $p \times q$
IV. $\frac{q}{r}$

Explanation:

I. We have: $-9 < y < -5$
$=> -8 \leq y \leq -6$
$=> -16 \leq 2y <= -12$ ... (i)

Again: $-5 < x < 3$
$=> -4 \leq x \leq 2$ ... (ii)
$=> -20 \leq x + 2y \leq -10$; adding (i) and (ii)

II. We have: $-5 < x < 3$
$=> -4 \leq x \leq 2$
$=> -8 \leq 2x \leq 4$ ... (i)

Again: $-9 < y < -5$
$=> 9 > -y > 5$
$=> 5 < -y < 9$
$=> 6 \leq -y \leq 8$... (ii)
$=> -8 + 6 \leq 2x - y \leq 8 + 4$; adding (i) and (ii)
$=> -2 \leq 2x - y \leq 12$

III. We have: $-13 < p < 8$ and $-4 < q < 6$
$=> -12 \leq p \leq 7$ and $-3 \leq q \leq 5$
Thus, the maximum value of $p \times q$

= The greater among $\{(-12) \times (-3)\}$ and $\{(7) \times (5)\} = 36$
Also, the minimum value of $p \times q$

= The smaller among $\{(-12) \times (5)\}$ and $\{(7) \times (-3)\} = -60$
Thus, we have: $-60 \leq pq \leq 36$

IV. We have: $-4 < q < 6 => -3 \leq q \leq 5$ and $1 \leq r \leq 8$
To determine the maximum value of $\frac{q}{r}$, we must keep the ratio positive. Since r is entirely positive, we must take the positive value of $q = 5$ and the least positive value of $r = 1$.
Thus, the maximum value of $\frac{q}{r} = \frac{5}{1} = 5$
Again, to determine the minimum value of $\frac{q}{r}$, we must keep the ratio negative. Since r is entirely positive, we must take the negative value of $q = -3$ and the least positive value of $r = 1$.

(Note: the minimum value of $\frac{q}{r}$ is the negative number with the largest magnitude)

Thus, the minimum value of $\frac{q}{r} = \frac{-3}{1} = -3$

Thus, we have: $-3 \leq \frac{q}{r} \leq 5$

3.5 Functions and graphs of functions

If y is expressed in terms of x, such that, for any x, a unique value of y is obtained, the expression of y in terms of x is called a function of x.

Functions of x are usually denoted by symbols of the form $f(x)$, $g(x)$, $h(x)$, etc.

Some examples of functions:

- $f(x) = 2x + 5$
- $f(x) = x^2 - 2x + 1$
- $f(x) = \frac{1}{x}$
- $f(x) = \sqrt{x} + 1$
- $g(x) = \frac{x-2}{x^2+3}$
- $g(x) = 1$

The idea of a function can be represented by a simple diagram as shown below:

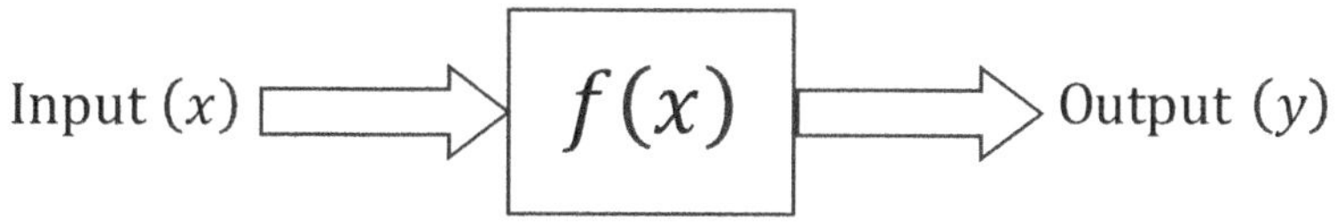

Method of substitution in functions:

For any given $f(x)$, the value of $f(a)$ is determined by substituting $x = a$ on the right side of the equality.

Let us take an example:

$f(x) = 2x + 1$

Thus, we have:

- $f(1) = 2(1) + 1$, i.e. we replace x with 1 on the right side of the equality
  $\Rightarrow f(1) = 3$
- $f(x^2) = 2(x^2) + 1 = 2x^2 + 1$
- $f(2x + 1) = 2(2x + 1) + 1 = 4x + 3$

Modulus function: $f(x) = |x|$:

- $f(x) = x$, if $x \geq 0$,
- $f(x) = -x$, if $x < 0$

The graph of $f(x) = |x|$ is shown:

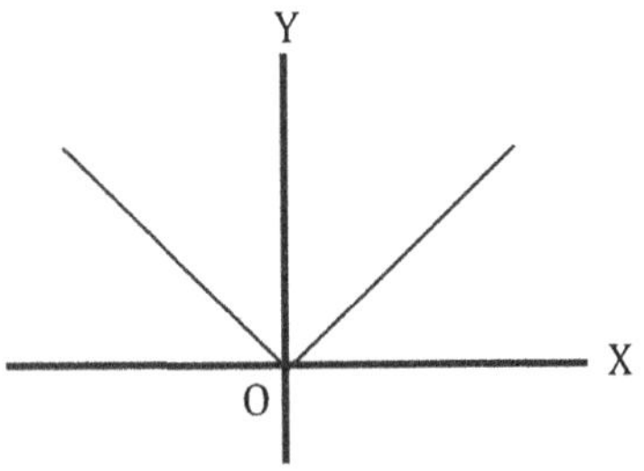

Additional solved problems:

(1) $f(x)$ is a function such that $f(2x+1) = x^2 - 2x$. What is the value of $f(3)$?
Explanation:

We have:
$f(2x+1) = x^2 - 2x \ldots$(i)

Since we need the value of $f(3)$, we need to use a suitable value of x so that we have:

$2x + 1 = 3$

$=> x = 1$

Thus, substituting $x = 1$ in (i):

$f(2 \times 1 + 1) = 1^2 - 2 \times 1$

$=> f(3) = -1$

(2) The function f is such that $f(x+1) = x(x+1)$. What is the value of $f(x)$?
Explanation:

We have:

$f(x+1) = x(x+1)$

Since we need the value of $f(x)$, we would replace $(x+1)$ with x.

In $f(x+1) = x(x+1)$, we have the term of $(x+1)$ on each side, and we need to write the single x as $[(x+1) - 1]$ so that we can replace $(x+1)$ with x.

Thus, $f(x+1) = [(x+1) - 1](x+1)$

Now, by replacing $(x+1)$ with x, we get,

Thus, $f(x) = [(x) - 1](x)$

$f(x) = (x-1)(x) = x(x-1)$

3.6 The Co-ordinate plane

Rectangular axes: The figure shows the XY Cartesian plane. The line XOX′ is called the X axis and YOY′ the Y axis. A point P (x, y) in the plane denotes the X value or *abscissa* of P and the Y value or the *ordinate* of P.

The plane is divided into four equal parts by the two axes, called the Quadrants (I, II, III, and IV).

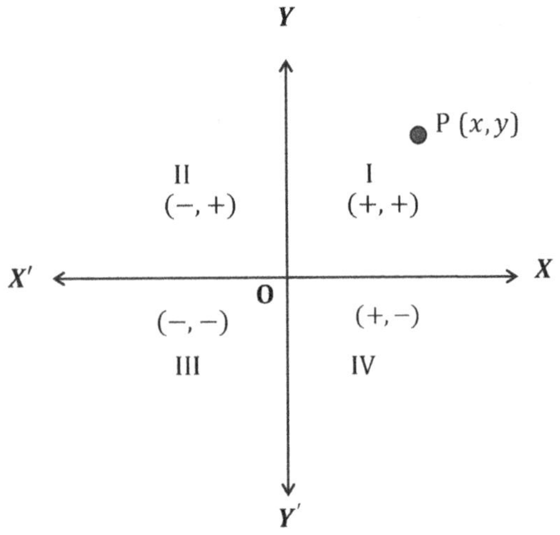

Linear function

The linear function is a straight line represented by the first-degree equation $ax + by = k$, where a, b, c are constants and a and b are not simultaneously zero.

The above equation can also be represented as:

$ax + by = k$

$=> by = -ax + k$

$=> y = -\left(\frac{a}{b}\right)x + \left(\frac{k}{b}\right)$

$=> y = mx + c$

Here, m is termed the slope and c is termed the Y-intercept, i.e. the value of y when the value of $x = 0$.

For a linear equation passing through a set of given points (x_1, y_1) and (x_2, y_2), we can substitute the values in the above equation and solve for the values of m and c.

Thus, we have:

$y_1 = mx_1 + c \ldots$ (i)

$y_2 = mx_2 + c \ldots$ (ii)

Subtracting: $y_1 - y_2 = mx_1 - mx_2$

$$=> m = \frac{y_1 - y_2}{x_1 - x_2}$$

Substituting the above in (i), we can determine the value of c.

For any linear relation between x and y, the slope, i.e. m represents the 'rate of change', or, in other words, the change in the value of y for unit change in the value of x.

For example, if x represents time in years and y represents the population of a country and it is observed that x and y follow a linear relationship, the slope defines the rate of change of population per year.

A relation between x and y where the y value remains a constant for all values of x is termed a 'constant relation.' Such a line is parallel to the X-axis and has a slope $m = 0$.

Polynomial function

A polynomial function is typically of the form:

- $y = ax^2 + bx + c$ (quadratic equation)
- $y = ax^3 + bx^2 + cx + d$ (cubic equation); and so on

They can be easily identified to be one where the exponent on x is an integer greater than 1.

The graphs of polynomial functions comprise of wavy-curves and are shown below:

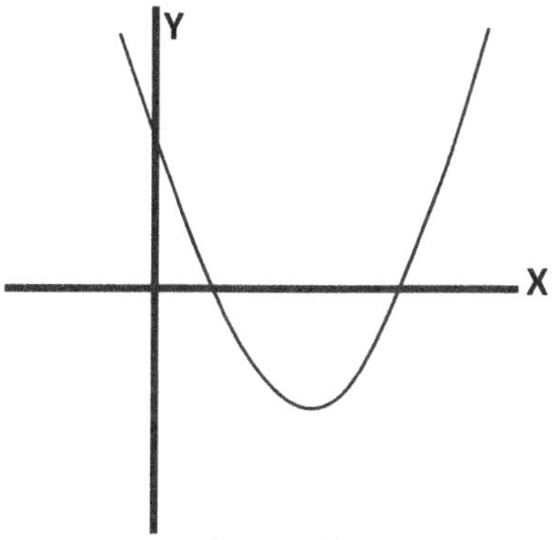

Quadratic function

$y = ax^2 + bx + c$

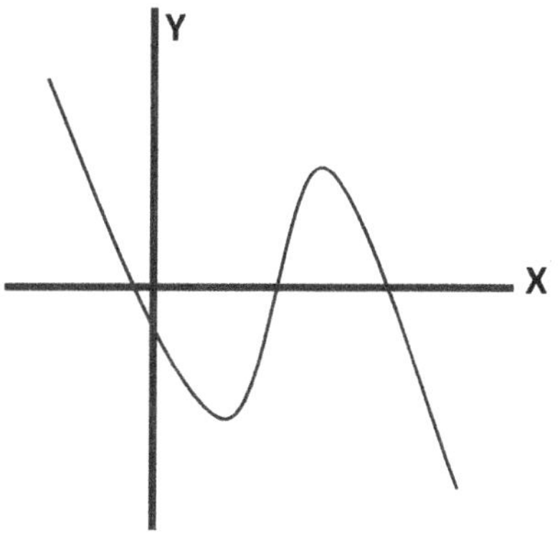

Cubic function

$y = ax^3 + bx^2 + cx + d$

Exponential function

An exponential function is typically of the form:

- $y = p^x$; where $p > 1$:

  An increasing exponential function since with increasing values of x, the values of y increase very rapidly.

For example: $y = 2^x$

x	y
1	2
2	4
3	8
4	16

- $y = p^x$; where $0 < p < 1$:
  A decreasing exponential function since with increasing values of x, the values of y keep decreasing.

  For example: $y = (0.5)^x$

x	y
1	0.5
2	0.25
3	0.125

The graphs of exponential functions are shown below:

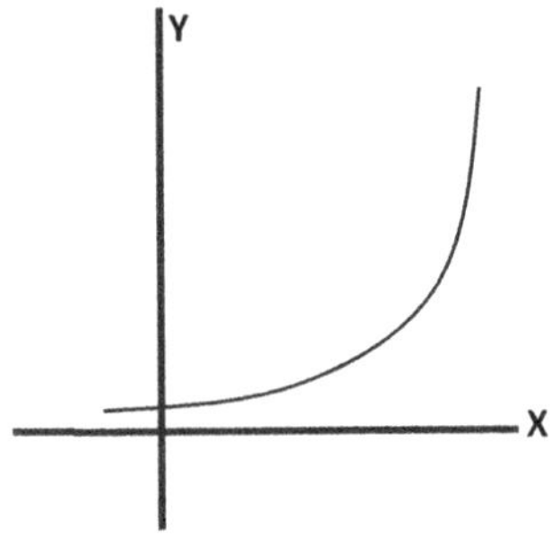

Increasing Exponential function

$y = 2^x$

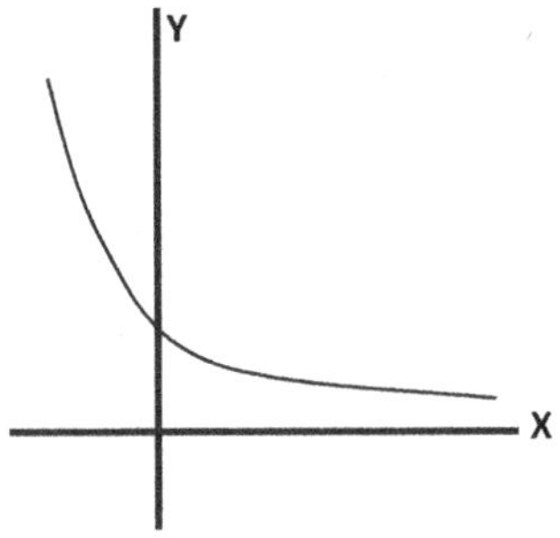

Decreasing Exponential function

$y = (0.5)^x$

In the next chapter, you will find exam-like questions. Best of luck!

Chapter 4

Algebra: Practice Questions

4.1 Algebraic Expressions

1. **SPR:**

$$\left(\frac{5}{x+1}\right)(x^2-1)=\left(3-\frac{1}{x+1}\right)(x+1)$$

In the equation above, what is the value of x?

2. Which of the following is equivalent to the expression $\frac{x^5-x^3}{x^4+x^3}$?

(A) $x+1$

(B) $x-1$

(C) x^2

(D) x

3. If $\frac{3}{x+1}=\frac{15}{2w-3}$, such that $x \neq -1$ and $w \neq \frac{3}{2}$, which is the correct expression of w in terms of x?

(A) $w=\frac{5}{2}x+8$

(B) $w=\frac{2}{5}x+\frac{1}{4}$

(C) $w=\frac{2}{5}x-\frac{8}{5}$

(D) $w=\frac{5}{2}x+4$

4. If $x \neq 0$, what quantity can be added to $\left(\frac{x+1}{x}\right)$ or multiplied with $\left(\frac{x-1}{x}\right)$ to obtain the same result?

(A) $\frac{x}{x+1}$

(B) $x+1$

(C) $\frac{x}{x-1}$

(D) $-(x+1)$

5. A store purchases goods in boxes that weigh either 40 pounds or 60 pounds each. Let x be the number of 40-pound boxes and y be the number of 60-pound boxes. If the total weight of goods purchased in all the boxes combined is 1,200 pounds, which of the following is the correct representation of x in terms of y?

(A) $60-3y$

(B) $30-y$

(C) $30-\frac{3}{2}y$

(D) $20-\frac{2}{3}y$

6. If $x = 1 + \sqrt{3}$, what is the value of $(x^2 - 2x - 2)$?

(A) -2

(B) $-\sqrt{3}$

(C) 0

(D) 1

7. If a percent of $(a - 2b)$ when added to b percent of b, the value obtained is 0.64, then which of the following statements could be true?

(A) $a = b$

(B) $a + b = -8$

(C) $a - b = 8$

(D) $a - 8 = -b$

8. If $a > 1$, which of the following is equivalent to $\left(\frac{1}{\sqrt{a+1} + \sqrt{a-1}}\right)$?

(A) $\sqrt{\frac{a+1}{8}} - \sqrt{\frac{a-1}{8}}$

(B) $\sqrt{\frac{a+1}{2}} - \sqrt{\frac{a-1}{2}}$

(C) $\frac{\sqrt{a+1} - \sqrt{a-1}}{2}$

(D) $\sqrt{a+1} - \sqrt{a-1}$

9. The terms P, Q, R, and S are connected by the relation $P^2Q = \frac{3R}{\sqrt{2S}}$. Which of the following is the correct expression for S?

(A) $\frac{3R^2}{2P^4Q^2}$

(B) $\frac{9R^2}{2P^4Q^2}$

(C) $\frac{9R^2}{2P^2Q}$

(D) $\frac{3R}{2P^4Q^2}$

10. In order to rent an office space, an initial deposit of \$1,200 has to be paid. The rent payable each month is \$200. Apart from that, there is a semi-annual maintenance charge of \$60. Which of the following represents the total amount paid for rent and maintenance changes, including the initial deposit, if the office is rented for t years?

(A) $\$(3,600 + 120t)$

(B) $\$(1,200 + 2,460t)$

(C) $\$(1,200 + 2,520t)$

(D) $\$(1,200 + 3,120t)$

11. The half-life of a radioactive substance is the number of years it takes for the mass of the substance to reach half its original value. The initial weight of such a radioactive substance is 10.6 grams. If the half-life of the substance is 30 minutes, which of the following represents the mass of the substance after t minutes (t is a multiple of 30)?

(A) $10.6 \times \left(\frac{1}{2}\right)^{30t}$

(B) $10.6 \times \left(\frac{1}{2}\right)^{\frac{t}{30}}$

(C) $10.6 \times \left(\frac{1}{2}\right)^{\frac{t}{2}}$

(D) $10.6 \times \left(\frac{1}{2}\right)^{2t}$

12. In an experiment, it was observed that the number of bacteria triples p hours. If the number of bacteria at the start of the experiment was N, which of the following equations best predicts the number of bacteria d days after the start of the experiment?

(A) $3N \times \left(\frac{d}{p}\right)$

(B) $3N \times \left(\frac{24d}{p}\right)$

(C) $3^{\left(\frac{d}{p}\right)}N$

(D) $3^{\left(\frac{24d}{p}\right)}N$

13. John's car has x gallons of gas. He drove his car to a gas station to get its tank filled up to its capacity. The car has a fuel efficiency of f miles per gallon. John drove d miles to reach the gas station, where he filled y gallons. If the tank capacity is c gallons, which of the following represents the correct relation between x, y, f, d, and c?

(A) $c - x + df - y = 0$

(B) $c - x - \frac{d}{f} + y = 0$

(C) $c + x - \frac{d}{f} + y = 0$

(D) $c - x + \frac{d}{f} - y = 0$

14. A man wishes that his Facebook page generates 1 million followers over a period of 30 days. He finds that the page generates 20,000 followers for each of the first 10 days and 15,000 followers for each of the next d days ($d < 20$). How many more followers does he need on average per day to reach his target of 1 million followers?

(A) $\frac{1,000\,(800 - 150d)}{30 - d}$

(B) $\frac{1000\,(800 - 15d)}{20 - d}$

(C) $\frac{100\,(800 - 15d)}{30 - d}$

(D) $\frac{100\,(800 - 15d)}{20 - d}$

15. A bodybuilder consumes 1,200 calories every day. However, to participate in a competition, he increases his calorie consumption by 5% every day until the day of the competition. If the competition is d days away, what is the total number of calories he consumes on the day of the competition?

(A) $1,200 \times 1.05d$

(B) $\frac{1200}{1.05^d}$

(C) $\frac{1200}{1.05d}$

(D) $1,200 \times 1.05^d$

16. The cost of manufacturing a shirt is $\$s$. The shirt, originally marked at $\$m$, is sold after giving a $d\%$ discount. If 10% of the gross profit is to be paid in tax, what is the net profit obtained from selling the shirt?

(A) $\$\left(\frac{9dm}{100} - s\right)$

(B) $\$\left(\frac{dm}{100} - s\right)$

(C) $\$\left\{\frac{1}{10}\left(\frac{(100-d)\,m}{100} - s\right)\right\}$

(D) $\$\left\{\frac{9}{10}\left(\frac{(100-d)\,m}{100} - s\right)\right\}$

17. **SPR:**

A farmer sells oranges by packing them in small, identical boxes, which hold 10 oranges each. He packs these boxes in identical cartons that can hold a fixed number of such boxes. The number of cartons needed to pack n oranges is given by the function $f(n) = \frac{n}{300}$. What is the number of boxes that one carton can hold? Assume that no box and no carton are partially filled.

18. The variables $A, B, p, q,$ and $\varnothing$ are related as $A = \frac{12pq}{\cos \varnothing}$ and $B = \frac{p \sin \varnothing}{14q}$. Which of the following is the correct expression of A in terms of $B, q,$ and $\varnothing$?

(A) $A = \frac{1}{B}\left(\frac{7 \sin \varnothing}{6q^2 \cos \varnothing}\right)$

(B) $A = B\left(\frac{6q^2 \sin \varnothing}{7 \cos \varnothing}\right)$

(C) $A = \frac{1}{B}\left(\frac{6q^2}{7 \sin \varnothing \ \cos \varnothing}\right)$

(D) $A = B\left(\frac{168q^2}{\sin \varnothing \ \cos \varnothing}\right)$

19. **SPR:**

If $a + 6b = 15$ and $4a - b = 10$, what is the value of $(a + b)$?

20. If $x = \frac{9x^2}{81}$, where $x \neq 0$, which of the following options is equivalent to x?

(A) $3x$

(B) $9\sqrt{x}$

(C) $\sqrt{3x}$

(D) $3\sqrt{x}$

4.2 Linear Equations

21. The sum of three integers is 40. The largest integer is 3 times the middle integer, and the smallest integer is 23 less than the largest integer. What is the product of the three integers?

(A) 1,102

(B) 972

(C) 672

(D) 292

22. The price of 19 chocolates and 21 pens is \$29, while the price of 21 chocolates and 19 pens is \$31. What is the price of one chocolate?

(A) \$0.50

(B) \$1.00

(C) \$1.25

(D) \$1.50

23. **SPR:**

$\frac{x+6}{3} = \frac{x\sqrt{3}}{\sqrt{27}} + k$ In the equation above, what is the value of k?

24. A piece of chalk, 8 cm long, is broken into three pieces whose lengths, in cm, are integers. The length of the longest piece is equal to the sum of the lengths of the other two pieces. What could be the product of the lengths of the three pieces?

I. 8

II. 12

III. 16

(A) Only I

(B) Only I and II

(C) Only II and III

(D) I, II and III

25. In a fraction $\frac{x}{y}$, where x and y are positive integers, if 4 is added to both the numerator and the denominator, the value of the fraction becomes $\frac{3}{4}$. If the denominator of the original fraction is 12 greater than the numerator, what is the value of $(x + y)$?

(A) 12

(B) 19

(C) 32

(D) 76

26. On January 1, 1999, there were x tons of trash in a landfill that had a capacity of y tons $(y > x)$. Every year since then, it has been found that the amount of trash in the landfill at the end of the year was z tons more than that in the previous year. Which of the following gives the value of the year when the landfill is at full capacity? (Assume that both x and y are multiples of z).

(A) $1998 + \frac{y - x}{z}$

(B) $1999 + \frac{y - x}{z}$

(C) $1999 + \frac{y}{z}$

(D) $2000 + \frac{y - x}{z}$

27. Abe's age is equal to the sum of the ages of his son and a 12-year-old daughter. If Abe's son is older than Abe's daughter, and the average age of Abe and his two children is 18 years, what is Abe's present age?

(A) 12 years

(B) 15 years

(C) 21 years

(D) 27 years

28. Elena purchased brand X pens for \$4.00 apiece and brand Y pens for \$2.80 apiece. If Elena purchased a total of n of these pens for \$42.00, what is the correct expression for the number of pens of brand X purchased?

(A) $\frac{7(15 - n)}{3}$

(B) $\frac{5(15 - 2n)}{3}$

(C) $25 - 3.33n$

(D) $\frac{5(2n - 21)}{3}$

29. In a pro-pool competition, the score, s, of the contestants is determined by the formula $s = \frac{3}{t}(5b - 2m)$, where t represents the total number of hits, b represents the points when a ball is potted, and m represents points when a ball is not potted. If the score of a contestant after 12 hits is 1, which of the following could be the value of b?

(A) 8

(B) 7

(C) 5

(D) 4

30. **SPR:**

A boy was asked to purchase some pens and pencils such that the cost of a pen, $\$p$, and the cost of a pencil, $\$c$, are related by the equation $2p + c = 7$. If the boy wants to buy an even number of pens for a total cost of \$6, what is the minimum cost, in dollars, of each pencil that he buys?

31. If $p = 2x - y$ and $q = 3x + 4y$, what is x in terms of p and q?

(A) $2q - 3p$

(B) $4p + q$

(C) $\frac{1}{11}(4p + q)$

(D) $\frac{1}{11}(2q - 3p)$

32. **SPR:**

$$\frac{7(2d-1)}{11} = \frac{8-(5-d)}{6}$$

What is the value of d, to the nearest hundredth digit, in the above equation?

33. **SPR:**

A store charges a base price of $\$x$ for one bottle of soft drink. A sales tax of a certain percent is applied to the base price, and an untaxed deposit for the bottle is also added. If the total amount, in dollars, paid at the time of purchase for one bottle of soft drink is given by the function $f(x) = 1.05x + 0.35$, what is the sales tax, expressed as a percentage of the base price? While writing the answer, ignore the 'percent' sign.

34. **SPR:**

In a store, the price of a shirt is \$20 and that of a pair of trousers is \$30. If a total of 40 shirts and trousers were sold on a day and the total sales were \$1,100, how many shirts were sold on that day?

35. **SPR:**

$3x + 11y = 34$

$2x - y = 6$

If $x = ky$, what is the value of k?

36. In a store, the price of three laptops and two tablets is \$3,100, and the price of one laptop and four tablets is \$3,700. If the store offers a discount of 10% on both laptops and tablets, what will be the combined price of two laptops and 3 tablets?

(A) \$3,060

(B) \$3,400

(C) \$6,120

(D) \$6,800

37. In a game of darts, a player gets 7.5 points for each dart that hits the bull's eye and -1.5 points for each dart that falls outside the bull's eye. If, out of 12 darts thrown by Bob, x hit the bull's eye, which of the following represents his total score?

(A) $6x + 18$

(B) $9x - 18$

(C) $9x + 18$

(D) $12x - 24$

38. The price of a car is \$13,500. It is expected to depreciate at a constant rate, and after 5 years, the depreciated value of the car is expected to be \$4,850. Which of the following equations correctly represents the value of the car after n years? It is given that $n \leq 5$.

(A) $\$\,(13,500 - 1,442n)$

(B) $\$\,(13,500 - 1,730n)$

(C) $\$\,(13,500 - 2,163n)$

(D) $\$\,(13,500 - 4,850n)$

39. A man wants to reduce his weight and thus follows an exercise regimen for four hours every day. Out of the four hours, he runs for some time and cycles for some time. For every hour he runs, he loses 400 calories, and for every hour he cycles, he loses 300 calories. If r and c represent the number of hours the man runs and cycles, respectively, and C represents the total number of calories he loses by exercising every day, which of the following set of equations represents the above information correctly?

(A) $r + c = 4$
$3r + 4c = 100C$

(B) $r + c = C$
$300r + 400c = 4$

(C) $r + c = 4$
$300r + 400c = C$

(D) $r + c = 4$
$400r + 300c = C$

4.3 Inequalities

40. In a shop, chocolates are sold in boxes with 20 pieces each. Joe has 20 pieces of chocolate, and he needs to have a minimum of 140 pieces. Which of the following shows all possible values for the number of boxes, x, that Joe needs to buy to meet his minimum requirement?

(A) $x \geq 3$

(B) $x \geq 4$

(C) $x \geq 5$

(D) $x \geq 6$

41. Gift boxes, to be distributed among school children, contain pencils and erasers. Each gift box can hold a maximum of 12 items. If each box must have at least 3 more pencils than the number of erasers, what is the maximum number of erasers in a gift box?

(A) 2

(B) 3

(C) 4

(D) 5

42. **SPR:**

David is planning to borrow books from a library. The charge for borrowing each book is \$5. However, if he takes a membership, then he can borrow each book for \$1 and pay a one-time membership fee of \$125. What is the minimum number of books David should borrow if taking the membership is a comparatively cost-effective option?

43. A contractor needs to color a wall with dimensions of 12 feet by 8 feet. The wall has a glass door with dimensions of 3 feet by 5 feet. The wall also has a glass window with an area of between 5 and 10 square feet. If the contractor charges \$0.60 per square feet, which of the following inequalities shows all possible values for p, the amount in dollars required to paint the wall?

(A) $71 \leq p \leq 76$

(B) $45.70 \leq p \leq 46.70$

(C) $42.60 \leq p \leq 45.60$

(D) $40 \leq p \leq 41.20$

44. If $-1 < h < 0$, which of the following has the greatest value?

(A) $1 - h$

(B) $1 + h$

(C) $1 + h^2$

(D) $1 - \frac{1}{h}$

45. Amy is planning to organize a picnic. The one-time charge for booking the venue is \$120, and charges per guest are \$24. If n represents the number of people Amy is planning to take along on the picnic, and she has a budget of \$350, what is the maximum value of n?

(A) 7

(B) 8

(C) 9

(D) 10

46. Ann wants to score an average of at least 70 in her five tests. She obtained an average score of 65 in the first four tests. If s is her score in the final test, which of the inequalities below shows all values of s that she could score to achieve the required average? All tests are scored out of 100.

(A) $s \leq 100$

(B) $s \leq 90$

(C) $90 < s < 100$

(D) $90 \leq s \leq 100$

47. SPR:

If $-\frac{5}{3} \leq 2a + 3 \leq \frac{7}{2}$, how many integer values are possible for $\left(\frac{a}{2} + 1\right)$?

48. SPR:

$$2x - \frac{7}{4} < \frac{11}{12}$$
$$-3x + \frac{1}{5} < \frac{1}{20}$$

What is the integer value of x which satisfies the above inequalities?

49. SPR:

A zoo has two elephants, one weighing 8,560 pounds and the other weighing 9,450 pounds. If elephants per day drink 2 to 3 liters of water for every 100 pounds of body weight, what is the difference, to the nearest integer, between the maximum and minimum amount of total water, w, that the two elephants drink in a day?

50. John follows a weight loss program for a month; he exercises daily for at least an hour up to a maximum of two hours every day. For each minute he exercises, he loses at least 15 calories and at most 20 calories. If he consumes at least 600 calories, up to a maximum of 800 calories in a day, which of the following inequalities represents the total number of calories, c, that John loses in a day?

(A) $100 \leq c \leq 1,800$

(B) $100 \leq c \leq 1,600$

(C) $300 \leq c \leq 1,600$

(D) $300 \leq c \leq 1,800$

51.

Profession	Hourly wage
Accountant	\$16
Journalist	\$20
Engineer	\$25
Architect	\$35

The table above shows the hourly wages for four professionals. The average working hours are 32 to 38 hours per week. If the difference between the weekly wages of any two people is \$$x$, which of the following inequalities is correct?

(A) $x \geq 512$

(B) $x \leq 818$

(C) $512 \leq x \leq 818$

(D) $512 \leq x \leq 1,330$

52. SPR:

A city-bus service charges \$17 for a trip from the airport of a city to the city center. However, there is a deal for people who frequently use the given route. People can purchase a discount card for \$350, after which they have to pay only \$9.50 per trip. How many minimum trips, t, must a person make for the discount card to be a better deal?

53. A cab service charges \$5 for the first $\frac{1}{2}$ mile. Thereafter, it charges \$1.20 for each additional $\frac{1}{3}$ miles. If a rider doesn't want to pay more than \$20 and m represents the number of miles he rode, which of the following inequalities, when solved, would give the maximum possible value of m?

(A) $3.6m + 3.2 \le 20$

(B) $3.6m + 5 \le 20$

(C) $1.2m + 4.4 < 20$

(D) $1.2m + 5 \le 20$

54. In the drilling process, a hole has to be drilled in a number of steel plates. The required radius of the hole is 2.5 cm. However, there is an error of up to 0.1 mm in the radius in the drilling process. If d represents the diameter of the hole in the steel plates, which of the following represents all possible values of d?

(A) $|d - 5| < 0.02$

(B) $|d - 5| \le 0.02$

(C) $|d - 5| \le 0.2$

(D) $|d - 0.02| < 5$

55. A man mixes two solutions of milk in water, one with 20% milk and the other with 60% milk, to form a 50-liter solution with 50% milk. If he took x liters of the first solution and y liters of the second solution, solving which of the following systems of equations yields the volumes of the two solutions that he took?

(A) $x + y = 50$
$x + 3y = \frac{5}{3}(x + y)$

(B) $x + y = 50$
$x + 3y = \frac{2}{5}(x + y)$

(C) $x + y = 50$
$x + 3y = \frac{5}{2}(x + y)$

(D) $x + y = 50$
$x + 3y = \frac{1}{2}(x + y)$

56. Several desks and chairs are to be purchased for a classroom. The number of desks and chairs should be at least 15 and 25, respectively. Each desk costs \$120, and each chair costs \$40. If the budget is not more than \$4,000, which of the following sets of inequalities can be used to determine the number of desks, d, and chairs, c, that can be purchased?

(A) $d > 25$
$c > 15$
$3d + c \le 100$

(B) $d > 15$
$c > 25$
$3d + c = 100$

(C) $d \geq 25$
$c \geq 15$
$d + 3c \leq 100$

(D) $d \geq 15$
$c \geq 25$
$3d + c \leq 100$

4.4 Absolute Value

57. If $|x| = |y|$, which of the following could be true?

(A) $x < y$
(B) $x = y$
(C) $x > y$
(D) All the above

58. **SPR:**

What is the greatest possible solution for $|x| = |2x - 3|$?

59. If $\left|x - \frac{4}{5}\right| \leq \frac{11}{5}$, what is the greatest possible value of $|x - 1|$?

(A) 1.5
(B) 2
(C) 2.4
(D) 3

60. In the drilling process, a hole has to be drilled in a number of steel plates. It was found that the maximum diameter of a hole was 12.6 cm, and the minimum diameter was 7.4 cm. If d represents the diameter of the hole in the steel plates, which of the following represents all possible values of d?

(A) $|d - 10| \leq 2.6$
(B) $|d - 10| < 2.6$
(C) $|d - 10| \geq 2.6$
(D) $|d - 10| > 2.6$

4.5 Functions

61. If $f(p, q) = \frac{p}{p + q}$ and $g(p, q) = \frac{p + q}{q}$, which of the following options is correct?

(A) $\frac{f(p, q)}{g(p, q)} = \frac{q}{p}$

(B) $\dfrac{f(p,\ q)}{g(p,\ q)} = \dfrac{pq}{p+q}$

(C) $f(p,\ q) \times g(p,\ q) = p^2q^2$

(D) $f(p,\ q) \times g(p,\ q) = \dfrac{p}{q}$

62. $f(x)$ is a function such that $f(2x-1) = x^2 + 3x$. What is the value of $f(5)$?

(A) 3

(B) 15

(C) 18

(D) 25

63. The function f is defined by $f(x) = -\dfrac{1}{x}$ for all non-zero numbers x. If $f(a) = -\dfrac{1}{2}$ and $f(ab) = \dfrac{1}{6}$, then what is the value of b?

(A) 3

(B) $\dfrac{1}{3}$

(C) $-\dfrac{1}{3}$

(D) -3

64. Which of the following functions $f(x)$ is equivalent to $f(1-x)$ for all x?

(A) $f(x) = 1 - x$

(B) $f(x) = 1 - x^2$

(C) $f(x) = x^2 - (1-x)^2$

(D) $f(x) = |x(1-x)|$

65. The function $f(x)$ is such that $\frac{1}{2}f(x) = f\left(\dfrac{x}{2}\right)$. Which of the following could be possible expressions for $f(x)$?

(A) $f(x) = 2x - 2$

(B) $f(x) = 3x + 1$

(C) $f(x) = x^2$

(D) $f(x) = (x+2)^2 - (x-2)^2$

66. The gravitational force, F, between two bodies having masses S and s, separated by a distance d is given by the equation $F = \dfrac{GSs}{d^2}$, where G is the gravitational constant. Which of the following gives the value of d in terms of G, S, s and F?

(A) $\left(\dfrac{GSs}{F}\right)^2$

(B) $\dfrac{FSs}{G}$

(C) $\sqrt{\dfrac{GSs}{F}}$

(D) $\sqrt{\frac{FSs}{G}}$

67. SPR:

Joe wants to convert x dollars to yen at a bank. The bank offers a particular exchange rate and charges a fee equal to 2% of the amount converted. If the function $f(x) = 0.98 \times 100x$ represents the amount of money received by Joe in yen, how many cents make a yen?

68. SPR:

The number of 'rewards points' received for making purchases worth x dollars using a credit card is given by the function $R(x) = \frac{x}{100} + 1$. If the purchases using the credit card increase by \$1,200, what is the increase in the number of reward points?

69. SPR:

x	$f(x)$
1	240
2	480
3	960

If the function $f(x)$ is defined as $f(x) = f(0) \times k^x$, what is the value of $f(0) + k$?

70. Joe puts \$100 in a bank and assumes that the rate of interest will be 3%, compounded semi-annually. However, after a year, he finds that the total amount of money is \$1 less than what it should have been. If the function $f(t) = 100\left(1 + \frac{r}{200}\right)^{2t}$ represents the total amount of money Joe would have after t years, and r represents the rate of interest, what is the value of r?

(A) 0.5%

(B) 1%

(C) 1.5%

(D) 2%

71. The cost, $c(n)$, in dollars, of manufacturing n LED bulbs is given as $c(n) = 12.8n + 132$. Such LED bulbs are sold at \$16 apiece. Which of the following inequalities gives the number of bulbs that have to be sold in order to make a profit?

(A) $n \geq 41$

(B) $n \geq 42$

(C) $n > 42$

(D) $n > 43$

72. The value of a commodity depreciates by \$10 every year. It was valued \$180 in 2009. M, the value of the commodity in any year, is represented by the equation $M = 230 - 10t$, where t is the number of years since it was produced. Which of the following explains the meaning of the number 230 in the equation?

(A) The value of the commodity in 2004

(B) The value of the commodity in 2006

(C) The value of the commodity in 2011

(D) The value of the commodity in 2014

73. A teacher wanted to purchase some pencils to distribute among the children in his class. Pencils can be purchased in boxes such that each box contains a definite number of pencils. He found that there were already a few pencils in stock with him. The number of pencils with him after the purchase was made is given by the function $p(b) = 10b + 40$, where b represents the number of boxes of pencils purchased. In the above function, what does the number '10' denote?

(A) The number of boxes of pencils already with the teacher

(B) The number of pencils already with the teacher

(C) The number of pencils in each box

(D) The number of boxes with the teacher after the purchase

4.6 Interpreting Equations

74. If one of the solutions of $x^2 - px + 12 = 0$ is $x = \frac{3}{2}$, what is the value of p?

(A) $\frac{3}{2}$

(B) 8

(C) $\frac{19}{2}$

(D) $\frac{57}{4}$

75. For what value of n, the equation $4(2x - 1) + 3(x - 2) = n(x + 2) - 3(x + 1)$ has no possible solution?

(A) -4

(B) $-\frac{7}{2}$

(C) $\frac{7}{2}$

(D) 14

76. $\frac{8d + 10}{4} = \frac{d\left(3.5^2 - 1.5^2\right) + x}{5}$

If the above equation has infinite values of d, what is the value of x?

(A) 2.5

(B) 12.5

(C) 50

(D) No such value of x exists

77. **SPR:**

$3x - (12 - x) = c(2x - 5) - p$

If the equation above has no possible solution, what is NOT a possible value of p?

78. $Ax + \frac{1}{2}y = -6$

$-\frac{1}{3}x - 2y = 24$

If the above set of equations has infinite possible solutions, what is the value of A?

79. **SPR:**

x	y
−5	11
−1	5
1	2
7	k

If the values in the table above represent a linear relationship, what is the value of k^2?

4.7 Graphic Equations and Inequalities

80. $Ax + 4y = 6$

$Bx + 2Ay = 4$

The graphs of the above system of equations intersect at $(2, -1)$, what is the value of $(A + B)$?

81. $3x - (5 - 4x) \square x + 3(x + 1) + 1$

Which of the following symbols given below, when placed in the square box above, satisfies the inequality $x > 3$?

(A) $>$

(B) $\geq$

(C) $<$

(D) $\leq$

82.

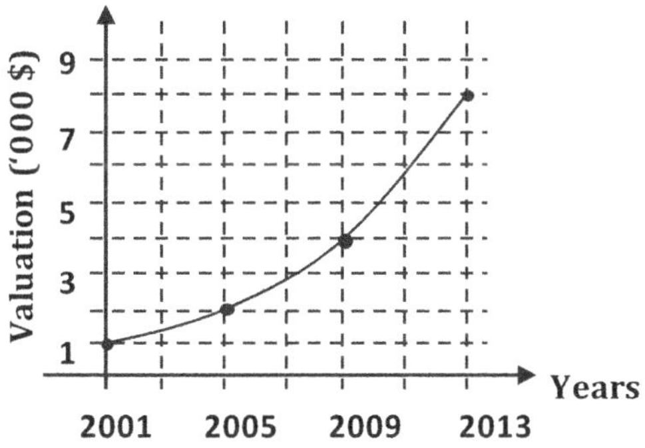

The graph above shows the value of an antique over the years. If t represents the number of years starting from 2001, which of the following best describes the value of the antique over the years?

(A) $\$\left(1,000 \times 2^{\frac{4}{t}}\right)$

(B) $\$ \left(1,000 \times 2^{4t}\right)$

(C) $\$ \left(1,000 \times 2^{t}\right)$

(D) $\$ \left(1,000 \times 2^{\frac{t}{4}}\right)$

83.

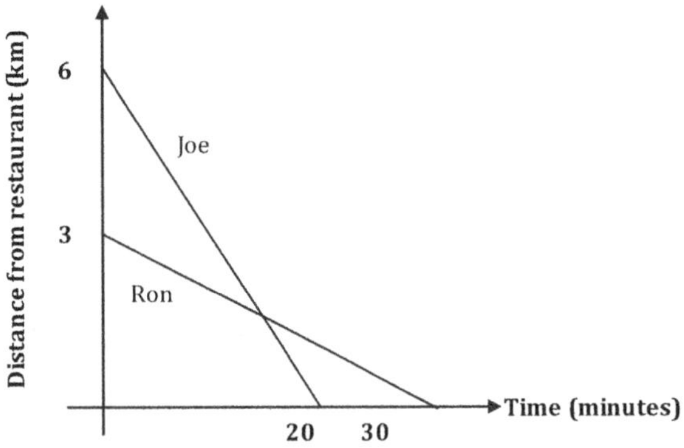

Joe and Ron plan to meet at a restaurant. The graph above represents their journey from their homes to the restaurant. Based on the above graph, which of the following statements is true?

(A) Joe traveled half as fast as Ron did but reached before Ron.

(B) Joe traveled one-third as fast as Ron did but arrived late since he stayed farther from the restaurant.

(C) Joe traveled three times as fast as Ron did and before Ron, despite staying farther from the restaurant.

(D) Both Joe and Ron traveled at the same speeds, but Joe arrived before Ron.

Chapter 5

Algebra Answer key

(1) $\frac{7}{2}$ or 3.5

(2) B

(3) D

(4) D

(5) C

(6) C

(7) C

(8) C

(9) B

(10) C

(11) B

(12) D

(13) D

(14) B

(15) D

(16) D

(17) 30

(18) D

(19) 5

(20) D

(21) B

(22) C

(23) 2

(24) C

(25) D

(26) A

(27) D

(28) A

(29) D

(30) 1

(31) C

(32) 0.96

(33) 5

(34) 10

(35) 2

(36) A

(37) B

(38) B

(39) D

(40) D

(41) C

(42) 32

(43) C

(44) D

(45) B

(46) D

(47) 2

(48) 1

(49) 180

(50) A

(51) A

(52) 47

(53) A

(54) D

(55) B

(56) C

(57) D

(58) 3

(59) C

(60) A

(61) D

(62) C

(63) D

(64) D

(65) D

(66) C

(67) 1

(68) 12

(69) 122

(70) D

(71) B

(72) A

(73) C

(74) C

(75) D

(76) B

(77) 2

(78) $\frac{1}{12}$

(79) 49

(80) 12

(81) A

(82) D

(83) C

Chapter 6

Algebra: Solutions to Practice Questions

6.1 Algebraic Expressions

1. We have:

$\left(\frac{5}{x+1}\right)(x^2-1) = \left(3 - \frac{1}{x+1}\right)(x+1)$

$=> \left(\frac{5}{x+1}\right)(x+1)(x-1) = \left(3 - \frac{1}{x+1}\right)(x+1)$; splitting (x^2-1) into $(x+1)(x-1)$

$=> 5(x-1) = 3(x+1) - \left(\frac{1}{x+1}\right)(x+1)$; cancelling $(x-1)$

$=> 5x - 5 = 3x + 3 - 1$; cancelling $(x+1)$

$=> 2x = 7$

$=> x = \frac{7}{2} = 3.5$

The correct answer is $\frac{7}{2}$ or 3.5.

2. We have:

$\frac{x^5 - x^3}{x^4 + x^3}$

$= \frac{x^3(x^2-1)}{x^3(x+1)}$

Cancelling x^3 and breaking $x^2 - 1$ as $(x+1)(x-1)$

$= \frac{(x+1)(x-1)}{x+1}$

$= x - 1$

The correct answer is option B.

3. We have:

$\frac{3}{x+1} = \frac{15}{2w-3}$

$=> 3\left(\frac{1}{x+1}\right) = 15\left(\frac{1}{2w-3}\right)$

$=> \left(\frac{1}{x+1}\right) = \frac{15}{3}\left(\frac{1}{2w-3}\right)$

$=> x + 1 = \frac{1}{5}(2w-3)$

$=> 5(x+1) = 2w - 3$

$=> 5x + 5 = 2w - 3$

$=> 2w = 5x + 8$

$=> w = \frac{5}{2}x + 4$

The correct answer is option D.

Alternate approach:

Let $x = 2$:

$\frac{3}{2+1} = \frac{15}{2w-3}$

$=> 2w - 3 = 15$

$=> w = 9$

Plugging in $x = 2$ in each option, we find that only option D results in $w = 9$.

4. Let the required quantity be k.

Thus, according to the problem, we have:

$k + \frac{x+1}{x} = k\left(\frac{x-1}{x}\right)$

$=> k\left(\frac{x-1}{x}\right) - k = \frac{x+1}{x}$

$=> k\left(\frac{x-1}{x} - 1\right) = \frac{x+1}{x}$

$=> k\left(-\frac{1}{x}\right) = \frac{x+1}{x}$

$=> k = -x\left(\frac{x+1}{x}\right)$

$=> k = -(x+1)$

The correct answer is option D.

5. We know that the store purchases goods in boxes that weigh either 40 pounds or 60 pounds each.

Also, x represents the number of 40-pound boxes and y represents the number of 65-pound boxes.

Thus, the total weight of all boxes = $(40x + 60y)$ pounds.

Since the total weight of goods purchased in all the boxes combined is 1200 pounds, we have:

$40x + 60y = 1,200$

$=> 40x = 1,200 - 60y$

$=> x = \frac{1,200 - 60y}{40} = 30 - \frac{3}{2}y$

The correct answer is option C.

6. We have:

$x = 1 + \sqrt{3}$

$=> x - 1 = \sqrt{3}$

Squaring both sides:

$(x-1)^2 = (\sqrt{3})^2$

$=> x^2 - 2x + 1 = 3$

$=> x^2 - 2x - 2 = 0$

The correct answer is option C.

Alternate approach:

Substituting $x = 1 + \sqrt{3}$ in the expression, we have:

$x^2 - 2x - 2$

$= (1 + \sqrt{3})^2 - 2(1 + \sqrt{3}) - 2$

$= 1 + 2\sqrt{3} + 3 - 2 - 2\sqrt{3} - 2$

$= 0$

7. $\frac{a}{100} \times (a - 2b) + \frac{b}{100} \times b = 0.64$

$=> \frac{a^2 - 2ab}{100} + \frac{b^2}{100} = 0.64$

$=> \frac{a^2 - 2ab + b^2}{100} = 0.64$

$=> a^2 - 2ab + b^2 = 64$

$=> (a - b)^2 = 64$

$=> a - b = \pm 8$

$=> a = b \pm 8$

$=> a = b + 8 => a - b = 8$

The correct answer is option C.

8. We have:

$\frac{1}{\sqrt{a+1} + \sqrt{a-1}}$

$= \frac{\sqrt{a+1} - \sqrt{a-1}}{(\sqrt{a+1} + \sqrt{a-1})(\sqrt{a+1} - \sqrt{a-1})}$

$= \frac{\sqrt{a+1} - \sqrt{a-1}}{(\sqrt{a+1})^2 - (\sqrt{a-1})^2}$

$= \frac{\sqrt{a+1} - \sqrt{a-1}}{(a+1) - (a-1)}$

$= \frac{\sqrt{a+1} - \sqrt{a-1}}{2}$

The correct answers is option C.

9. We have:

$P^2Q = \frac{3R}{\sqrt{2S}}$

Cross-multiplying, we have:

$=> \sqrt{2S} = \frac{3R}{P^2Q}$

Squaring both sides:

$=> 2S = \left(\frac{3R}{P^2Q}\right)^2$

$=> 2S = \frac{9R^2}{P^4Q^2}$

$=> S = \frac{9R^2}{2P^4Q^2}$

The correct answer is option B.

10. Initial deposit = \$1,200

Rent per month = \$200

Thus, rent per year = \$200 × 12 = \$2,400

Thus, total rent for t years = $\$2,400t$

Semi-annual maintenance charge = \$60

Thus, annual maintenance charge = $\$60 \times 2 = \$120$

Thus, maintenance charge for t years = $\$120t$

Thus, total amount = $\$\,(1,200 + 2,400t + 120t) = \$\,(1,200 + 2,520t)$

The correct answer is option C.

11. We know that the mass of the substance becomes half its initial value in every 30 minutes.

Thus, in t minutes, the number of times the mass of the substance will become half = $\frac{t}{30}$

Thus, the final weight of the substance after t minutes

$$= 10.6 \times \left\{\left(\frac{1}{2}\right) \times \left(\frac{1}{2}\right) \times \left(\frac{1}{2}\right) \ldots\ldots\ldots \left(\frac{t}{30} \text{ times}\right)\right\}$$

$$= 10.6 \times \left(\frac{1}{2}\right)^{\frac{t}{30}}$$

The correct answer is option B.

12. We know that the number of bacteria triples in every p hours.

Number of hours in d days = $24d$ hours.

Thus, number of times the number of bacteria will triple = $\left(\frac{24d}{p}\right)$ times.

Thus, the final number of bacteria

$$= N \times 3 \times 3 \times 3 \ldots \left(\frac{24d}{p}\right) \text{ times} = 3^{\left(\frac{24d}{p}\right)}N$$

The correct answer is option D.

13. Number of gallons already present in the tank = x gallons.

Distance travelled = d miles.

Fuel efficiency = f miles per gallon.

Thus, fuel consumed to travel d miles = $\frac{d}{f}$ gallons.

Thus, fuel remaining = $\left(x - \frac{d}{f}\right)$ gallons.

Amount of fuel filled = y gallons.

Thus, final amount of fuel = $\left(x - \frac{d}{f} + y\right)$ gallons.

Since John filled his tank to full capacity and the tank can hold c gallons, we have:

$$x - \frac{d}{f} + y = c$$

$$=> c - x + \frac{d}{f} - y = 0$$

The correct answer is option D.

14. Total number of followers required in 30 days = 1 million = 1,000,000

Number of followers in the first 10 days = $20,000 \times 10 = 200,000$

Number of followers in the next d days = 15,000 × d = $15,000d$

Number of days left = 30 - 10 - d = $20 - d$

Number of followers still required = 1,000,000 - 200,000 - $15,000d$ = $800,000 - 15,000d$

Thus, number of followers required on an average per day

$$= \frac{800,000 - 15,000d}{20 - d}$$

$$= \frac{1,000\,(800 - 15d)}{20 - d}$$

The correct answer is option B.

15. Number of calories consumed on the first day = $1,200$

Number of calories consumed on the second day = $1,200 \times 1.05$

Number of calories consumed on the third day = $1,200 \times 1.05^2$

Continuing in the above manner, we have:

Number of calories consumed on the dth day (the day of the competition) = $1,200 \times 1.05^d$

The correct answer is option D.

Alternate approach:

Assuming $d = 2$, we have:

For the second day, calorie consumption should be = $1,200 \times 1.05 \times 1.05$

This is satisfied only by option D.

16. Selling price of the shirt = \$ $\{(100 - d)\,\%\text{ of } m\}$ = \$ $\left\{\frac{(100 - d)\,m}{100}\right\}$

Cost of manufacturing the shirt = \$$s$

Thus, the gross profit = \$ $\left\{\frac{(100 - d)\,m}{100} - s\right\}$

Since 10% of the profit is paid as tax, the net profit obtained

$$= \$\left\{(100 - 10)\,\%\text{ of }\left(\frac{(100 - d)\,m}{100} - s\right)\right\}$$

$$= \$\left\{\frac{9}{10}\left(\frac{(100 - d)\,m}{100} - s\right)\right\}$$

The correct answer is option D.

Alternate approach:

Let us assume some values:

$s = \$100, m = \$150, d = 20\%$:

Selling price of the shirt = 80% of \$150 = \$120

Profit = \$120 - \$100 = \$20

Profit after tax = 90% of \$20 = \$18

Plugging in the above values in the options, we find that only option D satisfies.

17. Total number of oranges = n

Number of oranges in each box = 10

Thus, number of boxes required = $\frac{n}{10}$

Let the number of boxes in each carton be b

Thus, the number of cartons required for $\left(\frac{n}{10}\right)$ boxes

$= \frac{\left(\frac{n}{10}\right)}{b} = \frac{n}{10b}$

Since the number of cartons needed to pack n oranges is given by the function $f(n) = \frac{n}{300}$, we have:

$\frac{n}{300} = \frac{n}{10b}$

$\Rightarrow b = 30$

The correct answer is 30.

18. We have:

$A = \frac{12pq}{\cos \varnothing} \ldots$ (i)

$B = \frac{p \sin \varnothing}{14q} \ldots$ (ii)

We need the expression of A in terms of B, q and $\varnothing$. Thus, we need to eliminate p from (i) and (ii). We observe that p is in the numerator in both (i) and (ii).

Thus, to eliminate p, we need to divide (i) by (ii):

$\frac{A}{B} = \left(\frac{12pq}{\cos \varnothing}\right) \div \left(\frac{p \sin \varnothing}{14q}\right)$

$\Rightarrow \frac{A}{B} = \left(\frac{12pq}{\cos \varnothing}\right) \times \left(\frac{14q}{p \sin \varnothing}\right)$

$\Rightarrow \frac{A}{B} = \frac{12 \times 14 \times q^2}{\sin \varnothing \times \cos \varnothing}$

$\Rightarrow A = B\left(\frac{168q^2}{\sin \varnothing \cos \varnothing}\right)$

The correct answer is option D.

19. We have:

$a + 6b = 15 \ldots$ (i)

$4a - b = 10 \ldots$ (ii)

Multiplying (ii) by 6 and adding to (i), we have:

$a(1 + 4 \times 6) = 15 + 10 \times 6$

$\Rightarrow 25a = 75$

$\Rightarrow a = 3$

Substituting the above in (i), we have:

$3 + 6b = 15$

$\Rightarrow b = 2$

Thus, we have:

$a + b = 3 + 2 = 5$

The correct answer is 5.

Alternate approach:

These types of questions can often be solved by observing the coefficients of x and y.

Adding the two given equations, we have:

$a(1+4) + b(6-1) = 15 + 10$

$\Rightarrow 5a + 5b = 25$

$\Rightarrow a + b = 5$

20. We have:

$x = \frac{9x^2}{81}$

Since $x \neq 0$, we can cancel x from both sides:

$1 = \frac{9x}{81}$

$\Rightarrow x = 9$

Working with the options, we have:

(A) $3x = 3 \times 9 = 27 \neq x$ - Does not satisfy

(B) $9\sqrt{x} = 9 \times \sqrt{9} = 27 \neq x$ - Does not satisfy

(C) $\sqrt{3x} = \sqrt{3 \times 9} = 3\sqrt{3} \neq x$ - Does not satisfy

(D) $3\sqrt{x} = 3\sqrt{9} = 9 = x$ - Satisfies

The correct answer is option D.

6.2 Linear Equations

21. Let the middle integer be x.

Since the largest integer is 3 times the middle integer, we have:

The largest integer = $3x$.

Since the smallest integer is 23 less than the largest integer, we have:

The smallest integer = $(3x - 23)$.

Since the sum of the integers is 40, we have:

$(3x - 23) + x + 3x = 40$

$\Rightarrow x = 9$

$\Rightarrow$ The three integers are: $3 \times 9 - 23 = 4$, 9 and $3 \times 9 = 27$.

Thus, the product of the integers

$= 4 \times 9 \times 27$

$= 4 \times 243$

$= 972$

The correct answer is option B.

22. Let the price of 1 chocolate and 1 pen be $\$x$ and $\$y$, respectively.

Thus, we have:

$19x + 21y = 29 \ldots$ (i)

$21x + 19y = 31 \ldots$ (ii)

Adding (i) and (ii):

$40x + 40y = 60$

$=> x + y = \frac{3}{2} \ldots$ (iii)

Subtracting (i) from (ii):

$2x - 2y = 2$

$=> x - y = 1 \ldots$ (iv)

Adding (iii) and (iv):

$2x = 1 + \frac{3}{2} = \frac{5}{2}$

$=> x = \frac{5}{4} = 1.25$

Thus, the price of a chocolate is \$1.25

The correct answer is option C.

23. We have:

$\frac{x+6}{3} = \frac{x\sqrt{3}}{\sqrt{27}} + k$

$=> \frac{x+6}{3} = \frac{x\sqrt{3}}{3\sqrt{3}} + k$

$=> \frac{x+6}{3} = \frac{x}{3} + k$

Multiplying throughout by 3, we have:

$=> x + 6 = x + 3k$

$=> k = 2$

The correct answer is 2.

Alternate approach:

Since the value of k has to be determined (since it is a grid-in question, there must be a numeric answer), we can merely substitute any suitable value of x.

Let $x = -6$: We chose $x = -6$ so that LHS would become 0, easing the calculation.

$=> 0 = -\frac{6\sqrt{3}}{3\sqrt{3}} + k$

$=> 0 = -2 + k$

$=> k = 2$

24. Let the lengths of the three pieces, in cm, be a, b and c, where a is greatest.

Thus, we have:

$a + b + c = 8 \ldots$ (i)

We know that:

$a = b + c$

Thus, from (i):

$a + (b + c) = 8$

$=> a + a = 8$

$=> a = 4$

$=> b + c = 4$

$=> b = 3, c = 1 => b = 2, c = 2$

For => $b = 3, c = 1$, the product of the lengths

$= a \times b \times c$

$= 4 \times 3 \times 1$

$= 12$ – option II is correct

Similarly, for => $b = 2, c = 2$, the product of the lengths

$= a \times b \times c$

$= 4 \times 2 \times 2$

$= 16$ – option III is correct

The correct answer is C.

25. We have the fraction as $\frac{x}{y}$.

Since on adding 4 to both x and y, the fraction becomes $\frac{3}{4}$, we have:

$\frac{x+4}{y+4} = \frac{3}{4}$

$=> 4x + 16 = 3y + 12$

$=> 4x = 3y - 4 \ldots$ (i)

Also, the denominator of the fraction $\frac{x}{y}$ is 12 greater than the numerator.

Thus, we have:

$y = x + 12 \ldots$ (ii)

Substituting the value of y in (i):

$4x = 3(x + 12) - 4$

$=> x = 32$

$=> y = 32 + 12 = 44$

$=> x + y = 32 + 44 = 76$

The correct answer is option D.

26. Capacity of the landfill = y tons

Initial quantity of trash present = x tons

Thus, quantity of trash that can be accommodated = $(y - x)$ tons

Quantity of trash added each year = z tons

Thus, number of years it would take to fill the landfill to capacity = $\left(\frac{y-x}{z}\right)$ years

Since January 1, 1999 is the beginning of the year, the year when the landfill will be at full capacity would be $\left\{\left(\frac{y-x}{z}\right) - 1\right\}$ years after 1999, ie. December 31 of the year $\left\{1998 + \left(\frac{y-x}{z}\right)\right\}$.

Note: Since both x and y are multiples of z, the value $\left(\frac{y-x}{z}\right)$ would represent an integer number of years.

For example:

Let $x = 20,\ y = 50$ and $z = 10$

Thus, the number of years required to fill the landfill = $\frac{50-20}{10} = 3$ years.

Let us see the year when the landfill would be full:

- January 1, 1999 to December 31, 1999 – 1 year
- January 1, 2000 to December 31, 2000 – 1 year
- January 1, 2001 to December 31, 2001 – 1 year

Thus, on December 31, 2001, the landfill would be completely filled.

Thus, the required year is 1999 + 3 – 1 = 2001

The correct answer is option A.

27. We know that the average age of the Abe and his two children is 18 years.

Thus, the present total age of Abe and his two children = $18 \times 3 = 54$ years.

Since Abe's age is equal to the sum of the ages of his two children, we have:

Abe's age + Sum of ages of his two children = 54 years

=> 2× (Abe's age) = 54 years

=> Abe's age = $\frac{54}{2} = 27$ years.

The correct answer is option D.

Alternate approach:

Let the present age of Abe's son = x years.

Present age of Abe's daughter = 12 years.

Thus, Abe's present age = $(x + 12)$ years.

Thus, total present age of Abe and his two children = $(2x + 24)$ years.

Since their average age is 18 years, their total age = $18 \times 3 = 54$ years.

Thus, we have:

$2x + 24 = 54 \Rightarrow x = 15$

Thus, Abe's present age = $(x + 12) = 27$ years.

28. Let the number of brand X pens be x and the number of brand Y pens be y.

Thus, we have:

$x + y = n$

$\Rightarrow y = n - x \ldots$ (i)

Since brand X pens cost \$4.00 apiece and brand Y pens cost \$2.80 apiece, and the total cost was \$42.00, we have:

$4x + 2.8y = 42$

$\Rightarrow 40x + 28y = 420$ (Multiplying the above equation by 10 to avoid decimals)

$\Rightarrow 10x + 7y = 105 \ldots$ (ii) (Dividing the above equation by 4)

Substituting the value of y from (i) in equation (ii) above, we have:

$10x + 7(n - x) = 105$

$\Rightarrow 3x = 105 - 7n$

$\Rightarrow x = \frac{(105 - 7n)}{3}$

$\Rightarrow x = \frac{7(15 - n)}{3} \ldots$

The correct answer is option A.

29. We have:

$s = \frac{3}{t}(5b - 2m)$

Since $t = 12$ and $s = 1$, we have:

$1 = \frac{3}{12}(5b - 2m)$

$1 = \frac{1}{4}(5b - 2m)$

$=> 5b - 2m = 4$

Since $t = b + m$, we have $m = t - b => m = 12 - b$

$=> 5b - 2(12 - b) = 4$

$=> 5b - 24 + 2b = 4$

$=> 7b = 28$

$b = 4$

The correct answer is option D.

30. Total cost of pens = \$6

Since he needs to buy an even number of pens, he must buy either 2 or 4 or 6 etc. number of pens.

Since the cost of a pencil has to be minimized (given by $c = 7 - 2p$), the cost of a pen must be the maximum possible. This would happen only if he purchased the least possible number of pens, i.e. 2 pens.

Thus, the cost of each pen = \$ $\left(\frac{6}{2}\right)$ = \$3 (Maximum possible cost price of a pen)

$=> p = 3$

We know that:

$c = 7 - 2p$

$c = 7 - 2(3) = 1$

Thus, the cost of each pencil = \$1.

The correct answer is 1.

31. We have:

$p = 2x - y \ldots$ (i)

$q = 3x + 4y \ldots$ (ii)

Since we need to determine x, we need to eliminate y.

Multiplying (i) by 4:

$4p = 8x - 4y \ldots$ (iii)

Adding (iii) to (ii), we have

$4p + q = 11x$

$=> x = \frac{1}{11}(4p + q)$

The correct answer is option C.

32. We have:

$$\frac{7(2d-1)}{11} = \frac{8-(5-d)}{6}$$

Cross-multiplying, we have:

$6 \times 7(2d-1) = 11(8-(5-d))$

$=> 84d - 42 = 11(3+d)$

$=> 84 - 42d = 33 + 11d$

$=> 53d = 51$

$=> d = \frac{51}{53} = 0.962 \approx= 0.96$ (to the nearest hundredth digit)

The correct answer is 0.96

33. Base price = \$$x$

Let the tax as a percent of the base price be t%.

Since the tax is applicable on the base price, the total price

= \$ (x + t% of x)

$= \$\left(x + \frac{xt}{100}\right)$

$= \$\left\{x\left(1 + \frac{t}{100}\right)\right\}$

Since an untaxed amount, say \$$k$ is also added, the final price of the bottle of soft drink = $\$\left\{x\left(1 + \frac{t}{100}\right) + k\right\}$

Since the final price is given as \$ (1.05x + 0.35), we have:

$1.05x + 0.35 = x\left(1 + \frac{t}{100}\right) + k$

Thus, comparing the LHS and RHS, we have:

$1 + \frac{t}{100} = 1.05 => t = 5 \ldots$ (i)

$k = 0.35 \ldots$ (ii)

Thus, the tax applicable is 5% of the base price.

The correct answer is 5.

34. Let the number of shirts be x and the number of trousers be y.

Since 40 items were sold, we have:

$x + y = 40 \ldots$ (i)

Since the total sales were \$1,100, we have:

$20x + 30y = 1,100 \ldots$ (ii)

Multiplying (i) by 30 and subtracting (ii), we have:

$(30-20)x = 1,200 - 1,100$

$=> x = 10$

The correct answer is 10.

35. We have:

$3x + 11y = 34 \ldots$ (i)

$2x - y = 6 \ldots$ (ii)

Multiplying (ii) by 11 and adding to (i), we have:

$x(3 + 11 \times 2) = 34 + 6 \times 11$

$\Rightarrow x = 4$

Substituting the above in (ii), we have:

$8 - y = 6$

$\Rightarrow y = 2$

Thus, we have:

$\frac{x}{y} = \frac{4}{2} = 2$

$\Rightarrow x = 2y$

$\Rightarrow k = 2$

The correct answer is 2.

36. Let the price of a laptop and a tablet be $\$l$ and $\$t$, respectively.

Thus, we have:

$3l + 2t = \$3,100 \ldots$ (i)

$l + 4t = \$3,700 \ldots$ (ii)

We need to determine the price of 2 laptops and 3 tablets after 10% discount is applicable on both.

Let us first determine the price of 2 laptops and 3 tablets before any discount is applied, i.e. the value of $(2l + 3t)$.

From (i) and (ii), we observe that adding the equations results in:

$4l + 6t = \$6,800$

$\Rightarrow 2l + 3t = \$3,400$

Since both laptops and tablets have the same discount of 10%, overall, the discount will also be 10%.

Thus, the final price of 2 laptops and 3 tablets after 10% discount

$= (100 - 10)\%$ of \$3,400

= 90% of \$3,400

= \$3,060

The correct answer is option A.

37. Total number of darts = 12

Number of darts that hit the bull's eye = x

Number of points scored @ 7.5 points per dart = $7.5x$

Number of darts that miss the bull's eye = $12 - x$

Number of points scored @ (-1.5) points per dart = $-1.5(12 - x)$

Thus, total points

$= 7.5x - 1.5(12 - x)$

$= 9x - 18$

The correct answer is option B.

38. Initial price of the car = \$13,500

Expected price of the car after 5 years = \$4,850

Decrease in value of the car = $\$ (13,500 - 4,850)$ = \$8,650

Since the depreciation is constant per year, the decrease in value of the car per year

$= \$ \left(\frac{8,650}{5}\right) = \$1,730$

Thus, the decrease in value after n years = $\$1,730n$

Thus, the value of the car after n years = $\$ (13,500 - 1,730n)$

The correct answer is option B.

39. We know that r and s represent the number of hours the man runs and cycles, respectively.

Since the man exercises for 4 hours every day, we must have:

$r + c = 4 \ldots$ (i)

Thus, the answer must be among A, C or D.

Total calories lost by running @ 400 calories per hour = $400r$

Total calories lost by cycling @ 300 calories per hour = $300c$

Thus, total calories lost per day

$= 400r + 300c = C \ldots$ (ii)

The correct answer is option D.

6.3 Inequalities

40. Minimum number of chocolates required by Joe = 140

Number of chocolates already with Joe = 20

Thus, minimum number of additional pieces of chocolates required = 140 - 20 = 120

Number of boxes Joe needs to buy = x

Number of chocolates in those boxes = $20x$

Thus, we have:

$20x \geq 120$

$\Rightarrow x \geq 6$

The correct answer is option D.

41. Let the number of pencils and the number of erasers in a gift box be x and y, respectively.

Thus, we have:

$x + y \leq 12 \ldots$(i)

$x \geq y + 3 \ldots$(ii)

From (ii): Assuming $x = y + 3$ and substituting in (i), we have: $(y + 3) + y \leq 12$

$\Rightarrow 2y \leq 9$

$\Rightarrow y \leq 4.5$

If x is greater than $(y + 3)$, the value of y will reduce even further to accommodate the higher value of x.

Thus, the maximum value of y is 4 (Since y is an integer).

The correct answer is option C.

42. Let the number of books David borrows be x.

Cost of borrowing the books without membership = \$$5x$

Cost of borrowing the books with membership = \$ $(125 + x)$

Since taking the membership must be comparatively cost effective option, we must have:

$125 + x < 5x$

$\Rightarrow 4x > 125$

$\Rightarrow x > 31.25$

Thus, the minimum number of books is 32.

The correct answer is 32.

43. Total area of the wall = $12 \times 8 = 96$ square feet.

Area left out for the door = $3 \times 5 = 15$ square feet.

Area left out for the window = 5 square feet to 10 square feet.

Thus, minimum area to be colored = 96 - 15 - 10 = 71 square feet

Thus, minimum charge = \$ (71×0.6) = \$42.60

Maximum area to be colored = 96 - 15 - 5 = 76 square feet

Thus, maximum charge = \$ (76×0.6) = \$45.60

Thus, the charge, p, satisfies:

$42.60 \leq p \leq 45.60$

The correct answer is option C.

44. We have:

$-1 < h < 0$

Thus, h is negative.

We need to find the greatest value among the given options.

Thus, we need to either find the option which is positive and has the largest magnitude OR the option which is negative and has the minimum magnitude.

Working with the options:

(A) $-1 < h < 0$
Multiplying (-1) throughout:
$\Rightarrow 0 < -h < 1$
Adding 1 throughout:
$1 < 1 - h < 2 \ldots$ (i)

(B) $-1 < h < 0$
Adding 1 throughout:
$0 < 1 + h < 1 \ldots$ (ii)

(C) $-1 < h < 0$
Squaring throughout:
$0 < h^2 < 1$
Adding 1 to both the sides:
$1 < 1 + h^2 < 2 \ldots$ (iii)

(D) $-1 < h < 0$

Taking reciprocal throughout:

$-\infty < \frac{1}{h} < -1$

Multiplying by (-1) throughout:

$1 < -\frac{1}{h} < \infty$

Adding 1 throughout (and realizing that: $\infty + 1 \equiv \infty$):

$2 < 1 - \frac{1}{h} < \infty \ldots \text{(iv)}$

Thus, comparing (i) to (iv), we find that (iv) represents the largest set of values.

The correct answer is option D.

Alternate approach:

Let us assume a convenient, smart value of h and evaluate each of the answer options.

Since $-1 < h < 0$, let us assume $h = -\frac{1}{2}$.

(A) $1 - h = 1 - \left(-\frac{1}{2}\right) = \frac{3}{2}$

(B) $1 + h = 1 + \left(-\frac{1}{2}\right) = \frac{1}{2}$

(C) $1 + h^2 = 1 + \left(-\frac{1}{2}\right)^2 = 1 + \frac{1}{4} = \frac{5}{4}$

(D) $1 - \frac{1}{h} = 1 - \frac{1}{\left(-\frac{1}{2}\right)} = 3$ (greatest value)

45. One-time charge for booking = \$120

Cost per guest = \$24

Since Amy plans to bring along n people, there would be a total of $(n + 1)$ people in the picnic.

Thus, total cost = \$ $\{120 + 24(n + 1)\}$

Since the budget is \$350, we have:

$120 + 24(n + 1) \leq 350$

$\Rightarrow 24(n + 1) \leq 230$

$\Rightarrow n + 1 \leq \frac{230}{24}$

$\Rightarrow n + 1 \leq 9.58$

$\Rightarrow n \leq 8.58$

Thus, the maximum value of n is 8.

The correct answer is option B.

46. Minimum total score required in five tests = $70 \times 5 = 350$

Total score obtained in the first four tests = $65 \times 4 = 260$

Thus, minimum score required in the fifth test = $s = 350 - 260 = 90$

Since all tests are scored out of 100, the range of possible values of s is:

$90 \leq s \leq 100$

The correct answer is option D.

47. We have:

$-\frac{5}{3} \le 2a + 3 \le \frac{7}{2}$

$=> -\frac{5}{3} - 3 \le 2a \le \frac{7}{2} - 3$

$=> -\frac{14}{3} \le 2a \le \frac{1}{2}$

$=> -\frac{7}{3} \le a \le \frac{1}{4}$

$=> \frac{1}{2}\left(-\frac{7}{3}\right) \le \frac{a}{2} \le \frac{1}{2}\left(\frac{1}{4}\right)$

$=> -\frac{7}{6} \le \frac{a}{2} \le \frac{1}{8}$

$=> -\frac{7}{6} + 1 \le \frac{a}{2} + 1 \le \frac{1}{8} + 1$

$=> -\frac{1}{6} \le \left(\frac{a}{2} + 1\right) \le \frac{9}{8}$

Thus, the possible integer values of $\left(\frac{a}{2} + 1\right)$ are 0 and 1.

The correct answer is 2.

48. We have:

$2x - \frac{7}{4} < \frac{11}{12} \ldots$ (i)

$-3x + \frac{1}{5} < \frac{1}{20} \ldots$ (ii)

From (i):

$2x - \frac{7}{4} < \frac{11}{12}$

$=> 2x < \frac{11}{12} + \frac{7}{4}$

$=> 2x < \frac{32}{12}$

$=> x < \frac{4}{3} \ldots$ (iii)

From (ii):

$-3x + \frac{1}{5} < \frac{1}{20}$

$=> -3x < -\frac{3}{20}$

Dividing the inequality by -3. Notice the reversal of $<$ to $>$.

$=> x > \frac{1}{20} \ldots$ (iv)

Thus, from (iii) and (iv), we have:

$\frac{1}{20} < x < \frac{4}{3}$

Thus, the only integer value of $x = 1$

The correct answer is 1.

49. First elephant:

Minimum water it drinks in a day = $\frac{2}{100} \times 8,560 = 171.2$ liters

Maximum water it drinks in a day = $\frac{3}{100} \times 8,560 = 256.8$ liters

Second elephant:

Minimum water it drinks in a day = $\frac{2}{100} \times 9,450 = 189$ liters

Maximum water it drinks in a day = $\frac{3}{100} \times 9,450 = 283.5$ liters

Thus, total minimum water the two elephants drink in a day = 171.2 + 189 = 360.2 liters

Total maximum water the two elephants drink in a day = 256.8 + 283.5 = 540.3 liters

Thus, we have:

$360.2 \le w \le 540.3$

Thus, the required difference = 540.3 – 360.2 = 180.1 ≈= 180 (to the nearest integer)

The correct answer is 180.

Alternate approach:

Total weight of the two elephants = 18,010 pounds

Difference between the maximum and the minimum value of the amount of water an elephant drinks per 100 pound of body weight = $\frac{3}{100} - \frac{2}{100} = \frac{1}{100} = 0.01$ liters

Thus, the required difference = 18,010 × 0.01 = 180.1 ≈= 180 liters

50. Minimum number of hours John exercises in a day = 1 hour = 60 minutes

Minimum number of calories he can lose per minute = 15 calories

Thus, minimum number of calories he can lose in a day = 15 × 60 = 900 calories

Maximum number of hours John exercises in a day = 2 hours = 120 minutes

Maximum number of calories he can lose per minute = 20 calories

Thus, maximum number of calories he can lose in a day = 20 × 120 = 2,400 calories

Minimum number of calories he consumes daily = 600 calories

Maximum number of calories he consumes daily = 800 calories

Thus, we have:

Net minimum number of calories he loses

= Minimum number of calories he loses in a day - Maximum number of calories he consumes in a day

= 900 - 800 = 100 calories ... (i)

Net maximum number of calories he loses

= Maximum number of calories he loses in a day - Minimum number of calories he consumes in a day

= 2400 - 600 = 1,800 calories ... (ii)

Thus, from (i) and (ii), we have:

$100 \le c \le 1,800$

The correct answer is option A.

51. Maximum hourly wage earned (by the architect) = \$35

Minimum hourly wage earned (by the accountant) = \$16

Maximum number of working hours in a week = 38 hours

Minimum number of working hours in a week = 32 hours

Thus, we have:

Maximum possible weekly earnings of the professional who earns the most per hour

$= \$ (35 \times 38) = \$1,330$

Minimum possible weekly earnings of the professional who earns the least per hour

$= \$ (16 \times 32) = \$512$

Thus, maximum possible difference in earning $= \$ (1,330 - 512) = \$818 \ldots$ (i)

Similarly, we have:

Minimum possible weekly earnings of the professional who earns the most per hour

$= \$ (35 \times 32) = \$1,120$

Maximum possible weekly earnings of the professional who earns the least per hour

$= \$ (16 \times 38) = \$608$

Thus, minimum possible difference in earning $= \$ (1,120 - 608) = \$512 \ldots$ (ii)

Thus, we have:

$512 \le x \le 818$

The correct answer is option A.

52. Number of trips = t

Total charges for t trips without using the discount card @\$17 per trip = \$17t

Price of the discount card = \$350

Total charges for t trips using the discount card @ \$9.50 per trip = \$ $(350 + 9.5t)$

If the discount card has to be a better deal, the total charge using the card should be less than the normal charge.

Thus, we have:

$350 + 9.5t < 17t$

$=> 350 < t\,(17 - 9.5)$

$=> t > \frac{350}{7.5}$

$=> t > 46.67$

Since t represents the number of trips, an integer, the minimum value of $t = 47$

The correct answer is 47.

53. Charge for the first $\frac{1}{2}$ mile = \$5

Total distance travelled = m miles.

Distance left = $\left(m - \frac{1}{2}\right)$ miles.

Additional charge every $\frac{1}{3}$ miles = \$1.20

Thus, additional charge for $\left(m - \frac{1}{2}\right)$ miles

$= \$\left\{\frac{1.2}{\left(\frac{1}{3}\right)} \times \left(m - \frac{1}{2}\right)\right\}$

$= \$\left\{3.6 \times \left(m - \frac{1}{2}\right)\right\}$

Thus, total charge for m miles $= \$\left\{5 + 3.6 \times \left(m - \frac{1}{2}\right)\right\}$

Since the rider doesn't want to pay more than \$20, we have:

$5 + 3.6 \times \left(m - \frac{1}{2}\right) \leq 20$

$=> 5 + 3.6m - 1.8 \leq 20$

$=> 3.6m + 3.2 \leq 20$

The correct answer is option A.

54. Required radius = 2.5 cm

Thus, required diameter = 5 cm

Maximum error in the radius = 0.1 mm = 0.01 cm

Thus, maximum error in the diameter = 0.02 cm

Note that due to the error, the actual diameter may be less/more than the required value.

Thus, the positive difference between the actual diameter, d, and the required diameter should be less than or equal to 0.02 cm.

The positive difference between any two terms a and b is given as $|a - b|$. For example, if $a = 1$ and $b = 3$, then the positive difference is 3 - 1 = 2, similarly, $|a - b| = |1 - 3| = 2$.

Thus, we have:

$|d - 5| \leq 0.02$

The correct answer is option B.

55. Volume of milk in the first solution = 20% of $x = \frac{x}{5}$ liters

Volume of milk in the second solution = 60% of $y = \frac{3y}{5}$ liters

Total volume of milk in the two solutions $= \left(\frac{x}{5} + \frac{3y}{5}\right)$ liters

Total volume of the two solutions $= (x + y)$ liters

Thus, we have:

$x + y = 50 \ldots$ (i)

Since overall there is 50% milk, we have:

$\frac{x}{5} + \frac{3y}{5} = 50\% \text{ of } (x + y)$

$=> \frac{x}{5} + \frac{3y}{5} = \frac{1}{2}(x + y)$

$=> \frac{1}{5}(x + 3y) = \frac{1}{2}(x + y)$

$=> x + 3y = \frac{5}{2}(x + y) \ldots$ (ii)

The correct answer is option C.

56. Since we need at least 15 desks and 25 chairs, we have:

$d \geq 15 \ldots$ (i)

$c \geq 25 \ldots$ (ii)

Total cost of d desks @ \$120 each and c chairs @ \$40 each = \$ $(120d + 40c)$

Since the budget is \$4,000, we have:

$120d + 40c \leq 4,000$

$=> 3d + c \leq 100 \ldots$ (iii)

The correct answer is option D.

6.4 Absolute Value

57. We have:

$|x| = |y|$

$|x| = |y|$ mean that the absolute value of x equals the absolute value of y; the direction of the numbers do not matter.

Thus, if say, for example $x = 4$, y may be 4, or −4. In this case, either $x = y$ or $x > y$. Thus, option B and C are correct.

However, if say, for example $x = -4$, y can still be 4, or −4. In this case, either $x = y$ or $x < y$. Thus, option A and B are correct.

The correct answer is option D.

58. We have:

$|x| = |2x - 3|$

$=> \pm x = \pm(2x - 3)$; removing the absolute value signs and replacing with '±'

$=> x = \pm(2x - 3)$

Case I:

$x = 2x - 3 => x = 3$

Case II:

$x = -(2x - 3) => x = 1$

Thus, the greatest possible solution is 3.

The correct answer is 3.

59. We have:

$\left|x - \frac{4}{5}\right| \leq \frac{11}{5}$

$=> \pm\left(x - \frac{4}{5}\right) \leq \frac{11}{5}$

Case I:

$x - \frac{4}{5} \leq \frac{11}{5}$

$=> x \leq \frac{11}{5} + \frac{4}{5}$

$=> x \leq 3$

Case II:

$-\left(x-\frac{4}{5}\right) \leq \frac{11}{5}$

Multiplying the inequality by -1.

$=> x-\frac{4}{5} \geq -\frac{11}{5}$; notice the reversal of the inequality

$=> x \geq \frac{4}{5}-\frac{11}{5}$

$=> x \geq -\frac{7}{5}$

Thus, we have:

$-\frac{7}{5} \leq x \leq 3$

$=> -\frac{7}{5}-1 \leq (x-1) \leq 3-1$

$=> -\frac{12}{5} \leq (x-1) \leq 2$

Since $\left(-\frac{12}{5}\right)$ is greater in magnitude than 2, the maximum value of $|x-1|$ would be $\left|-\frac{12}{5}\right| = \frac{12}{5} = 2.4$

The correct answer is option C.

60. From the options, we find that all of them are inequalities involving absolute values.

Maximum diameter = 12.6 cm

Minimum diameter = 7.4 cm

Thus, average of the two values = (12.6 + 7.4)/2 = 10 cm

The difference from the maximum/minimum value and their mean

= 12.6 - 10 = 10 - 7.4

= 2.6

Thus, d, which represents the values of the diameters on the steel plates, has a maximum difference of 2.6 cm from the maximum/minimum diameter values

Thus, we have:

$|d-10| \leq 2.6$

The correct answer is option A.

6.5 Functions

61. Working with the options:

- Option A/B:

  $\frac{f(p, q)}{g(p, q)} = \left(\frac{p}{p+q}\right) \times \frac{1}{\left(\frac{p+q}{q}\right)} = \left(\frac{p}{p+q}\right) \times \left(\frac{q}{p+q}\right) = \frac{pq}{(p+q)^2}$

  Thus, options A and B are incorrect.

- Option C/D:

  $f(p, q) \times g(p, q) = \left(\frac{p}{p+q}\right) \times \left(\frac{p+q}{q}\right) = \frac{p}{q}$

Thus, option C is incorrect, and option D is correct.

The correct answer is option D.

62. We have:

$f(2x-1) = x^2 + 3x \ldots$(i)

Since we need the value of $f(5)$, we need to use a suitable value of x so that we have:

$2x - 1 = 5$

$\Rightarrow x = 3$

Thus, substituting $x = 3$ in (i):

$f(2 \times 3 - 1) = 3^2 + 3 \times 3$

$\Rightarrow f(5) = 18$

The correct answer is option C.

63. $f(x) = -\frac{1}{x}$

$\Rightarrow f(a) = -\frac{1}{a} = -\frac{1}{2}$

$\Rightarrow a = 2$

Also, we have:

$f(ab) = -\frac{1}{ab} = \frac{1}{6}$

$\Rightarrow ab = -6$

$\Rightarrow b = -\frac{6}{a} = -\frac{6}{2}$

$\Rightarrow b = -3$

The correct answer is option D.

64. Working with the options one at a time:

(A) $f(x) = 1 - x$
$\Rightarrow f(1-x) = 1 - (1-x) = x \neq f(x)$ - Incorrect

(B) $f(x) = 1 - x^2$
$\Rightarrow f(1-x) = 1 - (1-x)^2 = 1 - (1 - 2x + x^2) = 2x - x^2 \neq f(x)$ - Incorrect

(C) $f(x) = x^2 - (1-x)^2$
$\Rightarrow f(1-x) = (1-x)^2 - (1-(1-x))^2 = (1-x)^2 - x^2 \neq f(x)$ - Incorrect

(D) $f(x) = |x(1-x)|$
$\Rightarrow f(1-x) = |(1-x)(1-(1-x))| = |(1-x)x| = f(x)$ - Correct

The correct answer is option D.

Alternate approach:

Let us take a suitable value of x, say $x = 0.2$

We need $f(x) = f(1-x) \Rightarrow f(0.2) = f(0.8)$

Working with the options:

(A) $f(0.2) = 1 - 0.2 = 0.8; f(0.8) = 1 - 0.8 = 0.2$
$f(x) \neq f(1-x)$ - Incorrect

(B) $f(0.2) = 1 - 0.2^2 = 0.96; f(0.8) = 1 - 0.8^2 = 0.36$
$f(x) \neq f(1-x)$ - Incorrect

(C) $f(0.2) = 0.2^2 - 0.8^2 = -0.6; f(0.8) = 0.8^2 - 0.2^2 = 0.6$
$f(x) \neq f(1-x)$ - Incorrect

(D) $f(0.2) = |0.2 \times 0.8| = 0.16; f(0.8) = |0.8 \times 0.2| = 0.16$
$f(x) = f(1-x)$ - Correct

65. We have:

$$\frac{1}{2}f(x) = f\left(\frac{1}{2}x\right)$$

$$=> f(x) = 2f\left(\frac{1}{2}x\right)$$

Working with the options, we have:

(A) $f(x) = 2x - 2$

$$=> 2f\left(\frac{1}{2}x\right) = 2\left\{\left(2\left(\frac{1}{2}x\right) - 2\right)\right\} = 2x - 4 \neq f(x) \text{ - Does not satisfy}$$

(B) $f(x) = 3x + 1$

$$=> 2f\left(\frac{1}{2}x\right) = 2\left\{3\left(\frac{x}{2} + 1\right)\right\} = 3x + 2 \neq f(x) \text{ - Does not satisfy}$$

(C) $f(x) = x^2$

$$=> 2f\left(\frac{1}{2}x\right) = 2\left\{\left(\frac{1}{2}x\right)^2\right\} = \frac{1}{2}x^2 \neq f(x) \text{ - Does not satisfy}$$

(D) $f(x) = (x+2)^2 - (x-2)^2 = (x^2 + 4x + 4) - (x^2 - 4x + 4) = 8x$

$$=> 2f\left(\frac{1}{2}x\right) = 2\left\{8\left(\frac{1}{2}x\right)\right\} = 8x = f(x) \text{ - Satisfies}$$

The correct answer is option D.

66. We have:

$$F = \frac{GSs}{d^2}$$

Cross-multiplying, we have:

$$Fd^2 = GSs$$

Dividing throughout by F, we have:

$$=> d^2 = \frac{GSs}{F}$$

Taking square-root on both sides, we have:

$$=> d = \sqrt{\frac{GSs}{F}}$$

The correct answer is option C.

67. Let the exchange rate be R, i.e. \$1 = R Yen.

Thus, for $\$x$, the equivalent amount in Yen = $(R \times x)$ Yen.

Fee applicable is 2% of the amount converted.

Thus, the final amount after deducting fees

= (100 - 2)%, i.e. 98% of the amount converted

$= \left(\frac{98}{100} \times Rx\right)$ Yen

$= (0.98 \times Rx)$ Yen

The function representing the amount of money received by Joe in Yen is

$f(x) = 0.98 \times 100x$

Thus, we have:

$R = 100$

Thus, $1 = 100 Yen

$=> 1$ Yen $= \$\left(\frac{1}{100}\right) = \$0.01 = 1$ cent

The correct answer is 1.

68. We have:

$R(x) = \frac{x}{100} + 1$

Here, $R(x)$ represents the number of reward points and x represents purchases worth x dollars.

For a value of purchase worth $\$k$, the number of reward points

$= \left(\frac{k}{100} + 1\right)$

Thus, for a value of purchase worth $\$\,(k + 1200)$, the number of reward points

$= \left(\frac{k + 1,200}{100} + 1\right)$

Thus, the increase in the number of reward points

$= \left(\frac{k + 1,200}{100} + 1\right) - \left(\frac{k}{100} + 1\right)$

$= \frac{k + 1,200 - k}{100}$

$= 12$

The correct answer is 12.

69. We have:

$f(x) = f(0) \times k^x$

Using the data from the table, we have:

$f(1) = f(0) \times k^1$

$=> 240 = f(0) \times k \ldots$ (i)

$f(2) = f(0) \times k^2$

$=> 480 = f(0) \times k^2 \ldots$ (ii)

Dividing (ii) by (i), we have:

$\frac{480}{240} = \frac{f(0) \times k^2}{f(0) \times k}$

$=> k = 2$

Substituting the above in (i), we have:

$240 = f(0) \times 2$

$=> f(0) = 120$

Thus, we have:

$f(0) + k = 120 + 2 = 122$

The correct answer is 122.

70. We have:

$f(t) = 100\left(1 + \frac{r}{200}\right)^{2t}$

Here, $f(t)$ represents the total amount of money Joe would have after t years.

According to Joe, if $r = 3$, the total sum of money after 1 year

$= \$\left\{100\left(1 + \frac{3}{200}\right)^2\right\}$

$= \$103.02$

However, Joe found that the total amount of money is \$1 less than what it should have been, i.e. \$102.02

Thus, we have:

$102.02 = 100\left(1 + \frac{r}{200}\right)^2$

$=> \left(1 + \frac{r}{200}\right)^2 = \frac{102.02}{100} = 1.0202$

$=> 1 + \frac{r}{200} = \sqrt{1.0202} = 1.01$

$=> r = 2$

The correct answer is option D.

Alternate approach:

According to Joe, if $r = 3$, the total sum of money after 1 year

$= \$\left\{100\left(1 + \frac{3}{200}\right)^2\right\} = \$103.02$

However, Joe found that the total amount of money is \$1 less than what it should have been, i.e. \$102.02

Working with the options, we have:

Option D: Total sum of money $= \$\left\{100\left(1 + \frac{2}{200}\right)^2\right\} = \$102.01$

71. Total cost of manufacturing n bulbs = \$ $(12.8n + 132)$

Total selling price of n bulbs = \$$16n$

There would be profits only if the total selling price is strictly greater than the total cost price.

Thus, we have:

$16n > 12.8n + 132$

$=> n(16 - 12.8) > 132$

$=> 3.2n > 132$

$=> n > 41.25$

Since n represents the number of bulbs, an integer value, the minimum value of n must be 42 (Note: We should not round off 41.25 to 41).

Thus, we have:

$n \geq 42$

The correct answer is option B.

72. We have:

$M = 230 - 10t$

Price of the commodity in 2009 = \$180

Thus, we have:

$180 = 230 - 10t$

$=> t = 5$

Thus, the value of the commodity was \$180 after 5 years from when it was produced.

Thus, the commodity was produced in the year 2009 – 5 = 2004.

Thus, '230' signifies the value of the commodity in year 2004 and it depreciated at the rate of \$10 per year for 5 years, reaching to the value of \$180 in year 2009.

The correct answer is option A.

73. We have:

$p\,(b) = 10b + 40$

Here, b represents the number of boxes of pencils purchased and $p\,(b)$ represents the number of pencils with him after the purchase was made.

We know that there were already some pencils present with the teacher. Thus, the final number of pencils would be the sum of the number of pencils purchased and the number of pencils already present.

Since the pencils are purchased in boxes, the number of pencils purchased

= Number of boxes purchased × Number of pencils in each box

Thus, the total number of pencils after the purchase should be:

Number of boxes purchased × Number of pencils in each box + Number of pencils already present

Thus, comparing the above with the given function, we have:

Number of boxes purchased = b

Number of pencils in each box = 10

Number of pencils already present = 40

The correct answer is option C.

6.6 Interpreting Equations

74. We know that $x = \frac{3}{2}$ satisfies the equation $x^2 - px + 12 = 0$

Substituting $x = \frac{3}{2}$ in the equation, we have:

$\left(\frac{3}{2}\right)^2 - p\left(\frac{3}{2}\right) + 12 = 0$

$=> \frac{3}{2}p = \frac{9}{4} + 12 = \frac{57}{4}$

$=> p = \frac{2}{3} \times \frac{57}{4} = \frac{19}{2}$

The correct answer is option C.

75. The above is a linear equation. On simplifying the above, we would finally get a form:

$kx + l = nx + m$, where $k,\ l,\ n$ and m are constants

In the above, if $k = n$, the terms with x will cancel out.

Thus, if $l = m$, i.e. the constant terms are equal, it implies that the above equation is true for any value of x.

However, if $l \neq m$, the above equation would not be true for any value of x, i.e. there will be no solution.

We have:

$4(2x - 1) + 3(x - 2) = n(x + 2) - 3(x + 1)$

$=> n(x + 2) = 4(2x - 1) + 3(x - 2) + 3(x + 1)$

$=> nx + 2n = 14x - 7$

Thus, we should have:

$nx = 14x$ AND $2n \neq -7$

$=> n = 14$ (which satisfies $2n \neq -7$)

The correct answer is option D.

76. We have:

$$\frac{8d + 10}{4} = \frac{d\left(3.5^2 - 1.5^2\right) + x}{5}$$

Cross-multiplying, we have:

$=> 40d + 50 = 4\left\{d\left(3.5^2 - 1.5^2\right) + x\right\}$

$=> 40d + 50 = 4\{d(3.5 - 1.5)(3.5 + 1.5) + x\}$

$=> 40d + 50 = 4\{d(2)(5) + x\}$

$=> 40d + 50 = 40d + 4x$

We can observe that d cancels out completely. However, if the constant terms don't cancel out, we would end up with LHS $\neq$ RHS. This would imply no possible solution for the given equation.

Thus, in order to have infinite possible solutions, the constant terms must also cancel out. Thus, we have:

$4x = 50$

$=> x = 12.5$

The correct answer is option B.

77. If an equation of the form $ax + b = cx + d$ has no solution, we have:

$a = c$ and $b \neq d$

We have:

$3x - (12 - x) = c(2x - 5) - p$

$=> 3x - 12 + x = 2cx - 5c - p$

$=> 4x - 12 = 2cx - (5c + p)$

Since the above equation does not have any solution, we must have:

$4x = 2cx => c = 2 \ldots$ (i)

AND

$-(5c + p) \neq -12$

$=> 5c + p \neq 12$

$=> 5(2) + p \neq 12$

=> $p \neq 2$

Thus, the only value of p which is NOT possible is 2.

The correct answer is 2.

Alternate approach:

We have:

$3x - (12 - x) = c\,(2x - 5) - p$

=> $3x - 12 + x = 2cx - 5c - p$

=> $4x - 12 = 2cx - (5c + p)$

=> $4x - 2cx = 12 - 5c - p$

=> $2x(2 - c) = 12 - 5c - p$

Since the equation does not have any solution, we must make the co-efficient of $x = 0$.

Thus, $(2 - c) = 0$ => $c = 2$

At $c = 2$,

$2x\,(2 - c) = 12 - 5c - p$

=> $0 = 12 - 5 \times 2 - p$

=> $p = 2$

Since $p = 2$ renders a possible solution, $p \neq 2$.

78. We have:

$Ax + \frac{1}{2}y = -6 \ldots$ (i)

$-\frac{1}{3}x - 2y = 24$ (ii)

The set of equations $Px + Qy = K$ and $Mx + Ny = L$ have infinite solutions if $\frac{P}{M} = \frac{Q}{N} = \frac{K}{L}$.

Thus, we have:

$$\frac{A}{\left(-\frac{1}{3}\right)} = \frac{\left(\frac{1}{2}\right)}{-2} = -\frac{6}{24}$$

=> $-3A = -\frac{1}{4} = -\frac{1}{4}$

=> $A = \left(-\frac{1}{3}\right)\left(-\frac{1}{4}\right) = \frac{1}{12}$

The correct answer is $\frac{1}{12}$.

Alternate approach:

The equations: $Ax + \frac{1}{2}y = -6$ and $-\frac{1}{3}x - 2y = 24$ would have infinite number of solutions if both the equations are the same. Let's try to make the second equation same as the first.

We have:

$-\frac{1}{3}x - 2y = 24$

By dividing the equation by -4 to get -6 (as the constant term), equivalent to the RHS of the first equation:

$$\left(\frac{1}{-4}\right) \times \left(-\frac{1}{3}x\right) - (2y) \times \left(\frac{1}{-4}\right) = \left(\frac{1}{-4}\right) \times 24$$

$=> \frac{1}{12}x + \frac{1}{2}y = -6$

We notice that if $A = \frac{1}{12}$, the second equation would become the first equation, thus $A = \frac{1}{12}$.

79.

Change in x	x	y	Change in y
	-5	11	
4			-6
	-1	5	
2			-3
	1	2	
6			?
	7	k	

We can observe that, when the value of x increases by 4, the value of y decreases by 6; similarly, when the value of x increases by 2, the value of y decreases by 3.

Since the above represents a linear relationship, maintaining the same ratio as above, when the value of x increases by 6, the value of y should decrease by $6 \times \frac{3}{2} = 9$

Thus, the value of y should be $k = 2 - 9 = -7$

$=> k^2 = (-7)^2 = 49$

The correct answer is 49.

Alternate approach:

Let the equation of the line (linear relationship) be $y = mx + c$

Substituting the values from the table, we have:

$x = -1, y = 5:$

$5 = -m + c \ldots$ (i)

$x = 1,\ y = 2:$

$2 = m + c \ldots$ (ii)

Adding (i) and (ii):

$2c = 7 => c = \frac{7}{2}$

Substituting the above in (ii):

$2 = m + \frac{7}{2}$

$=> m = -\frac{3}{2}$

Thus, the equation is:

$y = -\frac{3}{2}x + \frac{7}{2}$

Thus, when $x = 7$, we have:

$y = -\frac{3}{2} \times 7 + \frac{7}{2} = -7$

$=> k = -7$

$=> k^2 = (-7)^2 = 49$

6.7 Graphic Equations and Inequalities

80. We have:

$Ax + 4y = 6 \ldots$ (i)

$Bx + 2Ay = 4 \ldots$ (ii)

Since the graphs of the above system of equations intersect at $(2, -1)$, $x = 2$ and $y = -1$ satisfy both the given equations.

From (i):

$A(2) + 4(-1) = 6$

$=> 2A = 10$

$=> A = 5$

From (ii):

$B(2) + 2(5)(-1) = 4$

$=> 2B = 14$

$=> B = 7$

Thus, we have:

$A + B = 5 + 7 = 12$

The correct answer is 12.

81. We have:

$3x - (5 - 4x) \,\square\, x + 3(x + 1) + 1$

$=> 3x - 5 + 4x \,\square\, x + 3x + 3 + 1$

$=> 7x - 5 \,\square\, 4x + 4$

$=> 7x - 4x \,\square\, 5 + 4$

$=> 3x \,\square\, 9$

$=> x \,\square\, 3$

Since the above inequality must be consistent with the given inequality $x > 3$, the correct symbol is '>.'

The correct answer is option A.

82. We can see that in 2001, the antique was valued at \$1,000.

Thereafter, every 4 years, the valuation of the antique doubled:

Value in 2005 = \$2,000

Value in 2009 = \$4,000

Value in 2013 = \$8,000

Thus, after t years, the number of times the valuation would double = $\left(\frac{t}{4}\right)$ times.

Thus, the valuation after t years

$= \$\left\{1,000 \times 2 \times 2 \ldots \left(\frac{t}{4} \text{ times}\right)\right\}$

$= \$\left(1,000 \times 2^{\frac{t}{4}}\right)$

The correct answer is option D.

Alternate approach:

Working with the options and using the value of the antique in 2005, we find that only option D gives the correct valuation of \$2,000.

83. In the above graph, the Y-axis shows the distance of Joe and Ron from the restaurant. At the beginning of the journey, Joe was 6 kilometers away from the restaurant, while Ron was 3 kilometers away.

However, Joe took 20 minutes to reach the restaurant (since at the 20-minute mark, his distance from the restaurant is '0'). Similarly, Ron took 30 minutes to reach the restaurant. So Joe reached before Ron.

Thus, Joe covered a larger distance in a shorter time and hence, he travelled faster than Ron.

Speed of Joe = $\frac{6}{20}$ = 0.3 kilometers per minute

Speed of Ron = $\frac{3}{30}$ = 0.1 kilometers per minute

We see that Joe's speed is $\frac{0.3}{0.1}$ = 3 times that of Ron.

The correct answer is option C.

Chapter 7

Advanced Math Concepts

7.1 Quadratic Equations

An equation is said to be quadratic when the variable contains the highest exponent of '2', thus ax^2+bx+c is a quadratic expression in one variable i.e. of x.

Solving a Quadratic Equation by Factorization:

A quadratic equation can be solved quickly if it can be written in the form of a product of two linear expressions.

Let us take an example:

$2x^2 - 7x + 6 = 0$

The given equation can be written as:

$2x^2 - 4x - 3x + 6 = 0$

While dividing the middle term in two parts, it should be seen that the product of the two terms equals the product of the first and last terms of the quadratic.

Thus, here we have: $(-4x) \times (-3x) = (2x^2) \times (6) = 12x^2$

$=> 2x(x - 2) - 3(x - 2) = 0$

$=> (2x - 3)(x - 2) = 0$

Here, $2x - 3$ and $x - 2$ are the two linear expressions. Since their product is 0, at least one of them must be 0.

This results in two separate linear equations:

$2x - 3 = 0$

OR

$x - 2 = 0$

$=> x = \frac{3}{2}$ OR $x = 2$

Thus, each quadratic equation can be written in the form $(x - p)(x - q) = 0$, where p and q are the roots of the quadratic equation.

Roots of a Quadratic Equation:

After suitable reduction, every quadratic equation can be written in the form:

$ax^2+bx+c = 0$

The solution or the roots of the above equation is given by:

$\boldsymbol{p} = \dfrac{-\boldsymbol{b}+\sqrt{\boldsymbol{b}^2-4\boldsymbol{ac}}}{2\boldsymbol{a}}$ and $\boldsymbol{q} = \dfrac{-\boldsymbol{b}-\sqrt{\boldsymbol{b}^2-4\boldsymbol{ac}}}{2\boldsymbol{a}}$; where $\boldsymbol{p}$ and $\boldsymbol{q}$ are the roots

Let us see how:

$ax^2 + bx + c = 0$

Dividing throughout by a:

$x^2 + \frac{b}{a}x + \frac{c}{a} = 0$

Let us combine the first two terms of the above equation to form a perfect square:

$$=> x^2 + 2x\left(\frac{b}{2a}\right) + \left(\frac{b}{2a}\right)^2 - \left(\frac{b}{2a}\right)^2 + \frac{c}{a} = 0$$

$$=> \left(x + \frac{b}{2a}\right)^2 = \left(\frac{b}{2a}\right)^2 - \frac{c}{a} = \frac{b^2 - 4ac}{4a^2}$$

Taking square roots on both sides:

$$=> x + \frac{b}{2a} = \pm\sqrt{\frac{b^2 - 4ac}{4a^2}} = \pm\frac{\sqrt{b^2 - 4ac}}{2a}$$

$$=> x = -\frac{b}{2a} \pm \frac{\sqrt{b^2 - 4ac}}{2a}$$

$$=> x = \frac{-b \pm \sqrt{b^2 - 4ac}}{2a}$$

Let us find the roots of the previous equation $2x^2 - 7x + 6 = 0$ using the above formula:

We have: $a = 2,\ b = -7,\ c = 6$

Thus, the roots are:

$$\frac{-b \pm \sqrt{b^2 - 4ac}}{2a} = \frac{-(-7) \pm \sqrt{(-7)^2 - 4(2)(6)}}{2(2)} = \frac{7 \pm 1}{4} = \frac{7-1}{4} \text{ OR } \frac{7+1}{4} = \frac{3}{2} \text{ OR } 2$$

Discriminant:

The value given by $\boldsymbol{D(\triangle)} = \boldsymbol{b^2} - \boldsymbol{4ac}$, i.e. the quantity under the square root, is called the discriminant and depending upon its value we can determine the nature of the roots of the quadratic equations:

- If $D \geq 0$: The roots are real
- If $D = 0$: The roots are real and equal
- If $D > 0$: The roots are real and unequal
- If $D < 0$: The roots are imaginary or complex
- If D is a perfect square: The roots are real and rational
- If D is not a perfect square: The roots are real and irrational (one root is conjugate of the other, i.e. the roots are of the form $a \pm \sqrt{b}$)

Let us take some example

- In the equation $x^2 + 2x + 1 = 0$, $D = b^2 - 4ac = 2^2 - 4.(1).(1) = 0$
  => The roots are real and equal.
- In the equation $x^2 + 2x + 2 = 0$, $D = b^2 - 4ac = 2^2 - 4.(1).(2) = -4 < 0$
  => The roots are complex.

Sum and Product of the roots:

If the roots of the quadratic equation $ax^2 + bx + c = 0$ are p and q, we have:

$ax^2 + bx + c \equiv a(x - p)(x - q)$

$=> ax^2 + bx + c \equiv a(x^2 - x(p + q) + pq)$

$=> ax^2 + bx + c \equiv ax^2 - a(p + q)x + apq$

Comparing the coefficient of x and the constant term, we have:

$b = -a(p + q)$ and $c = apq$

$=> p+q = -\frac{b}{a}$ and $pq = \frac{c}{a}$

Expressing a quadratic equation in terms of its roots:

Suppose we have to form the equation whose roots are p and q.

Thus, we have: $(x - p) = 0$ and $(x - q) = 0$

$=> (x - p)(x - q) = 0$

$=> x^2 - x(p + q) + pq = 0$

$=> \boldsymbol{x^2 - (\textbf{Sum of roots})\,x + (\textbf{Product of roots}) = 0}$

Let us take an example:

The equation having roots -2 and 5 is:

$x^2 - (-2 + 5)x + (-2)(5) = 0$

$=> x^2 - 3x - 10 = 0$

Expressing a quadratic equation in terms of its vertex:

Let the equation be

$y = ax^2 + bx + c$

$= a\left\{x^2 + \left(\frac{b}{a}\right)x + \frac{c}{a}\right\}$

$= a\left\{\left(x + \frac{b}{2a}\right)^2 - \left(\frac{b}{2a}\right)^2 + \frac{c}{a}\right\}$

$= a\left(x + \frac{b}{2a}\right)^2 - \frac{(b^2 - 4ac)}{4a}$

$= a(x - h)^2 - k$; here, $h = -\frac{b}{2a}$ and $k = \frac{(b^2 - 4ac)}{4a}$

The coordinates of the vertex is given by (h, k).

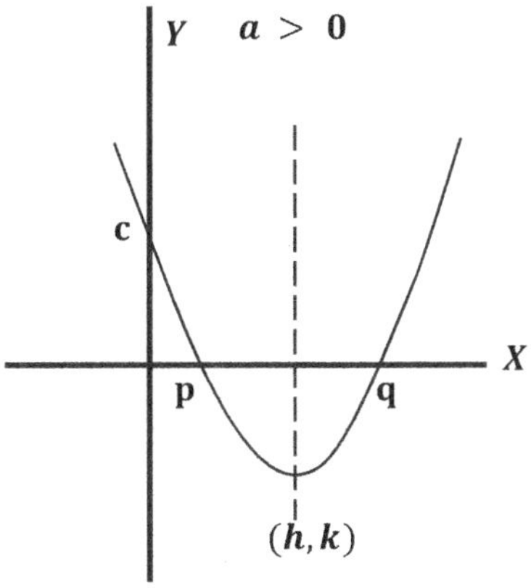

The line of symmetry: $x = h$

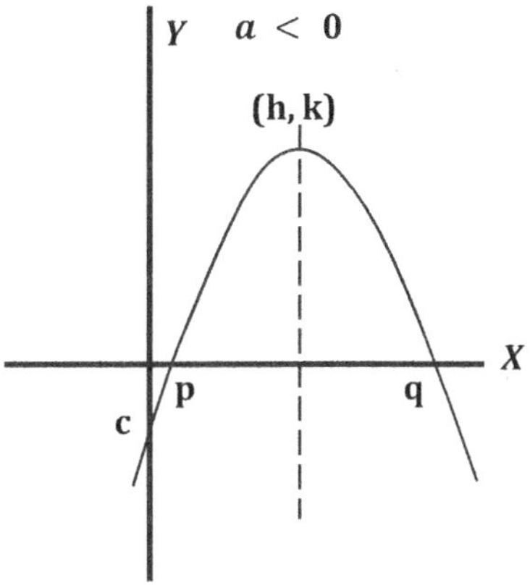

The line of symmetry: $x = h$

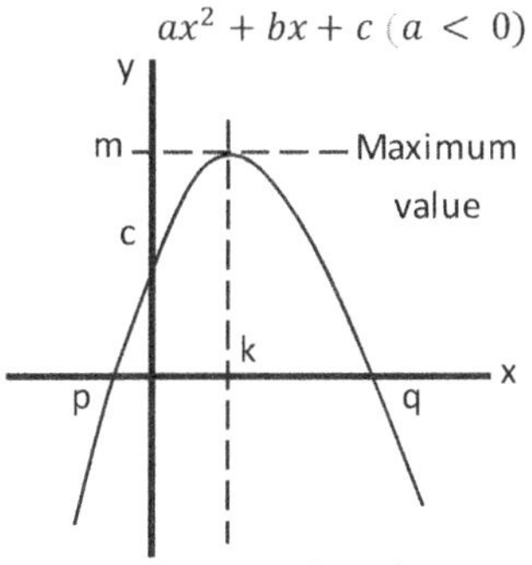

Real roots; value of m is positive, k is the mid-point of the roots p and q

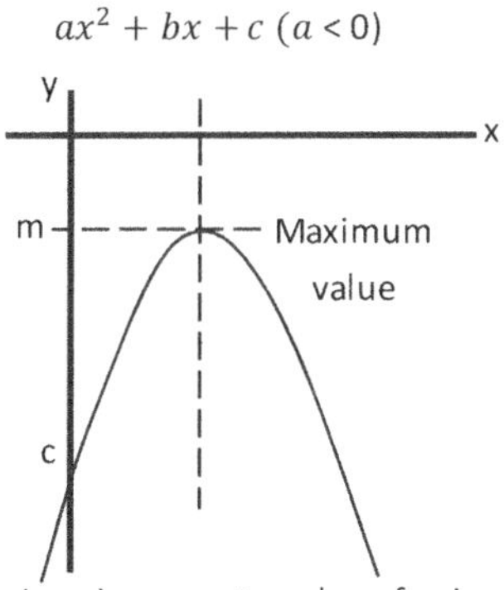

Imaginary roots; value of m is negative

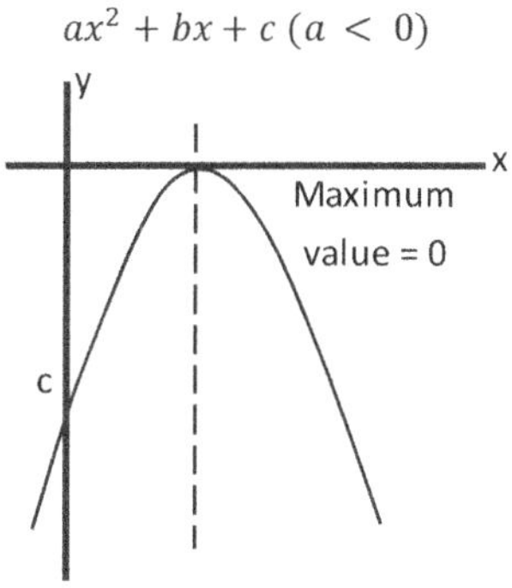

Single real root; value of m is zero

Additional solved problems:

(1) A man can swim in still water (without any current) at a speed of 4 miles per hour. He undertakes to swim from point A to point B and back in a river that has a current of its own. If the distance between points A and B is 15 miles and the total time he takes for the trip is 8 hours, what is the speed of the current in miles per hour?

Note: While swimming against the river, the man's normal swimming speed would be reduced by the speed of the current; similarly, while going along with the flow of the river, his speed would be increased by the speed of the current.

Explanation:

Let the speed of the river current be r miles per hour.

Since the man makes a round trip, one way he would be swimming against the river, while the other way, he would be swimming with the river.

Thus, while swimming against the river, the speed of the man = $(4 - r)$ miles per hour

Thus, time taken to cover 15 miles = $\left(\frac{15}{4 - r}\right)$ hours.

While swimming with the river, the speed of the man = $(4 + r)$ miles per hour

Thus, time taken to cover 15 miles = $\left(\frac{15}{4 + r}\right)$ hours.

Since the total time is 8 hours, we have:

$$\frac{15}{4 - r} + \frac{15}{4 + r} = 8$$

$$=> 15\,(4 + r + 4 - r) = 8\,(4^2 - r^2)$$

$$=> 8r^2 = 8 \times 4^2 - 15 \times 8 = 128 - 120$$

$$=> 8r^2 = 8 => r^2 = 1$$

$$=> r = 1$$

(2) A man buys a number of pieces of chocolate for \$24. If the price of a piece of chocolate increases by \$2, he can buy one piece of chocolate less for the same amount. What is the price of one piece of chocolate?

Explanation:

Let the price of a piece of chocolate be $\$x$.

Thus, the number of chocolate pieces the man can buy for \$24 = $\left(\frac{24}{x}\right)$

New price of a piece of chocolate = \$ $(x + 2)$.

Thus, the number of chocolate pieces the man can now buy for \$24 = $\left(\frac{24}{x+2}\right)$

Since the number of pieces of chocolate is 1 less than the previous occassion, we have:

$\frac{24}{x+2} = \frac{24}{x} - 1$

$=> \frac{24}{x} - \frac{24}{x+2} = 1$

$=> \frac{24(x + 2 - x)}{x(x+2)} = 1$

$=> x(x + 2) = 48$

Since $48 = 6 \times 8$, we have:

$x = 6$

Alternately, we can solve the quadratic:

$x(x + 2) = 48 => x^2 + 2x - 48 = 0$

$=> (x + 8)(x - 6) = 0$

$=> x = 6$

(3) The total cost, in dollars, of manufacturing n items of a product is given by $(2n^2 + 30)$. The selling price of each item is fixed at \$36. What is the number of items that must be sold so as to have maximum profit?

Explanation:

Total cost of manufacturing n items = \$ $(2n^2 + 30)$.

Selling price of each item = \$36.

Thus, total selling price of n items = \$ $(36n)$.

Thus, profit earned = Selling price − Cost price

$= 36n - (2n^2 + 30)$

$= -2\,(n^2 - 18n + 15)$

$= -2\,(n^2 - 18n + 81 - 66)$

$= -2(n - 9)^2 + 132$

The profit would be maximized if the negative square term, i.e. $-2(n - 9)^2$ becomes '0', which happens when $n = 9$.

The corresponding maximum value of the profit is \$132.

7.2 Inequalities

(1) **Number line based inequalities**: There are four important regions on a number line, as shown below:

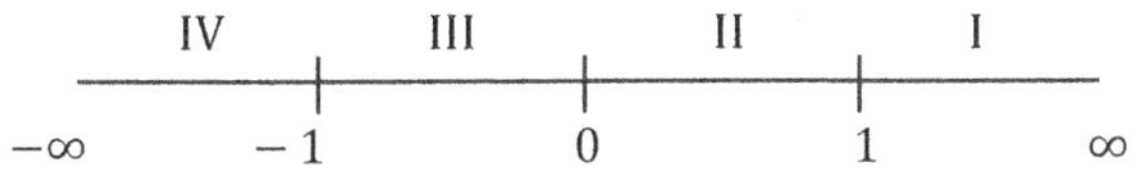

Some properties of the above four regions:

- **Region I**: A number x in the region satisfies $1 < x < \infty$

  Higher the exponent of x, higher is the value of the term and vice versa.
  Thus, we have:
  $x < x^2 < x^3 \dots$ For example: $2 < 2^2 < 2^3$
  $x > \sqrt{x} > \sqrt[3]{x} \dots$ For example: $2 > \sqrt{2} > \sqrt[3]{2}$

- **Region II**: A number x in the region satisfies $0 < x < 1$

  Higher the exponent of x, smaller is the value of the term and vice-versa.
  Thus, we have:
  $x > x^2 > x^3 \dots$ For example: $\frac{1}{2} > \left(\frac{1}{2}\right)^2 > \left(\frac{1}{2}\right)^3$
  $x < \sqrt{x} < \sqrt[3]{x} \dots$ For example: $\frac{1}{2} < \sqrt{\frac{1}{2}} < \sqrt[3]{\frac{1}{2}}$

- **Region III**: A number x in the region satisfies $-1 < x < 0$

  - For odd exponents (the values are always negative):
    Higher the exponent of x, higher is the value of the term and vice versa.
    Thus, we have:
    $x < x^3 < x^5 \dots$ For example: $-\frac{1}{2} < \left(-\frac{1}{2}\right)^3 < \left(-\frac{1}{2}\right)^5$
    $x > \sqrt[3]{x} > \sqrt[5]{x} \dots$ For example: $-\frac{1}{2} > \sqrt[3]{-\frac{1}{2}} > \sqrt[5]{-\frac{1}{2}}$

  - For even exponents (the values are always positive):
    Higher the exponent of x, smaller is the value of the term and vice versa.
    Thus, we have:
    $x^2 > x^4 > x^6 \dots$ For example: $\left(-\frac{1}{2}\right)^2 > \left(-\frac{1}{2}\right)^4 > \left(-\frac{1}{2}\right)^6$

    (Note: Square roots and fourth roots, etc. are not possible for negative numbers)

    An even exponent results in a positive value, which will always be greater than the value resulting from an odd exponent, which is always negative.

- **Region IV**: A number x in the region satisfies $-\infty < x < -1$
  - For odd exponents (the values are always negative):
    Higher the exponent of x, smaller is the value of the term and vice versa.

    Thus, we have:

    $x > x^3 > x^5 \ldots$ For example: $-2 > (-2)^3 > (-2)^5$

    $x < \sqrt[3]{x} < \sqrt[5]{x} \ldots$ For example: $-2 < \sqrt[3]{-2} < \sqrt[5]{-2}$

  - For even exponents (the values are always positive):
    Higher the exponent of x, higher is the value of the term and vice versa.

    Thus, we have:

    $x^2 < x^4 < x^6 \ldots$ For example: $(-2)^2 < (-2)^4 < (-2)^6$

    (Note: Square roots and fourth roots, etc. are not possible for negative numbers)

    An even exponent results in a positive value, which will always be greater than the value resulting from an odd exponent, which is always negative.

(2) **Quadratic inequalities**: We have the following two rules:

- $x^2 < k^2 => -k < x < k$
- $x^2 > k^2 => x < -k$ OR $x > k$

  A simple way to remember this is to take square root and add $\pm$ to the x term:

  - $x^2 < k^2 => \pm x < k => x < k$ or $x > -k => -k < x < k$
  - $x^2 > k^2 => \pm x > k => x > k$ or $x < -k$

Any quadratic can be converted to one of the above two forms.

Let us take an example:

$x^2 - 8x + 12 < 0$

$=> (x^2 - 2 \times x \times 4 + 4^2) - 4^2 + 12 < 0$

$=> (x - 4)^2 < 4$

$=> -2 < x - 4 < 2$

$=> 2 < x < 6$

Alternately, we can use the rules below: If $(x - k)$ and $(x - m)$ are the factors of a quadratic, and $k > m$, we have:

- $(x-k)(x-m) > 0 \Rightarrow x > k$ OR $x < m$
  Thus, x is greater than the greatest root OR smaller than the smallest root.

- $(x-k)(x-m) < 0 \Rightarrow m < x < k$
  Thus, x lies between the two roots.

  Let us take an example:

  $x^2 - 8x + 12 < 0$

  $\Rightarrow (x-2)(x-6) < 0$

  Since the roots are 2 and 6, with 6 being the greater root, we have:

  $\Rightarrow 2 < x < 6$

Additional solved problems:

(1) Which of the following is true about x if $x^2 > x > x^3$?

(A) $x > 1$
(B) $0 < x < 1$
(C) $-1 < x < 0$
(D) $x < -1$

Explanation:

We know that there are four major regions on the number line:

I. $x > 1$
II. $0 < x < 1$
III. $-1 < x < 0$
IV. $x < -1$

Let us pick 1 number from each region and check whether it satisfies the given inequality:

I. $x = 2 : 2^2 > 2 \ngtr 2^3$
II. $x = \frac{1}{2} : \left(\frac{1}{2}\right)^2 \ngtr \frac{1}{2} > \left(\frac{1}{2}\right)^3$
III. $x = -\frac{1}{2} : \left(-\frac{1}{2}\right)^2 > -\frac{1}{2} \ngtr \left(-\frac{1}{2}\right)^3$
IV. $x = -2 : (-2)^2 > -2 > (-2)^3$ – Satisfies

Hence, correct answer is option D.

Alternate approach:

Since x^2 is greater than both x, a term with a lower exponent and x^3, a term with a higher exponent, the value of x must be negative.

Since x, a term with a smaller odd exponent, is greater than x^3, a term with a higher odd exponent, the value of x must be less than -1.

7.3 Functions and graphs of functions

Domain and Range:

The Domain refers to the set of values of x that can be used in the function.

The Range refers to the set of values of $f(x)$ obtained using the above values of x.

Let us take an example:

$f(x) = x^2 - 1$

Let the set of values of x to be used be $\{-1,\ 0,\ 1\}$.

Thus, we have:

- $f(-1) = (-1)^2 - 1 = 0$
- $f(0) = 0^2 - 1 = -1$
- $f(1) = 1^2 - 1 = 0$

Here, Domain = $\{-1,\ 0,\ 1\}$ and Range = $\{-1,\ 0\}$

To determine the Domain for a function, two things need to be considered:

- For any term under a square-root or fourth-root, etc., the term should be non-negative.
  For example: $f(x) = \sqrt{x-1}$
  We have: $x - 1 \geq 0 => x \geq 1$
  Thus, the Domain is: $1 \leq x < \infty$

- For any term in the denominator, the term must be non-zero.
  For example: $f(x) = \dfrac{2x}{x-3}$
  We have: $x - 3 \neq 0 => x \neq 3$
  Thus, the Domain is: x is any real number except 3
  $=> -\infty < x < 3$ OR $3 < x < \infty$

Let us take an example:

$f(x) = \dfrac{x+3}{\sqrt{2x-4}-4}$

Thus, we have:

- $2x - 4 \geq 0 => x \geq 2 \ldots$(i)
- $\sqrt{2x-4}-4 \neq 0\ => \sqrt{2x-4} \neq 4\ => 2x-4 \neq 16 => x \neq 10 \ldots$(ii)

Thus, the Domain: $2 \leq x < 10$ OR $10 < x < \infty$

Note: The radical symbol for square root only gives the positive square root, i.e. $\sqrt{4} = 2$, not ± 2.

Composite functions:

For any two functions $f(x)$ and $g(x)$, the functions defined as $f(f(x))$, $f(g(x))$, $g(g(x))$ and $g(f(x))$ are composite functions.

Let us take an example:

$f(x) = 2x + 1$ and $g(x) = x^2 - 1$

Thus, we have:

- $f(f(x)) = 2(f(x)) + 1 = 2(2x + 1) + 1 = 4x + 3$
- $f(g(x)) = 2(g(x)) + 1 = 2(x^2 - 1) + 1 = 2x^2 - 1$
- $g(g(x)) = (g(x))^2 - 1 = (x^2 - 1)^2 - 1 = x^4 - 2x^2$
- $g(f(x)) = (f(x))^2 - 1 = (2x + 1)^2 - 1 = 4x^2 + 4x$

Special case in a composite function:

If $f(x)$ and $g(x)$ are two functions and it is observed that $f(g(x)) = g(f(x)) = x$, we have:

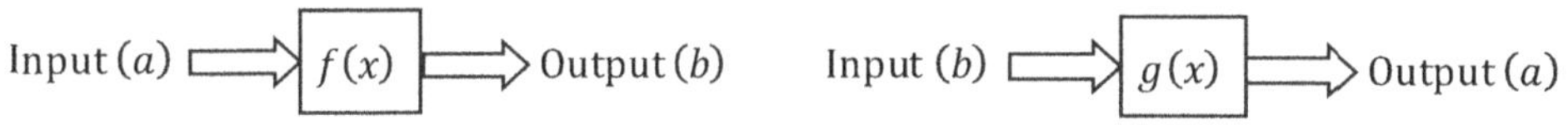

Thus, we have:

If $f(a) = b \Rightarrow g(b) = a$

Note: Such functions $f(x)$ and $g(x)$ are inverse functions of one another.

Let us take an example:

If $f(x) = (x + 1)^3 - 1$ and $g(x) = \sqrt[3]{x + 1} - k$, such that $f(g(x)) = g(f(x))$, what is the value of k?

The normal way of solving, by evaluating the composite functions $f(g(x))$ and $g(f(x))$, is complicated. Instead, we use the above method:

$f(1) = (1 + 1)^3 - 1 = 7$

$\Rightarrow g(7) = 1$

$=> \sqrt[3]{7+1} - k = 1 => 2 - k = 1$

$=> k = 1$

Periodic function:

A function $f(x)$ is periodic if there exists a number n so that $f(x+n) = f(x)$ for all x. Here, n is the period of the function.

Let us take an example:

If $f(x+3) = f(x+2) - f(x+1)$, what is the value of n if $f(1) = -f(1+n)$?

We have: $f(x+3) = f(x+2) - f(x+1)$

Substituting different values of x:

- $x = 0 : f(3) = f(2) - f(1) \ldots$ (i)
- $x = 1 : f(4) = f(3) - f(2) \ldots$ (ii)

Adding (i) and (ii):

$f(3) + f(4) = f(3) - f(1)$

$=> f(1) = -f(4)$

$=> f(1) = -f(1+3)$

$=> n = 3$

Properties of graphs of functions:

- The graph of $f(x+p)$ is obtained by shifting the graph of $f(x)$ by p units left
- The graph of $f(x-p)$ is obtained by shifting the graph of $f(x)$ by p units right
- The graph of $f(x)+p$ is obtained by shifting the graph of $f(x)$ by p units up
- The graph of $f(x)-p$ is obtained by shifting the graph of $f(x)$ by p units down
- The graph of $f(-x)$ is obtained by reflecting the graph of $f(x)$ about the Y-axis
- The graph of $-f(x)$ is obtained by reflecting the graph of $f(x)$ about the X-axis

Graphs of some quadratic functions:

- $f(x) = x^2$:

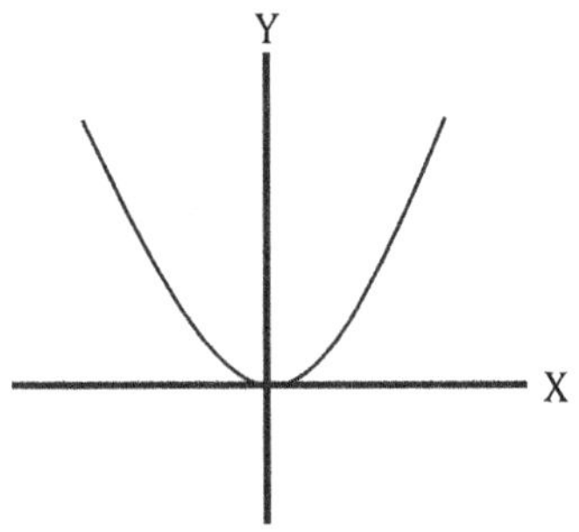

- $f(x) = x^2 + 1$:

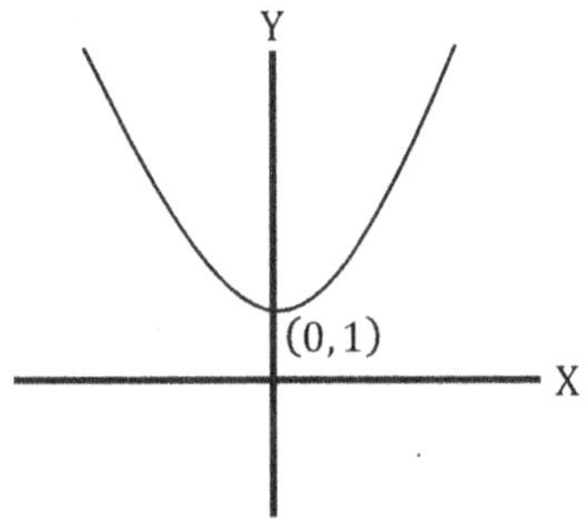

- $f(x) = (x - 1)^2$:

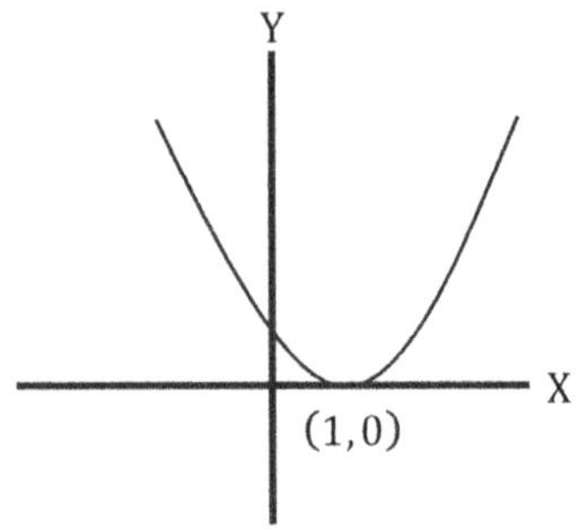

Additional solved problems:

(1) Are the following functions the same?

(A) $f(x) = x^x$

(B) $g(x) = x \times x \times x \times x \times \cdots \times x$ (x times)

Explanation:

(A) We have $f(x) = x^x$

For example:

If $x = 3 => f(x) = 3^3 = 27$

If $x = -1 => f(-1) = (-1)^{-1} = -1$

If $x = \frac{1}{2} => f\left(\frac{1}{2}\right) = \left(\frac{1}{2}\right)^{\frac{1}{2}} = \frac{1}{\sqrt{2}} =\approx \frac{1}{1.4} = 0.71$

Thus, $f(x) = x^x$ is valid for all real values of x except for reciprocals of negative even numbers like $x = -\frac{1}{2}$. (since, if $x = -\frac{1}{2} => x^x = \left(-\frac{1}{2}\right)^{\left(-\frac{1}{2}\right)} = \frac{1}{\left(-\frac{1}{2}\right)^{\left(\frac{1}{2}\right)}}$, an imaginary number.

(B) We have: $g(x) = x \times x \times x \times x \times \cdots \times x$ (x times)

If x is multiplied for x times, the result obtained is x^x.

Thus, apparently, $f(x)$ and $g(x)$ appear to be identical.

However, we are multiplying x for x times, which makes sense only when x is a positive integer. (Multiplying $\frac{1}{2}$ for $\frac{1}{2}$ times or multiplying -1 for -1 times makes no sense)

For example:

$g(3) = 3 \times 3 \times 3$ (3 is multiplied for 3 times)

$= 27$

Thus, $g(x)$ is valid only for positive integer values of x.

Thus, $f(x)$ is not the same as $g(x)$.

(2) If $(2, 1)$ are the coordinates of a point on the graph of $f(x)$, what would be the coordinates of that point for the function $-f(x) + 1$?

Explanation:

To modify $f(x)$ to $-f(x) + 1$, we follow the following steps:

- $f(x) \to -f(x)$: The graph is reflected about the X-axis. Thus, the Y-coordinate of the point would be negated.
  Thus, the coordinates of $-f(x) = (2, -1)$

- $-f(x) \to -f(x) + 1$: The graph is shifted 'up' by 1 unit. Thus, the Y-coordinate of th point would increase by '1'.
  Thus, the final coordinates of $-f(x) + 1 = [2, (-1 + 1)] = (2, 0)$

(3) If (2, 1) are the coordinates of a point on the graph of $f(x)$, what would be the coordinates of that point for the function $f(x - 1)$?

Explanation:

We need to modify $f(x) \rightarrow f(x-1)$, i.e. the graph should shift one unit left, hence, value of x should decrease by 1. Thus, (2, 1) gets modified to (1, 1).

Chapter 8

Advanced Math: Practice Questions

1. If $p(1+r^2)^3 = q$, and r is a positive quantity, which of the following gives the correct relation of r in terms of p and q?

(A) $r = \left\{\left(\frac{p}{q}\right)^{\frac{1}{3}} - 1\right\}^{\frac{1}{2}}$

(B) $r = \left\{\left(\frac{p}{q}\right)^{\frac{1}{2}} - 1\right\}^{\frac{1}{3}}$

(C) $r = \left\{\left(\frac{q}{p}\right)^{\frac{1}{3}} - 1\right\}^{\frac{1}{2}}$

(D) $r = \left\{\left(\frac{q}{p}\right)^{\frac{1}{2}} - 1\right\}^{\frac{1}{3}}$

2. A factory manufactures s units of a product X whose average manufacturing cost per unit, in dollars, is a function of s, given as $f(s) = 3 + 2.5s$. If each item is sold at a fixed price of $\$p$, which of the following represents the condition for a profit by selling all s units?

(A) $3s + 2.5s - ps > 0$

(B) $3 + 2.5s^2 - ps > 0$

(C) $3 + 2.5s^2 - p < 0$

(D) $3s + 2.5s^2 - ps < 0$

3. A man saves $\$x$ in a particular month. Thereafter, each month, he saves $\$y$ more than he saved in the respective previous months. If he saved $\$z$ in the tenth month from the beginning, what is the relationship between x, y, and z?

(A) $y = \frac{z+x}{10}$

(B) $y = \frac{z+x}{9}$

(C) $y = \frac{z-x}{10}$

(D) $y = \frac{z-x}{9}$

4. In a bacterial sample with n bacteria, it was found that each bacterium divides into two identical cells after 20 minutes. After t hours, the number of bacteria was found to be m. Which of the following represents the relationship between m, n, and t?

(A) $n = m \times 2^t$

(B) $n = m \times 2^{3t}$

(C) $n = m \times 2^{-3t}$

(D) $n = m \times 2^{-t}$

5. **SPR:**

The distance, in miles, covered by a rocket in t seconds after its launch is given by the function $d(t) = 3t^2$. What is the distance, in miles, covered by the rocket in the 5th second after its launch?

6. **SPR:**

If the roots of the equation $x^2 - 2x - 15 = 0$ are a and b, where $a > b$, what is the value of $a + b^2$?

7. If the roots of the equation $x^2 - (b + c)x + bc = 0$ are equal in magnitude and opposite in sign, which of the following statements are true?

(A) $b = -c$

(B) $b = c$

(C) $b = c^2$

(D) $b^2 = c$

8. If a and b are the roots of $x^2 - 10x + 16 = 0$, what is the equation whose roots are given by $(a + b)^2$ and $(a - b)^2$?

(A) $x^2 - 16x - 60 = 0$

(B) $x^2 + 4x + 60 = 0$

(C) $x^2 - 3600x + 136 = 0$

(D) $x^2 - 136x + 3600 = 0$

9. **SPR:**

In the quadratic equation $x^2 + (a + 1)x + (a + 5) = 0$, if the sum of the roots is -5, what is the product of the roots?

10. If $9^x - 3^{x+1} - 54 = 0$, what is the value of 9^x?

(A) -6

(B) 2

(C) 9

(D) 81

11. If $\sqrt{x+2} + \sqrt{x-1} = 3$, what is the value of x^2?

(A) 2

(B) 4

(C) 8

(D) 16

12. A quadratic equation $x^2 + 6x + 8 = 0$ has two roots p and q, while another quadratic equation has the roots $\frac{1}{p}$ and $\frac{1}{q}$. Which of the following could be that quadratic equation?

(A) $6x^2 + 8x + 1 = 0$

(B) $8x^2 + 6x + 1 = 0$

(C) $x^2 + 6x + 8 = 0$

(D) $8x^2 + 6x - 1 = 0$

13. If $x^2 - 5x + 6 = 0$ and $y^2 + 6y - 7 = 0$, which of the following is correct?

(A) Both values of x are greater than both values of y.

(B) Both values of y are greater than both values of x.

(C) Both values of x are equal to both values of y.

(D) One value of the values of x is greater than one value of of the values of y.

14. **SPR:**

If the breadth of a rectangle is 6 centimeters less than its length and the area of the rectangle is 16 square centimeters, what is the perimeter of the rectangle in centimeters?

15. $f(x) = x^2 + 6x + 8$ and $h(x) = 32 + 4x$. At how many points do the graphs of the two functions intersect?

(A) None

(B) One

(C) Two

(D) Three

16. **SPR:**

The sum of the squares of two consecutive positive odd numbers is 74. What is the sum of the two numbers?

17. A local club has a rectangular football field whose length is 16 meters more than its width. There is a running track surrounding the football field of width 2 meters. If the area of the track is $\frac{1}{3}$ of the football field, what is the length of the football field (in meters)?

(A) 18

(B) 20

(C) 24

(D) 36

18. Peter bought some oranges from the market for \$144. He later calculated that if each orange cost \$2 less, he could buy exactly 12 more oranges for the same money. How many oranges did Peter buy?

(A) 12

(B) 18

(C) 24

(D) 26

19. A man drove at a uniform speed of x miles per hour and reached his destination in $(x + 2)$ hours. Had he driven at y more miles per hour, he would have reached his destination in $\frac{3y}{4}$ hours less time. Which of the following is the correct relation between x and y?

(A) $x = 8y - 3$
(B) $x = 8y + 3$
(C) $x = 3y - 8$
(D) $x = 8 - 3y$

20. If $\frac{1}{2}\left(\frac{1}{x-4}\right)(x^2 - 16) = \left(1 + \frac{4}{x-4}\right)(x - 4)$, and $x \neq 4$, how many values of x exist?

(A) None
(B) One
(C) Two
(D) Three

21. A man deposits \$100 in a bank. The bank gives interest at the rate of $r\%$ every 3 years, compounded triennially (computed once in every three years). What will be the amount of money in the bank after n years?

(A) $\$\left\{100\left(1 + \frac{r}{100}\right)^{3n}\right\}$
(B) $\$\left\{100\left(1 + \frac{r}{100}\right)^{\left(\frac{n}{3}\right)}\right\}$
(C) $\$\left\{100\left(1 + \frac{r}{300}\right)^{n}\right\}$
(D) $\$\left\{100\left(1 + \frac{r}{300}\right)^{3n}\right\}$

22. **SPR:**

Initially, two strains of bacteria had 1,024 and 64 bacteria, respectively. It was observed that the first strain doubled every 9 hours, while the second strain doubled every 3 hours. In how many hours would the strains have the same number of bacteria?

23. What is the value of $(a + b)$ if the equations $x^2 - 6ax + 12 = 0$ and $x^2 - 4x + 9b = 0$ have both roots common?

(A) 1
(B) $\frac{2}{3}$
(C) $\frac{4}{3}$
(D) 2

24. What is the sum of all possible values of x if $(x - 3)^{(x-6)} = 1$?

(A) 6
(B) 8
(C) 10

(D) 12

25. The distance, in feet, covered by a ball in t seconds after dropped from the top of a building is $9t^2$. If the distance covered by the ball in the tth second after it was dropped is h feet, which of the following is the correct expression of t in terms of h?

(A) $\frac{1}{2}(h+9)$

(B) $\frac{1}{2}\left(\frac{h}{9}-1\right)$

(C) $\frac{1}{2}\left(\frac{h}{9}+1\right)$

(D) $2\left(\frac{h}{9}+1\right)$

26. **SPR:**

If one of the solutions of $x^2 - px + 12 = 0$ is $x = 3$, what is the value of p?

27. The curve $y = x^2 + 3x + 1$ intersects the line $y = x + 1$ at two points. What is the y coordinate of the intersecting point closer to the x-axis?

(A) -3

(B) -2

(C) 0

(D) 1

28. For $f(x) = x^2 + bx + c,\ f(1) = 2$. If $f(2) = 0$, what is the value of $f(0)$?

(A) -6

(B) -5

(C) 5

(D) 6

29. If $f(x) = x^3 - kx + 2$ and $f(2) = f(-2)$, what is the value of k?

(A) 0

(B) 1

(C) 2

(D) 4

30. The function $f(x)$ is defined according to the given table:

x	$f(x)$
1	2
2	5
3	8
4	11

Which of the following is a correct representation of the function $f(x)$?

(A) $x^2 + 1$

(B) $3x - 1$

(C) $x + 4$

(D) $x^3 + 1$

31. Functions $f(x)$ and $g(x)$ are defined such that their values for different values of x are as shown:

x	$f(x)$	$g(x)$
1	5	3
2	4	4
3	3	5
4	2	6
5	1	7

What is the value of $f(g(1)) + g(f(1))$?

(A) 2

(B) 3

(C) 7

(D) 10

32. If $f(x) = |x| + x$ and $g(x) + f(x) = 2x$, what is the value of $g(-2)$?

(A) -8

(B) -4

(C) 0

(D) 8

33. **SPR:**

If $f(x) = ax^2 + x$ such that $f(1) = f(0) + 2$, what is the value of $f(2)$?

34. If $x = \sqrt{3} - 1$, what is the value of $(x^2 + 2x)$?

(A) -2

(B) $-\sqrt{3}$

(C) $\sqrt{3}$

(D) 2

35. For all integers $x > 1$, let $f(x)$ be defined as $f(x) = (-1)^x \times f(x-1)$. If $f(1) = 1$, what is the value of $f(4)$?

(A) -3

(B) -1

(C) 1

(D) 3

36.

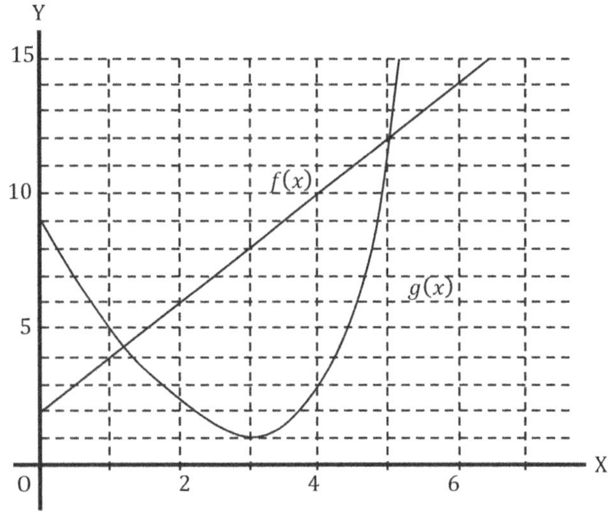

The given graph represents two functions $f(x)$ and $g(x)$. If $f(1) = k$, which of the following is the value of $g(k)$?

(A) 0

(B) 1

(C) 2

(D) 3

37. If $a > 1$, which of the following is an equivalent expression of $\left(\frac{1}{\sqrt{a+1}+\sqrt{a}}\right)$?

(A) $\sqrt{a-1}+\sqrt{a}$

(B) $\sqrt{a-1}-\sqrt{a}$

(C) $\sqrt{a+1}-\sqrt{a}$

(D) $\sqrt{a+1}+\sqrt{a}$

38. SPR:

$f(x) = \frac{x-k}{5}$ and $g(x) = 5x + 1$. If $f(g(x)) = x$, what is the value of k?

39. SPR:

A function $f(x)$ is defined such that:

- $f(x) = f(x-1) + 2$ if x is prime
- $f(x) = f(x-1) + 1$ if x is non-prime

If $f(1) = 3$, what is the value of (5)?

40. If $f(x) = \dfrac{2^x + 1}{2^x}$, which of the following is correct?

(A) $f(-1) = f(1)$
(B) $f(-1) = -f(1)$
(C) $f(-1) = -f(-1)$
(D) $f(-1) = 2f(1)$

41. If $f(x) = ax^2 + b$ and $f(x+1) = f(x) + x + \frac{1}{2}$, for all values of x, what is the value of a?

(A) -2
(B) -1
(C) $\frac{1}{2}$
(D) 1

42. If $f(x) = 2^{x-1}$, what is the value of $\dfrac{f(1) \times f(2)}{f(3)}$?

(A) 0.04
(B) 0.06
(C) 0.25
(D) 0.50

43. A man puts \$1,000 in a bank that offers r% interest compounded annually. After 3 years, the amount of money in the bank is \$$P$. Which of the following is the correct expression of r in terms of P?

(A) $100\left\{\frac{P^{\left(\frac{1}{3}\right)}}{1{,}000} - 1\right\}$
(B) $100\left\{\left(\frac{P}{1{,}000}\right)^3 - 1\right\}$
(C) $100\left\{\left(\frac{P}{1{,}000}\right)^{\left(\frac{1}{3}\right)} - 1\right\}$
(D) $100\left\{\left(\frac{P}{1{,}000}\right)^{\left(\frac{1}{3}\right)} + 1\right\}$

44. A man travels a distance of d miles at a speed of s miles per hour. Had he traveled the distance at $(s + 10)$ miles per hour, he would have reached his destination more than an hour earlier. Which of the following is the correct relation between s and d?

(A) $ds > \frac{s}{10} + 10$
(B) $ds > \frac{s}{10} - 1$
(C) $\frac{1}{10}s\,(s + 10) < d$
(D) $\frac{1}{10}s\,(s + 10) > d$

45. An athlete follows a strict training regimen: on the first day, he runs a distance of d units; thereafter, each day, he increases the distance by 20%. If, on the final day of his regimen, he runs more than D units, what is the relationship among the number of days, n, for which he follows the training regimen and the distances he ran on the first and last days?

(A) $\left(\frac{D}{d}\right)^{(n-1)} > \frac{6}{5}$

(B) $\left(\frac{D}{d}\right)^{(n-1)} < \frac{6}{5}$

(C) $\left(\frac{D}{d}\right)^{\left(\frac{1}{n-1}\right)} > \frac{6}{5}$

(D) $\left(\frac{D}{d}\right)^{\left(\frac{1}{n-1}\right)} < \frac{6}{5}$

46. If $m = px + qy$ and $n = qx + py$, and $q^2 \neq p^2$, what is x in terms of m, n, p, and q?

(A) $x = \dfrac{mq - np}{q^2 - p^2}$

(B) $x = \dfrac{mp - nq}{p^2 - q^2}$

(C) $x = \dfrac{mq - np}{p^2 - q^2}$

(D) $x = \dfrac{mp - nq}{q^2 - p^2}$

47. The terms A, B, C, and D are related by the equation $A \times \sqrt[3]{C} = \frac{B \times \sqrt{D}}{5}$. What is the value of C if $A = 5$, $B = 2$, and $D = 16$?

(A) $\frac{8}{25}$

(B) $\frac{64}{625}$

(C) $\frac{512}{15{,}625}$

(D) $\frac{256}{15{,}625}$

48. The half-life of a substance is the number of hours it takes for the mass of the substance to become half its original value. The initial weight of one such substance is 28 grams. If the half-life of the substance is 42 hours, which of the following represents the mass of the substance after 210 hours?

(A) $28 \times \left(\frac{1}{2}\right)^{109}$

(B) $28 \times \left(\frac{1}{2}\right)^{42}$

(C) $28 \times \left(\frac{1}{2}\right)^{5}$

(D) $28 \times \left(\frac{1}{2}\right)^{2}$

49. The force, F, between two bodies having masses M and m, placed at a distance of D, is given by the equation $F = \frac{GMm}{D^2}$, where G is a constant. If the distance between the masses and the mass of one of the bodies are doubled, what will be the effect on the force?

(A) The force is quadrupled.

(B) The force is doubled.

(C) The force remains unchanged.

(D) The force is halved.

50. A man types x pages on the first day, and thereafter, every day he increases his typing speed by $n\%$ from the previous day until he completes the task in D days. If he works for the same duration of time every day, which of the following represents the correct expression for n in terms of D, x, and the number of pages, p, that he typed on the last day?

(A) $n = \left\{\left(\frac{p}{x}\right)^{(D-1)} - 1\right\} \times 100$

(B) $n = \left\{\left(\frac{p}{x}\right)^{(D)} - 1\right\} \times 100$

(C) $n = \left\{\left(\frac{p}{x}\right)^{\left(\frac{1}{D}\right)} - 1\right\} \times 100$

(D) $n = \left\{\left(\frac{p}{x}\right)^{\left(\frac{1}{D-1}\right)} - 1\right\} \times 100$

51. A ball is thrown from the top of a building. The distance, in feet, covered by the ball in t seconds after it was dropped is $5t^2$ feet. If the distance covered by the ball in the tth second after it was dropped was 45 feet, what is the value of t?

(A) 2

(B) 3

(C) 5

(D) 6

52. Which of the following could be the value of p for which the quadratic equation $x^2 - px + 8p = 0$ has equal roots?

(A) 12

(B) 27

(C) 32

(D) 33

53. If one of the roots of $x^2 - px + 12 = 0$ is 4, what is the value of p?

(A) 3

(B) 7

(C) 12

(D) 15

54. If $y = x^2 + kx + l$ intersects the X-axis at (4, 0) and the Y-axis at (0, 4), what is the value of $(k + l)$?

(A) −20

(B) −5

(C) −4

(D) −1

55. **SPR:**

For $f(x) = x^2 + bx + 1$, if $f(1) = 1$, what is the value of $f(2)$?

56. **SPR:**

$f(x)$ is a function such that $f(2x-1) = x^2 + 3x$. What is the value of $f(3)$?

57. If $f(x) = x^2 + ax + b$ and $f(k) = f(k-1) + 1$, what is the value of k?

(A) $1 - \frac{a}{2}$

(B) $1 + \frac{a}{2}$

(C) $2 - a$

(D) $2 + a$

58. If a and b are non-zero integers such that $a^2 + b^2 - 2a - b = 0$ and $a^2 + b^2 = 2ab$, what is the correct expression of a in terms of b?

(A) $\frac{b}{2(b-1)}$

(B) $\frac{b}{(b-1)}$

(C) $\frac{b-1}{2b}$

(D) $\frac{b-1}{b}$

59. If $x + y = 4$ and $z = x^2 + 2y$, what is the expression of z in terms of y?

(A) $(y-3)^2$

(B) $(y-3)^2 + 7$

(C) $(y-3)^2 + 13$

(D) $(y-3)^2 + 16$

60. If $x = 1 + \sqrt{3}$, what is the value of $(x^2 - 2x)$?

(A) $-\sqrt{3}$

(B) 0

(C) $\sqrt{3}$

(D) 2

61. If $f(x) = \sqrt{x} + 1$ and $f(g(2)) = 4$, then which of the following could be a possible expression for $g(x)$?

(A) $5x - 1$

(B) $x^3 - 1$

(C) $6x + 3$

(D) $x^4 - 1$

62. SPR:

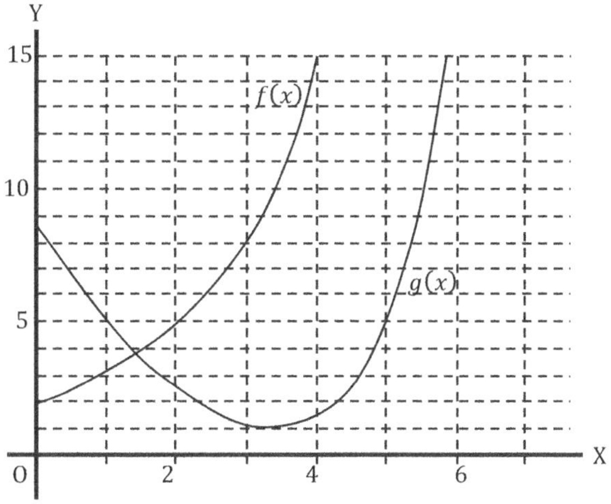

The given diagram represents two functions $f(x)$ and $g(x)$. If $f(2) = k$, what is the value of $g(k)$?

63. Let the function $f(x)$ be defined as $f(x) = x^2 + 1$ and $g(x)$ be a function such that $f(x) = g(x-2) + 1$. What is the expression for $g(x)$?

(A) $(x+1)^2$
(B) $(x-1)^2$
(C) $(x+2)^2$
(D) $(x-2)^2$

64. Which of the following is the x-value of a point of intersection of $f(x) = x + 2^x$ and $g(x) = 2 + x^2$?

(A) -1
(B) 0
(C) 2
(D) 4

65. The function $f(x)$ is such that $f\left(\frac{x}{2}\right) = 2x^2+3x$. Which of the following is the correct expression for $f(x)$?

(A) $f(x) = \dfrac{x^2 + 3x}{2}$
(B) $f(x) = 8x^2 + 3x$
(C) $f(x) = 4x^2 + 6x$
(D) $f(x) = 8x^2 + 6x$

66. Joe spends \$$x$ to purchase p gallons of gasoline every month. Joe calculated that if the price of gasoline increased by \$2 per gallon, he would be able to purchase 1 gallon less for the same amount of money. Which of the following is the correct relation between x and p?

(A) $x = (p-1)(2-p)$
(B) $x = 2p(p-1)$
(C) $x = \frac{p(2+p)}{p-1}$
(D) $x = \frac{p(p-2)}{p-1}$

67. A sum of money, \$$p$, invested at compound interest with $r\%$ annual rate of interest amounts to \$$q$ after 2 years. Which of the following gives the correct expression for r?

(A) $\sqrt{\frac{q}{p}} - 1$
(B) $100\left(\sqrt{\frac{q}{p}} - 1\right)$
(C) $100\left(\sqrt{\frac{p}{q}} - 1\right)$
(D) $100\left(1 - \sqrt{\frac{p}{q}}\right)$

68. The inside dimensions of a rectangular photo frame, which has a uniform width of x inches, are $2a$ inches by a inches, as shown in the diagram. If the area of the frame is equal to the area for the display of the photo, which of the following quadratic equations shows the relation between a and x?

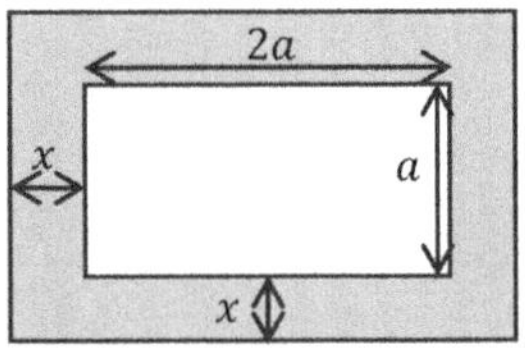

(A) $a^2 = 3ax + 2x^2$
(B) $a^2 = ax + 2x^2$
(C) $a^2 = 3ax - 2x^2$
(D) $a^2 + 3ax = 2x^2$

69. Which of the following values of x satisfies the inequality $\frac{x+6}{x^2} < 1$?

I. $x > 3$
II. $x < 3$
III. $x < -2$

(A) Only I
(B) Only II
(C) Both I and II
(D) Both I and III

70. What is the sum of all possible values of x that satisfy $(x^4 + 8x)(x^2 - 4x + 3) = 0$?

(A) 2
(B) 6

(C) 8

(D) 9

71. SPR:

If $f(x) = x^4 - 13x^2 + 36$, what is the sum of the roots for $f(x) = 0$?

72. For all positive integer values of x, if $f(x) = \frac{1}{x}$ and $g(x) = \frac{x}{x+1}$, what is the value of $f(g(x)) \times g(f(x))$?

(A) 0

(B) 1

(C) $\frac{1}{x}$

(D) $x + 1$

73. If $f(x) = x^2 - 8$ and $g(x) = x - 2$, for how many integer values of x is $f(x) \leq g(x)$?

(A) Three

(B) Four

(C) Seven

(D) Nine

74. The graph of a quadratic function $f(x)$ shown below is symmetric about the line $x = 3$. If $f(1) = 5$, what is the value of $f(5)$?

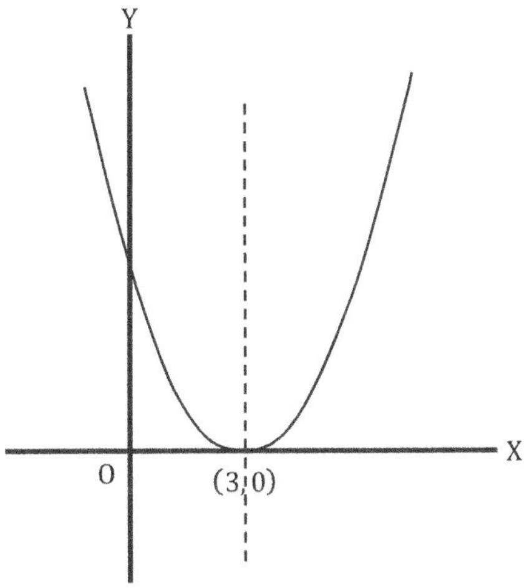

(A) 3

(B) 5

(C) 6

(D) Not enough information

75. In the graph shown below, the functions $f(x) = x^2 - 6x + 8$, which intersects the X-axis at $x = 2$ and $x = 4$, and $g(x) = x - 2$ are shown. For how many integer values of x is $g(x) \geq f(x)$?

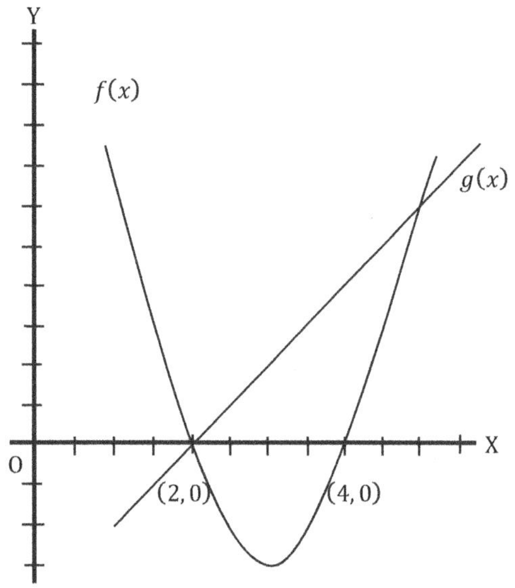

(A) One

(B) Two

(C) Three

(D) Four

76. The graphs of $f(x)$ and $g(x)$ are shown below. Which of the choices is correct?

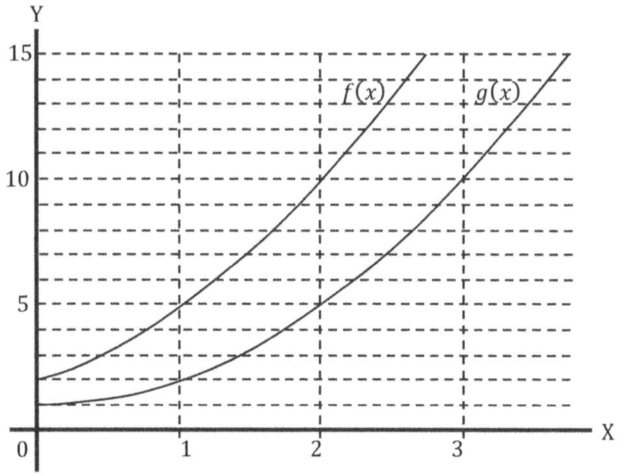

(A) $f(x) = 5x$

(B) $f(x) = (x+1)^2 + 1$

(C) $g(x) = (x+1)^2 - 4$

(D) $g(x) = (x-1)^2 + 4$

77. The graph of a quadratic function $f(x) = x^2 + ax + b$ shown below intersects the X-axis at $(2,\ 0)$ and $(4,\ 0)$. What is the value of $(a + b)$?

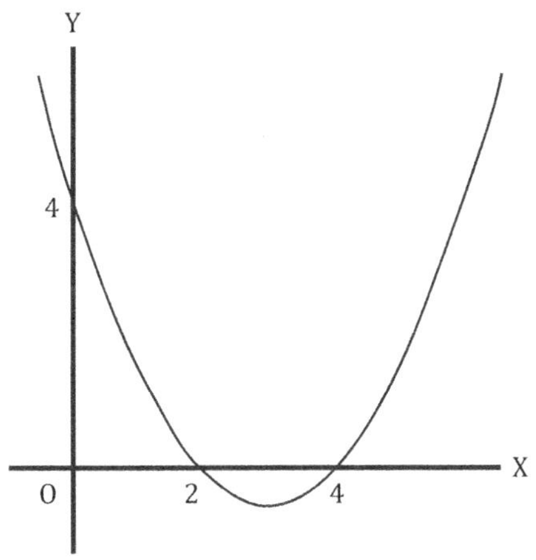

(A) 1
(B) 2
(C) 3
(D) 4

78. What is the value of $\left\{\dfrac{(a+b)^2 - (a^2+b^2)}{(a+b)^2 - (a-b)^2}\right\}$?

(A) 0.3
(B) 0.4
(C) 0.5
(D) 2.0

79. If $2x = a + \frac{1}{a}$, where a is a positive number, what is the value of $x + \sqrt{(x^2 - 1)}$?

(A) 0
(B) 1
(C) a
(D) $\frac{1}{2}\left(a + \frac{1}{a}\right)$

80. **SPR:**

If $x = \frac{1}{2}$ and $y = \frac{1}{3}$, what is the value of $\left(\frac{x^4 - y^4}{x^2 + y^2}\right)$?

81. **SPR:**

If $a + \dfrac{1}{a} = 2$, what is the value of $\left(a^4 + \dfrac{1}{a^4}\right)$?

82. **SPR:**

If $2x + 4y = 10$ and $x^2 + y^2 = 10$, what is the value of $(2x^3 + 4y^3 + 2xy^2 + 4x^2y)$?

83. The graph of $f(x)$ is shown below. Which of the following choices is the graph of $g(x)$, where $g(x) = f\left(\frac{x}{2} - 1\right) + 2$?

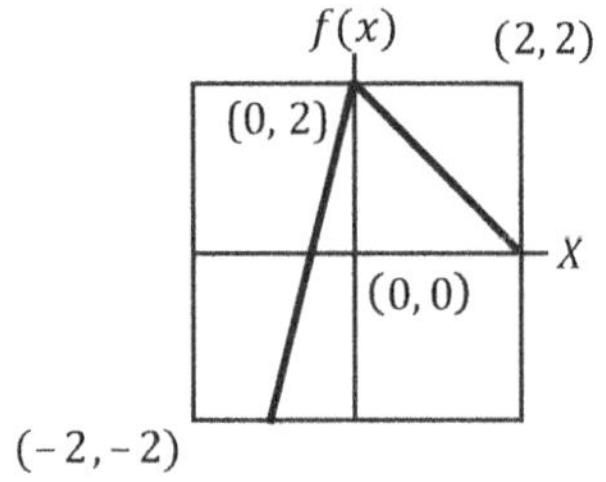

(A)

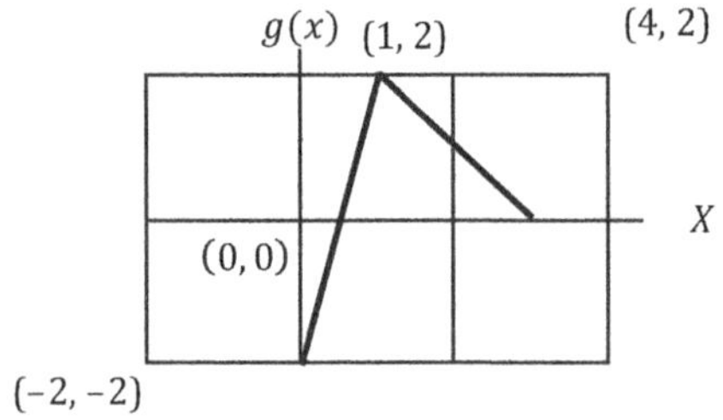

(B)

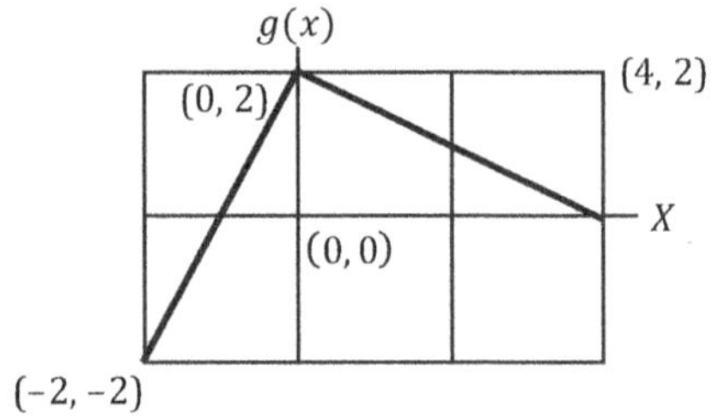

(C)

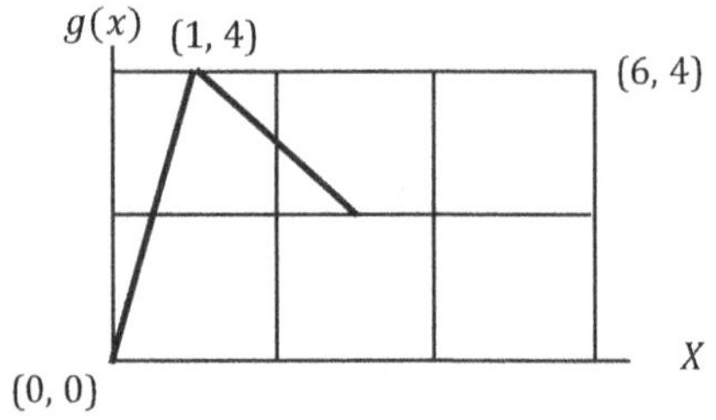

(D)

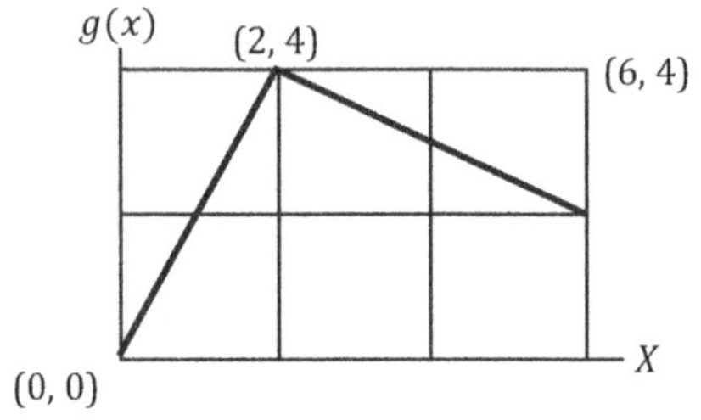

84. **SPR:**

If $\frac{2x+3}{(x+1)(x+2)} = \frac{A}{x+1} + \frac{B}{x+2}$ for all real values of x, what is the value of $(A + B)$?

85. The graph shown below shows the number of bacteria (y), in thousands, in an experiment over a period of time (x) (in hours):

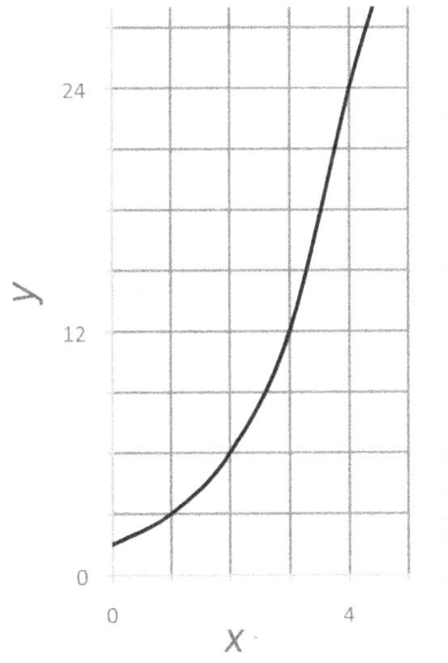

If the above graph represents $y = ab^x$, where y is the number of bacteria in thousands and x is the time in hours, what is the value of $(a + b)$?

(A) 1.5

(B) 2

(C) 3.5

(D) 4

86. The graph $y = \frac{3}{2} \times 2^x$ shown below shows the number of bacteria (y), in thousands, in an experiment over a period of time (x) (in hours):

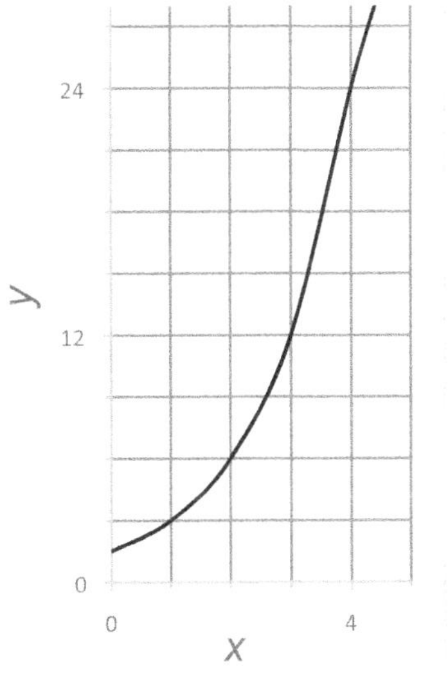

After minimum how many hours from the beginning of the experiment are the number of bacteria at least 10 times the number of bacteria present after 30 minutes from the start of the experiment?

(A) Between 1 and 2 hours

(B) Between 2 and 3 hours

(C) Between 3 and 4 hours

(D) Between 4 and 5 hours

Chapter 9

Advance Math Answer key

(1) C	(23) D	(45) D	(67) B
(2) D	(24) D	(46) B	(68) A
(3) D	(25) C	(47) C	(69) D
(4) C	(26) 7	(48) C	(70) A
(5) 27	(27) D	(49) D	(71) 0
(6) 14	(28) D	(50) D	(72) C
(7) A	(29) D	(51) C	(73) C
(8) D	(30) B	(52) C	(74) B
(9) 9	(31) D	(53) B	(75) D
(10) D	(32) B	(54) D	(76) B
(11) B	(33) 6	(55) 3	(77) B
(12) B	(34) D	(56) 10	(78) C
(13) A	(35) B	(57) A	(79) C
(14) 20cm	(36) D	(58) A	(80) '$\frac{5}{36}$' or '0.139'
(15) C	(37) C	(59) B	(81) 2
(16) 12	(38) 1	(60) D	(82) 100
(17) D	(39) 10	(61) A	(83) D
(18) C	(40) D	(62) 5	(84) 2
(19) C	(41) C	(63) C	(85) C
(20) A	(42) D	(64) C	(86) C
(21) B	(43) C	(65) D	
(22) 18	(44) C	(66) B	

Chapter 10

Advanced Math: Solutions to Practice Questions

1. We have

$$p(1 + r^2)^3 = q$$

$$=> (1 + r^2)^3 = \frac{q}{p}$$

$$=> 1 + r^2 = \left(\frac{q}{p}\right)^{\frac{1}{3}}$$

$$=> r^2 = \left(\frac{q}{p}\right)^{\frac{1}{3}} - 1$$

$=> r = \left\{\left(\frac{q}{p}\right)^{\frac{1}{3}} - 1\right\}^{\frac{1}{2}}$ (Since r is given to be a positive quantity, the negative square root of r^2 is rejected)

The correct answer is option C.

2. Number of units = s

Average cost per unit = \$ $(3 + 2.5s)$

Thus, total cost = \$ $(s \times (3 + 2.5s))$ = \$ $(3s + 2.5s^2)$

Selling price per unit = \$$p$

Thus, total selling price = \$ $(s \times p)$

Thus, for profit, we should have the selling price greater than the cost price:

$$s \times p > 3s + 2.5s^2$$

$$=> 3s + 2.5s^2 - ps < 0$$

The correct answer is option D.

3. Savings in the first month = \$$x$

Since in each month he increases his savings by \$$y$, the total increase in the 10th month, i.e. after 9 months of increasing savings = \$ $(9y)$

Thus:

$$z = x + 9y$$

$$=> y = \frac{z - x}{9}$$

The correct answer is option D.

4. Initial number of bacteria = n

The number of bacteria doubles in every 20 minutes, i.e. $\frac{1}{3}$ of an hour.

Thus, the number of times the bacteria doubles in a period of t hours $= \frac{t}{\left(\frac{1}{3}\right)} = 3t$

Thus, the final number of bacteria

$= n \times 2^{3t}$

$=> m = n \times 2^{3t}$

$=> n = m \times 2^{-3t}$

The correct answer is option C.

5. Distance covered by the rocket in t seconds = $3t^2$ miles.

Distance covered in the 5th second

= (Distance covered in 5 seconds) – (Distance covered in 4 seconds)

Distance covered by the rocket in 5 seconds = $3 \times 5^2 = 75$ miles.

Distance covered by the rocket in 4 seconds = $3 \times 4^2 = 48$ miles.

Thus, (required distance) = 75 – 48 = 27 miles.

The correct answer is 27.

6. $x^2 - 2x - 15 = 0$

$=> x^2 - 5x + 3x - 15 = 0$

$=> x(x - 5) + 3(x - 5) = 0$

$=> (x + 3)(x - 5) = 0$

$=> x = -3$, or 5

Now since it is given that the roots are a and b, where $a > b$, we have

$a = 5, \; b = -3$

$=> a + b^2 = 5 + 9 = 14$

The correct answer is 14.

7. $x^2 - (b + c)x + bc = 0$

$=> x^2 - bx - cx + bc = 0$

$=> x(x - b) - c(x - b) = 0$

$=> (x - c)(x - b) = 0$

$=> x = c$ or b

Since the roots are equal in magnitude but opposite in sign, the sum of the roots must be '0'.

$=> b + c = 0$

$=> b = -c$

The correct answer is option A.

8. $x^2 - 10x + 16 = 0$

$=> x^2 - 2x - 8x + 16 = 0$

$=> x(x - 2) - 8(x - 2) = 0$

$=> (x - 2)(x - 8) = 0$

$=> x = 2$ or 8

So, either $a = 2$ and $b = 8$ Or $a = 8$ and $b = 2$

Note that in both cases, $(a + b)^2 = 10^2 = 100$

And $(a - b)^2 =$ Either $(2 - 8)^2$ Or $(8 - 2)^2 = 36$

Thus, the new roots are 100 and 36.

Thus, the required equation:

$=> x^2 - (100 + 36)x + 100 \times 36 = 0$

$=> x^2 - 136x + 3600 = 0$

The correct answer is option D.

9. We know that in a quadratic equation $px^2 + qx + r = 0$, the sum of the roots is given by $\left(-\frac{q}{p}\right)$ and the product of the roots is given by $\left(\frac{r}{p}\right)$.

Thus, in the above equation, we have sum of the roots $= -\frac{(a+1)}{1} = -5 \Rightarrow a = 4$

The product of the roots in the above equation is given by $\frac{a+5}{1} = 4 + 5 = 9$.

The correct answer is 9.

10. Let us assume $3^x = k$.

Thus, $9^x = (3^x)^2 = k^2$ and $3^{x+1} = 3 \times 3^x$.

Hence, our equation becomes:

$k^2 - 3k - 54 = 0$

$\Rightarrow k^2 - 9k + 6k - 54 = 0$

$\Rightarrow k(k-9) + 6(k-9) = 0$

$\Rightarrow (k+6)(k-9) = 0$

$\Rightarrow k = -6$ or 9

Thus, we have $3^x = -6$ or $3^x = 9$.

Since 3^x is always positive, irrespective of whether x is positive, zero, or negative, we can reject the possibility of $3^x = -6$ and conclude that $3^x = 9$.

Thus, $9^x = 9^2 = 81$

The correct answer is option D.

11. We have

$\sqrt{x+2} + \sqrt{x-1} = 3$

$\Rightarrow \left(\sqrt{x+2} + \sqrt{x-1}\right)^2 = 3^2$

$\Rightarrow (x+2) + (x-1) + 2\left(\sqrt{x+2}\right)\left(\sqrt{x-1}\right) = 9$

$\Rightarrow 2x + 1 + 2\left(\sqrt{x+2}\right)\left(\sqrt{x-1}\right) = 9$

$\Rightarrow 2\sqrt{(x+2)(x-1)} = 8 - 2x$

$\Rightarrow \sqrt{(x+2)(x-1)} = 4 - x$

Again, squaring both sides, we get:

$(x + 2)(x - 1) = (4 - x)^2$

=> $(x^2 - x + 2x - 2) = 16 + x^2 - 8x$

=> $(x^2 + x - 2) = 16 + x^2 - 8x$

=> $9x = 18$

=> $x = 2$

=> $x^2 = 2^2 = 4$

Alternate approach:

$\sqrt{x+2} + \sqrt{x-1} = 3$

=> $\sqrt{x+2} = 3 - \sqrt{x-1}$

Squaring both sides:

=> $x + 2 = 9 + (x - 1) - 6\sqrt{x-1}$

=> $2 + 1 - 9 = -6\sqrt{x-1}$

=> $-6 = -6\sqrt{x-1}$

=> $1 = \sqrt{x-1}$

Squaring both sides again:

$1 = x - 1$

=> $x = 2$

=> $x^2 = 2^2 = 4$

The correct answer is option B.

12. It is given that the equation $x^2 + 6x + 8 = 0$ has two roots p and q.

Thus, we have

$p + q = -\frac{6}{1} = -6 \ldots$ (i)

$pq = \frac{8}{1} = 8$ ... (ii)

Say the another quadratic equation be $ax^2 + bx + c = 0$; so its roots are $\frac{1}{p}$ and $\frac{1}{q}$.

Sum of the roots $= -\frac{b}{a} = \left(\frac{1}{p} + \frac{1}{q}\right) = \frac{p+q}{pq} = -\frac{6}{8} = -\frac{3}{4}$

Product of the roots $= \frac{c}{a} = \left(\frac{1}{p} \times \frac{1}{q}\right) = \frac{1}{pq} = \frac{1}{8}$

Thus, the quadratic equation could be

$x^2 - \left(-\frac{3}{4}\right)x + \left(\frac{1}{8}\right) = 0$

$=> 8x^2 + 6x + 1 = 0$

The correct answer is option B.

13. Let us solve the equations and find out the roots.

$x^2 - 5x + 6 = 0$

$=> x^2 - 3x - 2x + 6 = 0$

$=> x(x - 3) - 2(x - 3) = 0$

$=> (x - 2)(x - 3) = 0$

$=> x = 3$ or 2

$y^2 + 6y - 7 = 0$

$=> y^2 + 7y - y - 7 = 0$

$=> y(y + 7) - 1(y + 7) = 0$

$=> (y - 1)(y + 7) = 0$

$=> y = 1$ or -7

On comparing the roots, we see that $x > y$ for both values of x and y.

The correct answer is option A.

14. Let us first find out the length of the rectangle, from which we would be able to calculate the breadth and, consequently, the perimeter of the rectangle.

Let the length of the rectangle be x cm

Therefore, the breadth of the rectangle is $(x - 6)$ cm

Area of a rectangle = (Length of the rectangle) × (Breadth of the rectangle)

=> $x(x - 6) = 16$

=> $x^2 - 6x = 16$

=> $x^2 - 6x - 16 = 0$

=> $x^2 - 8x + 2x - 16 = 0$

=> $x(x - 8) + 2(x - 8) = 0$

=> $(x - 8)(x + 2) = 0$

=> $x = 8$ or -2

Since x is the length of the rectangle, it cannot be negative.

Hence $x = 8$ cm.

Thus, the sides of the rectangle are $x = 8$ cm and $(x - 6) = 2$ cm.

Thus, the perimeter of the rectangle = $2(8 + 2) = 20$ cm

Hence, the correct answer is 20 cm.

15. At the points where $f(x)$ and $h(x)$ intersect, the values of the functions will be equal.

Thus the points where these two graphs meet will satisfy the condition $f(x) = h(x)$

=> $x^2 + 6x + 8 = 32 + 4x$

=> $x^2 + 2x - 24 = 0$

=> $x^2 + 6x - 4x - 24 = 0$

=> $x(x + 6) - 4(x + 6) = 0$

=> $(x - 4)(x + 6) = 0$

=> $x = 4$ or -6

Since x has two possible values, the two graphs intersect at two points.

The correct answer is option C.

16. Let one of the numbers be x. Therefore, the next odd number will be 2 greater than x, hence $(x + 2)$.

As per the problem, therefore we have $x^2 + (x + 2)^2 = 74$

$=> x^2 + x^2 + 4x + 4 = 74$

$=> 2x^2 + 4x - 70 = 0$

$=> x^2 + 2x - 35 = 0$

$=> x^2 + 7x - 5x - 35 = 0$

$=> x (x + 7) - 5 (x + 7) = 0$

$=> (x + 7) (x - 5) = 0$

$=> x = -7$ or 5

Since it is given that the numbers are positive, we have $x = 5$.

Hence, the two numbers are $x = 5$ and $x = 5 + 2 = 7$.

Thus, the sum of the two numbers is $5 + 7 = 12$.

Hence, the correct answer is 12.

17. Let the length of the field be x meters.

Therefore the width of the field = $(x - 16)$ meters.

Therefore, the area of the field is $x (x - 16)$ m^2.

In the figure here, the football field is represented by the inner rectangle. This rectangle is surrounded by a track 2-meter-wide on all sides.

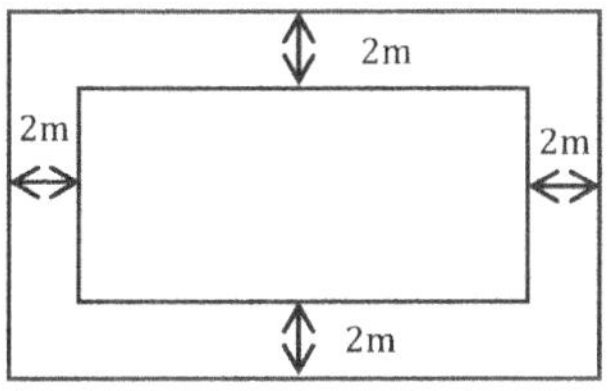

Thus, the length and width of the outer rectangle = $(x + 4)$ m and $(x - 16 + 4) = (x - 12)$ m.

Thus, the total area enclosed by the outer rectangle = $(x + 4) (x - 12)$ m^2

Thus, the area of the track

$= (x + 4)(x - 12) - x(x - 16)$

$= (x^2 - 8x - 48 - x^2 + 16x)$

$= 8(x - 6)$

Since this is $\frac{1}{3}$ of the football field's area, we have

$8(x - 6) = \frac{1}{3}x(x - 16)$

$=> 24x - 144 = x^2 - 16x$

$=> x^2 - 40x - 144 = 0$

$=> (x - 36)(x - 4) = 0$

$=> x = 4$ or 36

However, x must be greater than 16, else width $((x - 16))$ would be negative.

Therefore, the length of the field $= x = 36$ m

The correct answer is option D.

18. Let the cost of each orange be $\$x$.

Therefore, the number of oranges bought by Peter $= \left(\frac{144}{x}\right)$.

If the cost of each orange would have been \$2 less, he would have got $\left(\frac{144}{x-2}\right)$

Thus we have

$$\frac{144}{x - 2} - \frac{144}{x} = 12$$

$$=> \frac{1}{x - 2} - \frac{1}{x} = \frac{12}{144}$$

$$=> \frac{x - x + 2}{x(x - 2)} = \frac{1}{12}$$

$=> x(x - 2) = 24$

$=> x^2 - 2x - 24 = 0$

$=> x^2 - 6x + 4x - 24 = 0$

$=> x(x-6)+4(x-6)=0$

$=> (x+4)(x-6)=0$

$=> x=-4$ or 6

Since x is the price of an orange, it cannot be negative.

Thus, we have $x=6$

Thus the number of oranges bought by Peter $= \left(\frac{144}{6}\right) = 24$

The correct answer is option C.

19. If the man drives at x miles per hour, he takes $(x+2)$ hours.

Therefore, the total distance traveled

$= x(x+2)$ miles

Now it is given that if he increases his speed by y miles per hour, he would take $\frac{3y}{4}$ hours less than before.

Therefore, again, the total distance traveled

$= (x+y)\left(x+2-\frac{3y}{4}\right)$

Thus, we have

$x(x+2) = (x+y)\left(x+2-\frac{3y}{4}\right)$

$=> x^2+2x = x^2+2x-\frac{3xy}{4}+xy+2y-\frac{3y^2}{4}$

$=> \frac{3y^2}{4} - 2y = \frac{xy}{4}$

$=> y(3y-8) = xy$

$=> 3y-8 = x$ (Cancelling y from both sides)

$=> x = 3y-8$

The correct answer is option C.

20. $\frac{1}{2}\left(\frac{1}{x-4}\right)(x^2-16) = \left(1+\frac{4}{x-4}\right)(x-4)$

$=> \frac{1}{2(x-4)}(x+4)(x-4) = \frac{(x-4+4)}{(x-4)}(x-4)$

Thus, cancelling $(x - 4)$ from both sides (we can cancel it since $x \neq 4$).

$\frac{(x+4)}{2} = x$

$=> x = 4$

However, since $x \neq 4$, no real value of x satisfies the given equation.

The correct answer is option A.

21. We know that the deposit is \$100.

Since the interest is $r\%$ in 3 years, we have

Amount after 3 years = \$ $\left\{100\left(1 + \frac{r}{100}\right)^{1}\right\}$; note that the rate is $r\%$ per 3 years, not per anum.

In n years, number of 3-year periods = $\frac{n}{3}$

Thus, the amount after t years = \$ $\left\{100\left(1 + \frac{r}{100}\right)^{\frac{n}{3}}\right\}$

Hence, the correct answer is option B.

22. Let time taken be t hours when the strains have the same number of bacteria.

Number of 9-hour periods in t hours = $\frac{t}{9}$

Number of 3-hour periods in t hours = $\frac{t}{3}$

Final number of bacteria of the 1st strain = $1,024 \times 2^{\left(\frac{t}{9}\right)}$

Final number of bacteria of the 2nd strain = $64 \times 2^{\left(\frac{t}{3}\right)}$

Thus, $1,024 \times 2^{\left(\frac{t}{9}\right)} = 64 \times 2^{\left(\frac{t}{3}\right)}$

$=> 2^{\left(\frac{t}{9}+10\right)} = 2^{\left(6+\frac{t}{3}\right)}$

$=> \frac{t}{9} + 10 = 6 + \frac{t}{3}$

$=> \frac{2t}{9} = 4 => t = 18$ hours

The correct answer is 18.

23. The equations $x^2 + ax + b = 0$ and $x^2 + px + q = 0$ have both roots common if they represent the same equation, i.e. $a = p$ and $b = q$

Thus, we have

$-6a = -4 => a = \frac{2}{3}$

and $12 = 9b => b = \frac{4}{3}$

$=> a + b = \frac{2}{3} + \frac{4}{3} = 2$

The correct answer is option D.

24. Since $a^b = 1$, possible cases are:

- $a = 1 => x - 3 = 1 => x = 4$ ... (i)
- $b = 0$ such that $a \neq 0 => x - 6 = 0 => x = 6$ ... (ii)
- $a = -1$ and b is even $=> x - 3 = -1 => x = 2$ (Here, $x - 6 = -4$, which is even)

Thus, the possible values of x are 4, 6, and 2.

Required sum = 12

The correct answer is option D.

25. The distance covered by the ball in the tth second

= (Distance covered in t seconds) – (Distance covered in $(t - 1)$ seconds) = $9t^2 - 9(t-1)^2$

$= 9\left\{t^2 - (t-1)^2\right\}$

$= 9\,(t - (t-1))\,(t + (t-1))$

$= 9\,(2t - 1)$

Thus, we have

$9\,(2t - 1) = h$

$=> 2t - 1 = \dfrac{h}{9}$

$=> t = \frac{1}{2}\left(\frac{h}{9} + 1\right)$

The correct answer is option C.

26. We know that $x = 3$ satisfies the equation $x^2 - px + 12 = 0$.

Substituting $x = 3$ in the equation, we have

$(3)^2 - p(3) + 12 = 0$

$=> 3p = 21$

$=> p = 7$

The correct answer is 7.

27. To find the point of intersection, we need to equate the two equations.

$=> x^2 + 3x + 1 = x + 1$

$=> x^2 + 2x = 0$

$=> x(x + 2) = 0$

$=> x = 0$ or -2

The corresponding y values are obtained by substituting the x values in $y = x + 1$ (or in the other equation):

$x = 0 => y = 1$

$x = -2 => y = -1$

So, the given curve and line intersect at these two points $(0, 1)$ and $(-2, -1)$.

The point closer to the x-axis is (0, 1).

Thus, the y coordinate of the point, (0, 1), closer to the x-axis is 1.

The correct answer is option D.

28. We know that $f(1) = 2$

$=> (1)^2 + b(1) + c = 2$

$=> b + c = 1 \ldots$ (i)

We also know that $f(2) = 0$

$=> 2^2 + 2b + c = 0$

$\Rightarrow 2b + c = -4 \ldots$ (ii)

Thus, from (i) and (ii), we have

$b = -5$

$c = 6$

$\Rightarrow f(x) = x^2 - 5x + 6$

$\Rightarrow f(0) = 0^2 - 5 \times 0 + 6 = 6$

The correct answer is option D.

29. We have $f(2) = f(-2)$

$\Rightarrow 2^3 - k \times 2 + 2 = (-2)^3 - k \times (-2) + 2$

$\Rightarrow 8 - 2k + 2 = -8 + 2k + 2$

$\Rightarrow k = 4$

The correct answer is option D.

30. We know that $f(1) = 2$.

By substituting $x = 1$ in the four options, we observe that options A, B, and D satisfy this condition:

(A) $x^2 + 1 = 1^2 + 1 = 2$

(B) $3x - 1 = 3(1) - 1 = 2$

(C) $x + 4 = 1 + 4 = 5 \neq 2$

(D) $x^3 + 1 = 1^3 + 1 = 2$

By substituting $x = 2$ in the four options, we observe that options A and B satisfy this condition:

(A) $x^2 + 1 = 2^2 + 1 = 5$

(B) $3x - 1 = 3(2) - 1 = 5$

(C) Ignored

(D) $x^3 + 1 = 2^3 + 1 = 9$

By substituting $x = 3$ in the four options, we observe that only option B satisfies this condition:

(A) $x^2 + 1 = 3^2 + 1 = 10$

(B) $3x - 1 = 3(3) - 1 = 8$

The correct answer is option B.

31. We have

$g(1) = 3$ and $f(1) = 5$

$f(g(1)) = f(3) = 3$

$g(f(1)) = g(5) = 7$

Thus, we have

$f(g(1)) + g(f(1)) = 3 + 7 = 10$

The correct answer is option D.

32. We have

$g(-2) + f(-2) = 2 \times (-2) = -4$

$=> g(-2) = -4 - f(-2)$

Also, we have

$f(-2) = |-2| + (-2) = 2 - 2 = 0$

$=> g(-2) = -4 - 0 = -4$

The correct answer is option B.

33. We have $f(x) = ax^2 + x$

Since $f(1) = f(0) + 2$, we have

$a(1^2) + 1 = a(0)^2 + (0) + 2$

$=> a + 1 = 2$

$=> a = 1$

$=> f(x) = x^2 + x$

$=> f(2) = 2^2 + 2 = 6$

The correct answer is 6.

34. We have $x = \sqrt{3} - 1$

$=> x + 1 = \sqrt{3}$

Squaring both sides:

$(x + 1)^2 = (\sqrt{3})^2$

$=> x^2 + 2x + 1 = 3$

$=> x^2 + 2x = 2$

The correct answer is option D.

35. We know that $f(1) = 1$.

- $f(2) = (-1)^2 \times f(2 - 1) = (-1)^2 \times f(1) = 1 \times 1 = 1$
- $f(3) = (-1)^3 \times f(3 - 1) = (-1)^3 \times f(2) = (-1) \times 1 = -1$
- $f(4) = (-1)^4 \times f(4 - 1) = (-1)^4 \times f(3) = 1 \times (-1) = -1$

The correct answer is option B.

36.

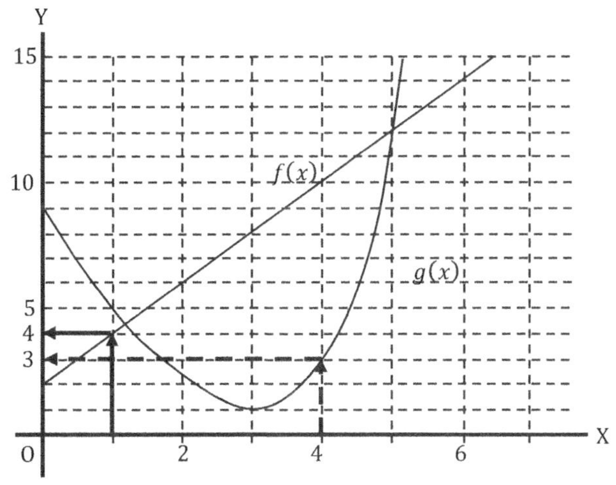

From the graph: $f(1) = 4$

Thus, we have $k = 4$

=> $g(k) = g(4) = 3$ (from the graph)

The correct answer is option D.

37. We have

$$\frac{1}{\sqrt{a+1}+\sqrt{a}}$$

$$= \frac{(\sqrt{a+1}-\sqrt{a})}{(\sqrt{a+1}+\sqrt{a})(\sqrt{a+1}-\sqrt{a})}$$

$$= \frac{\sqrt{a+1}-\sqrt{a}}{(\sqrt{a+1})^2-(\sqrt{a})^2}$$

$$= \frac{\sqrt{a+1}-\sqrt{a}}{(a+1)-(a)}$$

$$= \sqrt{a+1}-\sqrt{a}$$

The correct answer is option C.

38. $f(g(x)) = \frac{g(x)-k}{5} = \frac{(5x+1)-k}{5} = \frac{5x+1-k}{5}$

Since $f(g(x)) = x$

=> $\frac{5x+1-k}{5} = x$ => $5x + 1 - k = 5x$

=> $k = 1$

The correct answer is 1.

39. Given:

If x is prime, $f(x) = f(x-1) + 2 \ldots.$ 1st relation

If x is non-prime, $f(x) = f(x-1) + 1 \ldots.$ 2nd relation

- For $x = 2$ (prime) : $f(2) = f(2-1) + 2 = f(1) + 2 = 5$ from the 1st relation
- For $x = 3$ (prime) : $f(3) = f(3-1) + 2 = f(2) + 2 = 7$ from the 1st relation
- For $x = 4$ (non – prime) : $f(4) = f(4-1) + 1 = f(3) + 1 = 8$ from the 1st relation
- For $x = 5$ (prime) : $f(5) = f(5-1) + 2 = f(4) + 2 = 10$ from the 2nd relation

The correct answer is 10.

40. $f(x) = \frac{2^x+1}{2^x}$

$=> f(1) = \frac{2^1+1}{2^1} = \frac{2+1}{2} = \frac{3}{2}$

Also, $f(-1) = \frac{2^{-1}+1}{2^{-1}} = 3$

We find that $f(-1) = 2f(1)$.

The correct answer is option D.

41. $f(x+1) = f(x) + x + \frac{1}{2}$

$=> a(x+1)^2 + b = (ax^2 + b) + x + \frac{1}{2}$

$=> a(1+2x) = x + \frac{1}{2}$

$=> a = \frac{x+\frac{1}{2}}{1+2x}$

$=> a = \frac{\frac{1}{2}(2x+1)}{2x+1}$

$=> a = \frac{1}{2}$

The correct answer is option C.

Alternate approach:

$f(x+1) = f(x) + x + \frac{1}{2}$

Substituting $x = 0$: $f(0+1) = f(0) + 0 + \frac{1}{2}$

$=> f(1) = f(0) + 1$

$=> a + b = b + \frac{1}{2}$

$=> a = \frac{1}{2}$

42. $\frac{f(1) \times f(2)}{f(3)}$

$= \frac{(2^{1-1} \times 2^{2-1})}{2^{3-1}}$

$= \frac{2^0 \times 2^1}{2^2}$

$= 2^{-1} = 0.5$

The correct answer is option D.

43. Amount after 3 years = \$ $\left\{1,000\left(1+\frac{r}{100}\right)^3\right\}$

$=> P = 1,000\left(1+\frac{r}{100}\right)^3$

$=> \left(1+\frac{r}{100}\right)^3 = \frac{P}{1,000}$

Taking cube root:

$1+\frac{r}{100} = \left(\frac{P}{1,000}\right)^{\left(\frac{1}{3}\right)}$

$=> \frac{r}{100} = \left(\frac{P}{1,000}\right)^{\left(\frac{1}{3}\right)} - 1$

$=> r = 100\left\{\left(\frac{P}{1,000}\right)^{\left(\frac{1}{3}\right)} - 1\right\}$

The correct answer is option C.

44. Distance traveled = d miles

Speed = s miles per hour

Thus, time taken = $\frac{d}{s}$ hours

New speed = $(s+10)$ miles per hour

Thus, new time taken = $\frac{d}{(s+10)}$ hours

Since he would have reached more than 1 hour earlier, we have

$\frac{d}{s} - \frac{d}{s+10} > 1$

$=> d\left(\frac{1}{s} - \frac{1}{s+10}\right) > 1$

$=> \frac{d(s+10-s)}{s(s+10)} > 1$

$=> 10d > s\,(s+10)$

$=> \frac{1}{10}s\,(s+10) < d$

The correct answer is option C.

45. Distance on 1st day = d units

Distance on 2nd day = $d \times \left(\frac{120}{100}\right) = d \times \left(\frac{6}{5}\right)$ units

Distance on 3rd day = $d \times \left(\frac{6}{5}\right)^2$ units; and so on

Thus, distance on the nth day = $d \times \left(\frac{6}{5}\right)^{n-1}$

We have $d \times \left(\frac{6}{5}\right)^{n-1} > D$

$=> \left(\frac{6}{5}\right)^{n-1} > \frac{D}{d}$

$=> \left(\frac{D}{d}\right)^{\left(\frac{1}{n-1}\right)} < \frac{6}{5}$

The correct answer is option D.

46. From the given equations, we need to eliminate the variable y. Thus, we multiply the first equation by p and second equation by q.

Thus we get the following equations:

$mp = p^2x - pqy$...(1)

$nq = q^2x - pqy$...(2)

Subtracting the above two equations, we get

$p^2x - q^2x = mp - nq$

$$x = \frac{mp - nq}{p^2 - q^2}$$

The correct answer is option B.

47. From the given equation, let us express C in terms of A, B and D:

We have

$A \times \sqrt[3]{C} = \frac{B \times \sqrt{D}}{5}$

$=> \sqrt[3]{C} = \frac{B\sqrt{D}}{5A}$

$=> C = \left[\frac{B\sqrt{D}}{5A}\right]^3$

$=> C = \left[\frac{2 \times \sqrt{16}}{5 \times 5}\right]^3$

$=> C = \left(\frac{8}{25}\right)^3$

$=> C = \frac{512}{15{,}625}$

The correct answer is option C.

48. We know that the mass of the substance becomes half its initial value in every 42 hours, i.e., after 42 hours, the mass of the substance X will become $\frac{1}{2} \times 28 = 14$ grams.

After yet another 42 hours, i.e., after a total of 84 hours, the mass will be $\left(\frac{1}{2} \times \left(\frac{1}{2} \times 28\right)\right)$ grams, which is $\left\{\left(\frac{1}{2}\right)^2 \times 28\right\}$ grams. The power on $\left(\frac{1}{2}\right)$ is equal to $\left(\frac{84}{42}\right) = 2$.

Number of '42 hours' slots in 201 hours $= \frac{210}{42} = 5$.

Thus, the final weight of the substance after 210 hours

$= 28 \times \left\{\left(\frac{1}{2}\right) \times \left(\frac{1}{2}\right) \times \left(\frac{1}{2}\right) \times \left(\frac{1}{2}\right) \times \left(\frac{1}{2}\right)\right\}$

$= 28 \times \left(\frac{1}{2}\right)^5$ grams

The correct answer is option C.

49. We have

$F = \frac{GMm}{D^2} \ldots \text{(i)}$

The new distance, as per the given problem, is $2D$, and one of the masses, say, M becomes $2M$.

Therefore, the new force denoted by $F' = \frac{G(2M)m}{(2D)^2} = \frac{GMm}{2D^2} \ldots \text{(ii)}$

Therefore, from equations (i) and (ii), we can say

$F' = \frac{1}{2}\left(\frac{GMm}{D^2}\right)$

$=> F' = \frac{1}{2}F$

Thus, the force becomes half of the initial value.

The correct answer is option D.

50. Number of pages the man typed on the first day $= x$.

Percent increase in the number of pages typed each day $= n\%$

Thus, number of pages typed on the second day $= x\left(1 + \frac{n}{100}\right)$

Since the man typed for D days, this increase would occur for $(D - 1)$ days after the first day.

Thus, the number of he typed on the last day $= x\left(1 + \frac{n}{100}\right)^{(D-1)}$

Thus, we have

$p = x\left(1 + \frac{n}{100}\right)^{D-1}$

$=> \left(1 + \frac{n}{100}\right)^{(D-1)} = \frac{p}{x}$

$=> 1 + \frac{n}{100} = \left(\frac{p}{x}\right)^{\left(\frac{1}{D-1}\right)}$

$=> n = \left\{\left(\frac{p}{x}\right)^{\left(\frac{1}{D-1}\right)} - 1\right\} \times 100$

The correct answer is option D.

51. $\left(\text{The distance covered by the ball in the } t^{\text{th}} \text{ second}\right)$

$= \text{(Distance covered in } t \text{ seconds)} - \text{(Distance covered in } (t-1) \text{ seconds)}$

$= 5t^2 - 5(t-1)^2$

$= 5\left\{t^2 - (t-1)^2\right\}$

$= 5\,(t - (t-1))\,(t + (t-1))$

$= 5\,(2t - 1)$

Thus, we have

$5\,(2t - 1) = 45$

$=> 2t - 1 = 9$

$=> t = 5$

The correct answer is option C.

52. In a quadratic $ax^2 + bx + c = 0$, the roots are real and equal if the discriminant, i.e., $b^2 - 4ac = 0$.

Thus, for the quadratic $x^2 - px + 8p = 0$, we have

Discriminant $= (-p)^2 - 4\,(1)\,(8p) = 0$

$=> p^2 - 32p = 0$

$=> p\,(p - 32) = 0$

$=> p = 0$ or 32

Of the options, only 32 is the probable value of p.

The correct answer is option C.

53. We know that $x = 4$ satisfies the equation $x^2 - px + 12 = 0$

Substituting $x = 4$ in the equation, we have

$4^2 - 4p + 12 = 0$

$=> 4p = 28$

$=> p = 7$

The correct answer is option B.

54. To determine the point where $y = x^2 + kx + l$ intersects the Y-axis, we need to substitute $x = 0$.

Since the point of intersection on the Y-axis is given to be $(0, 4)$, i.e. $x = 0, \ y = 4$, we have

$0 + 0 + l = 4$
$=> l = 4 \ldots$ (i)

We also know that $y = x^2 + kx + l$ intersects the X-axis at $(4, 0)$, i.e.

$x = 4, y = 0$:
$4^2 + 4k + l = 0$
$=> 16 + 4k + 4 = 0$
$=> k = -5$
$=> k + l = -1$

The correct answer is option D.

55. We know that $f(1) = 1$

$=> (1)^2 + b(1) + 1 = 1$
$=> b = -1$
$=> f(x) = x^2 - x + 1$
$=> f(2) = 2^2 - 2 + 1 = 3$

The correct answer is '3.'

56. We have

$f(2x - 1) = x^2 + 3x$

Since we need the value of $f(3)$, we need to use a suitable value of x so that we have

$2x - 1 = 3$
$=> x = 2$

Thus, substituting $x = 2$:
$f(2 \times 2 - 1) = 2^2 + 3 \times 2 = 10$

The correct answer is '10.'

57. We have $f(x) = x^2 + ax + b$

Also, we have $f(k) = f(k-1) + 1$

$=> k^2 + ak + b = (k-1)^2 + a(k-1) + b + 1$
$=> k^2 + ak + b = k^2 - 2k + 1 + ak - a + b + 1$
$=> 2k = 2 - a$
$=> k = 1 - \frac{a}{2}$

The correct answer is option A.

58. We have

$a^2 + b^2 - 2a - b = 0 \ldots$ (i)

$a^2 + b^2 = 2ab \ldots$ (ii)

$=> 2ab - 2a - b = 0$
$=> 2ab - 2a = b$
$=> 2a(b-1) = b$
$=> 2a = \frac{b}{b-1}$
$=> a = \frac{b}{2(b-1)}$

The correct answer is option A.

59. We have
$x + y = 4$
$=> x = 4 - y$

Thus, we have
$z = x^2 + 2y$
$= (4-y)^2 + 2y$
$= y^2 - 8y + 16 + 2y$
$= y^2 - 6y + 16$
$= (y-3)^2 + 7$

The correct answer is option B.

60. We have
$x = 1 + \sqrt{3}$
$=> x - 1 = \sqrt{3}$

Squaring both sides:

$(x-1)^2 = (\sqrt{3})^2$
$\Rightarrow x^2 - 2x + 1 = 3$
$\Rightarrow x^2 - 2x = 2$

Alternate approach:

Given $x = 1 + \sqrt{3}$
$\Rightarrow x^2 - 2x = x(x-2) = (1+\sqrt{3})(1+\sqrt{3}-2) = (\sqrt{3}+1)(\sqrt{3}-1) = (\sqrt{3})^2 - 1^2 = 3 - 1 = 2$

The correct answer is option D.

61. We know that:
$f(x) = \sqrt{x} + 1$
$\Rightarrow f(g(2)) = \sqrt{g(2)} + 1$

Since $f(g(2)) = 4$, we have

$\sqrt{g(2)} + 1 = 4$
$\Rightarrow \sqrt{g(2)} = 3$
$\Rightarrow g(2) = 9$ ...(i)

Working with the options:
(A) $g(x) = 5x - 1 \Rightarrow g(2) = 10 - 1 = 9$ - Satisfies (i)
(B) $g(x) = x^3 - 1 \Rightarrow g(2) = 8 - 1 = 7$ - Does not satisfy (i)
(C) $g(x) = 6x + 3 \Rightarrow g(2) = 12 + 3 = 15$ - Does not satisfy (i)
(D) $g(x) = x^4 - 1 \Rightarrow g(2) = 16 - 1 = 15$ - Does not satisfy (i)

The correct answer is option A.

62. We need to determine the value of $f(2)$ from the graph.
To determine the value, we draw a (imaginary) vertical line from $x = 2$ and read the y value of the point where it intersects the line for $f(x)$.

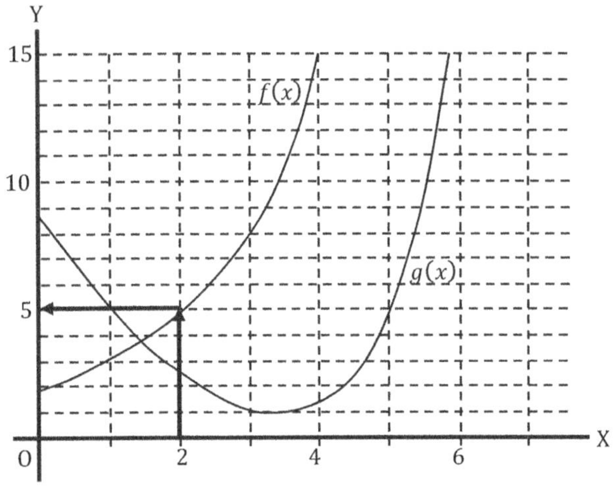

Thus, we have $f(2) = 5 \Rightarrow k = 5$

We need to determine the value of $g(k)$, i.e. the value of $g(5)$

To determine the value, we draw a (imaginary) vertical line from $x = 5$ and read the y value of the point where it intersects the line for $g(x)$.

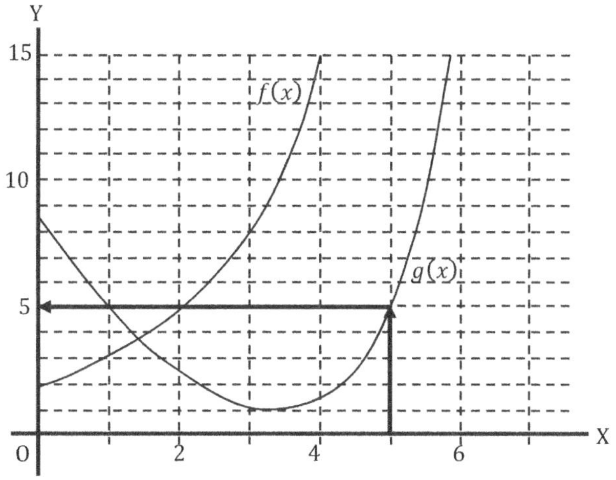

Thus, we have $g(5) = 5$

The correct answer is '5.'

63. We have

$f(x) = x^2 + 1$

Also, we have

$f(x) = g(x - 2) + 1$

Replacing x with $(x + 2)$, we have

$\Rightarrow f(x + 2) = g(x + 2 - 2) + 1$

$\Rightarrow g(x) + 1 = (x + 2)^2 + 1$

$\Rightarrow g(x) = (x + 2)^2$

The correct answer is option C.

64. To find the points of intersection of the functions $f(x)$ and $g(x)$, we need to solve for:

$f(x) = g(x)$

However, it is best to work with options since it is difficult to solve the above equation.

Thus, we have

(A) $f(-1) = -1 + 2^{-1} = -\frac{1}{2}$; $g(-1) = 2 + (-1)^2 = 3$. Thus, $f(x) \neq g(x)$

(B) $f(0) = 0 + 2^0 = 1$; $g(0) = 2 + (0)^2 = 2$. Thus, $f(x) \neq g(x)$

(C) $f(2) = 2 + 2^2 = 6$; $g(2) = 2 + 2^2 = 6$. Thus, $f(x) = g(x)$ - Satisfies

(D) $f(4) = 4 + 2^4 = 20$; $g(4) = 2 + 4^2 = 18$. Thus, $f(x) \neq g(x)$

The correct answer is option C.

65. We have

$$f\left(\frac{x}{2}\right) = 2x^2 + 3x$$

Replacing x with $2x$, we have

$$f\left(\frac{2x}{2}\right) = 2(2x)^2 + 3(2x)$$

$$\Rightarrow f(x) = 8x^2 + 6x$$

The correct answer is option D.

66. Price of gasoline per gallon = $\$\left(\frac{x}{p}\right)$

New price per gallon = $\$\left(\frac{x}{p} + 2\right)$

New quantity of gasoline = $(p - 1)$ gallons

Thus, the total amount he spends = $\$\left(\left(\frac{x}{p} + 2\right)(p - 1)\right)$

Since the amount he spends remains the same, we have

$$\left(\frac{x}{p} + 2\right)(p - 1) = x$$

$$\Rightarrow x + 2p - \frac{x}{p} - 2 = x$$

$$\Rightarrow 2p - \frac{x}{p} - 2 = 0$$

$$\Rightarrow \frac{x}{p} = 2p - 2$$

$$\Rightarrow x = 2p(p - 1)$$

The correct answer is option B.

67. Final amount after 2 years when $\$p$ is invested at $r\%$ rate under compound interest

$$= \$\left\{p\left(1 + \frac{r}{100}\right)^2\right\} = \$q$$

$$\Rightarrow \left(1 + \frac{r}{100}\right)^2 = \frac{q}{p}$$

$$\Rightarrow \left(1 + \frac{r}{100}\right) = \sqrt{\frac{q}{p}}$$

$$\Rightarrow \frac{r}{100} = \sqrt{\frac{q}{p}} - 1$$

$=> r = 100\left(\sqrt{\frac{q}{p}} - 1\right)$

The correct answer is option B.

68. We know that area of the frame is equal to the area for the display of the picture.

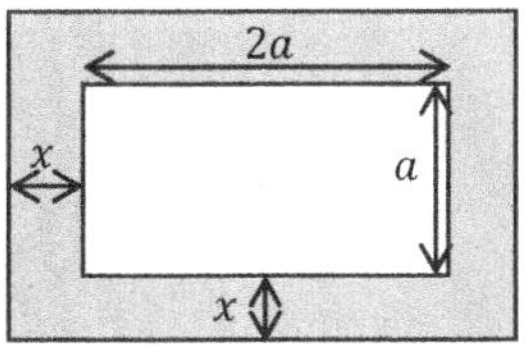

=> (Total Area of the outer rectangle) = (Area of the display) + (Area of the frame)

=> The area for the display = $\frac{1}{2}$ × Total area of the outer rectangle

$=> 2a \times a = \frac{1}{2}(2a + 2x)(a + 2x)$

$=> 2a^2 = a^2 + 3ax + 2x^2$

$=> a^2 = 3ax + 2x^2$

The correct answer is option A.

69. We have

$\frac{x + 6}{x^2} < 1$

Since $x^2 \geq 0$ for all x, we may cross-multiply without changing the inequality:

$x + 6 < x^2$

$=> x^2 - x - 6 > 0$

$=> (x - 3)(x + 2) > 0$

$=> x > 3$ OR $x < -2$

The correct answer is option D.

70. We have

$(x^4 + 8x)(x^2 - 4x + 3) = 0$

$=> x(x^3 + 8)(x^2 - 3x - x + 3) = 0$

$=> x(x^3 + 8)(x - 3)(x - 1) = 0$

Thus, we have

$x = 0$, OR $x^3 = -8$, OR $x = 3$, OR $x = 1$

=> Possible values of x are $0,\ -2,\ 3,\ \&\ 1$

Thus, the required sum = 2

The correct answer is option A.

71. $f(x) = x^4 - 13x^2 + 36$

Thus, we have
$x^4 - 13x^2 + 36 = 0$

Let $k = x^2$:
$=> k^2 - 13k + 36 = 0$
$=> k^2 - 4k - 9k + 36 = 0$
$=> (k - 4)(k - 9) = 0$
$=> k = 4$ or 9
$=> x^2 = 4$ or 9
$=> x = \pm 2$ or ± 3

So, there are four roots: $-3, -2, 2,$ &3.

Thus, the sum of the roots $= (-2) + (2) + (-3) + 3 = 0$

The correct answer is '0.'

72. $f(x) = \frac{1}{x}$
$g(x) = \frac{x}{x+1}$

$=> f(g(x)) = f\left(\frac{x}{x+1}\right)$
$= \frac{1}{\left(\frac{x}{x+1}\right)} = \frac{x+1}{x}$

Also, we have
$g(f(x)) = g\left(\frac{1}{x}\right)$
$= \frac{\left(\frac{1}{x}\right)}{\left(\frac{1}{x}\right)+1} = \frac{1}{x+1}$

Thus, we have
$f(g(x)) \times g(f(x)) = \frac{x+1}{x} \times \frac{1}{x+1} = \frac{1}{x}$

The correct answer is option C.

73. $f(x) = x^2 - 8$
$g(x) = x - 2$

We have
$f(x) \leq g(x)$
$=> x^2 - 8 \leq x - 2$
$=> x^2 - x - 6 \leq 0$
$=> (x + 2)(x - 3) \leq 0$

$=> -2 \leq x \leq 3$

Thus, the non-negative integer values of x are $-3, -2, -1, 0, 1, 2,$ & 3.
Thus, there are 'seven' possible values of x.

The correct answer is option C.

74.

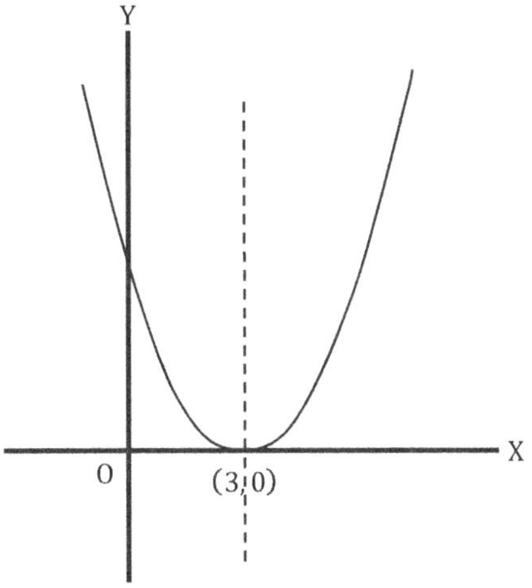

$f(1)$ refers to the value of y at $x = 1$, which is '2' units to the left of $x = 3$.

Since the graph is symmetric about the line $x = 3$, the value of y for a value of x, '2' units to the left of $x = 3$ is the same as the value of y for a value of x, '2' units to the right of $x = 3$, i.e. for $x = 3 + 2 = 5$.

Thus, we have
$f(5) = f(1) = 5$

The correct answer is option B.

75. From the graph, it is clear that the shaded region depicts $g(x) \geq f(x)$, since the line for $g(x)$ is above the line for $f(x)$.

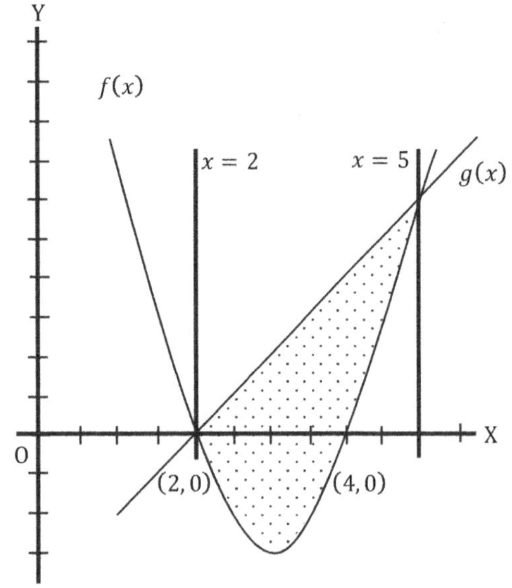

Solving $f(x) = g(x)$:

$x^2 - 6x + 8 = x - 2$
$=> x^2 - 7x + 10 = 0$
$=> (x - 2)(x - 5) = 0$
$=> x = 2$ OR 5

Thus, the shaded region, where $g(x) \geq f(x)$ is bounded by the lines $x = 2$ and $x = 5$.
Thus, the integer values of x are: 2, 3, 4, & 5.
Thus, there are 'four' possible values of x.

The correct answer is option D.

76.

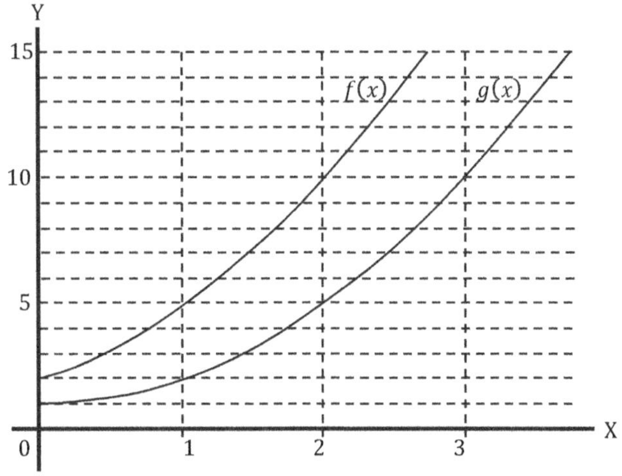

From the graphs, we observe that:

- $f(x)$ satisfies the coordinates $(0, 2)$, $(1, 5)$, & $(2, 10)$
- $g(x)$ satisfies the coordinates $(0, 1)$, $(1, 2)$, $(2, 5)$, & $(3, 10)$

Working with the choices:

(A) Does not satisfy the coordinate $(0, 2)$ for $f(x)$.

(B) Satisfies all the coordinates for $f(x)$.

(C) Does not satisfy the coordinates $(0, 1)$ or $(1, 2)$ or $(3, 10)$ for $g(x)$.

(D) Does not satisfy the coordinates $(0, 1)$ or $(1, 2)$ or $(3, 10)$ for $g(x)$.

The correct answer is option B.

77.

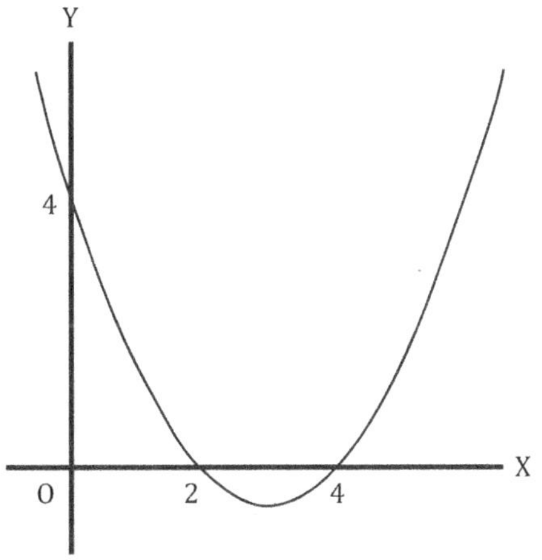

The above quadratic has roots 2 and 4.

Thus, we have

$f(x) = x^2 + ax + b = (x - 2)(x - 4) = x^2 - 6x + 8$

Thus, we have

$=> a = -6, \ b = 8$

$=> a + b = 2$

The correct answer is option B.

78. $\dfrac{(a+b)^2 - (a^2 + b^2)}{(a+b)^2 - (a-b)^2}$

$$= \frac{a^2 + 2ab + b^2 - (a^2 + b^2)}{(a^2 + 2ab + b^2) - (a^2 - 2ab + b^2)}$$

$$= \frac{2ab}{4ab} = 0.5$$

The correct answer is option C.

79. We have

$x = \frac{1}{2}\left(a + \frac{1}{a}\right)$

$=> x^2 - 1 = \left\{\frac{1}{2}\left(a + \frac{1}{a}\right)\right\}^2 - 1$

$= \frac{1}{4}\left(a^2 + \frac{1}{a^2} + 2\right) - 1$

$= \frac{1}{4}\left(a^2 + \frac{1}{a^2} - 2\right)$

$= \frac{1}{4}\left(a - \frac{1}{a}\right)^2$

$=> \sqrt{(x^2 - 1)} = \frac{1}{2}\left(a - \frac{1}{a}\right)$

Thus, we have

$x + \sqrt{(x^2 - 1)} = \frac{1}{2}\left(a + \frac{1}{a}\right) + \frac{1}{2}\left(a - \frac{1}{a}\right) = a$

The correct answer is option C.

80. $\frac{x^4 - y^4}{x^2 + y^2} = \frac{(x^2 - y^2)(x^2 + y^2)}{x^2 + y^2}$

$= x^2 - y^2$

$= \frac{1}{4} - \frac{1}{9} = \frac{5}{36} = 0.139$

The correct answer is '$\frac{5}{36}$' or '0.139.'

81. We have $a + \frac{1}{a} = 2$

Squaring, we have

$a^2 + \frac{1}{a^2} + 2 = 4$

$=> a^2 + \frac{1}{a^2} = 2$

Squaring again, we have

$a^4 + \frac{1}{a^4} + 2 = 4$

$=> a^4 + \frac{1}{a^4} = 2$

The correct answer is '2.'

82. We have

$2x^3 + 4y^3 + 2xy^2 + 4x^2y$

$= 2x^3 + 4x^2y + 2xy^2 + 4y^3$

$= 2x^2(x + 2y) + 2y^2(x + 2y)$

$= (x + 2y)(2x^2 + 2y^2)$

$= 2(x + 2y)(x^2 + y^2)$

$= (2x + 4y)(x^2 + y^2)$

$= 10 \times 10 = 100$

The correct answer is '100.'

Alternate approach:

The solution presented above is the more elegant one, and a student who carefully observes the expression $2x^3 + 4y^3 + 2xy^2 + 4x^2y$ before starting to solve the question will likely go as above.

A longer solution is as follows:

$2x + 4y = 10$

$=> x + 2y = 5$

$=> x = 5 - 2y$

Substituting this expression in the equation $x^2 + y^2 = 10$ and solving for y, one gets:

- Either $y = 3,\ x = -1$

  Putting these values in the expression $2x^3 + 4y^3 + 2xy^2 + 4x^2y$, its value comes out to be '100.'

- Or $y = 1,\ x = 3$

  Putting these values in the expression $2x^3 + 4y^3 + 2xy^2 + 4x^2y$, its value comes out to be '100.'

Thus, the value of the expression is the same in both the cases: '100.'

83. In the graph of $f(x)$, we see that $f(0) = 2$.

We have $g(x) = f\left(\frac{x}{2} - 1\right) + 2$

So, when $\frac{x}{2} - 1 = 0$, we have $x = 2$; thus, $g(x = 2) = f(0) + 2$

$=>$ When $x = 2,\ g(x = 2) = 2 + 2 = 4$

Thus, the graph of $g(x)$ should contain the point $(2,\ 4)$.

Only 'Option D' has such a point.

The correct answer is option D.

84. We have

$\frac{2x+3}{(x+1)(x+2)} = \frac{A}{x+1} + \frac{B}{x+2}$

Multiplying throughout by $(x + 1)(x + 2)$, we have

$2x + 3 = A(x + 2) + B(x + 1)$

Since the above is true for all values of x, it is an identity.

Thus, it is best to substitute some values of x and solve for A and B.

Substituting $x = -1$ (to eliminate B):

$=> 2(-1) + 3 = A(-1 + 2) + B(-1 + 1)$

$=> A = 1$

Similarly, substituting $x = -2$ (to eliminate A):

$=> 2(-2) + 3 = A(-2 + 2) + B(-2 + 1)$

$=> B = 1$

Thus, we have $A + B = 1 + 1 = 2$.

The correct answer is '2.'

Alternate approach:

$2x + 3 = A(x + 2) + B(x + 1)$

$=> 2x + 3 = (A + B)x + (2A + B)$

Since the above is true for all values of x, it is an identity.

$=> A + B = 2$ and $2A + B = 3$

Solving these two equations, we get $A = 1; B = 1 => A + B = 2$

85. We have $y = ab^x$

Observing values, we have $x = 1 => y = 3$

$=> 3 = ab \ldots$ (i)

Again, we have $x = 2 => y = 6$

$=> 6 = ab^2 \ldots$ (ii)

Dividing (ii) by (i):

$b = 2$

Thus, from (i):

$a = \frac{3}{2}$

Note: Thus, we have $y = \frac{3}{2} \times 2^x$

Thus: $a + b = \frac{3}{2} + 2 = 3.5$

The correct answer is option C.

86. We know that:

$y = \frac{3}{2} \times 2^x$

The number of bacteria present 30 minutes after the start, i.e. $x = \frac{1}{2}$ is:

$y = \frac{3}{2} \times 2^{\left(\frac{1}{2}\right)}$

Since we need at least 10 times the above value, we have

$\frac{3}{2} \times 2^x \geq 10 \times \frac{3}{2} \times 2^{\left(\frac{1}{2}\right)}$

$=> 2^x \geq 10 \times 2^{\left(\frac{1}{2}\right)}$

$=> 2^x \geq 10 \times 1.414$

$=> 2^x \geq 14.14$

At $t = 3$, we have $2^x = 8$, which doesn't satisfy

At $t = 4$, we have $2^x = 16 > 14.14$

Thus, the value of t is between 3 hours and 4 hours.

The correct answer is option C.

Chapter 11

Problem Solving & Data Analysis Concepts

11.1 Percents

Percent means 'out of hundred'. A fraction having 100 in the denominator is called a percent.

To convert a fraction into a percent, we need to multiply it with 100.

For example, the fraction $\frac{1}{4} = \frac{1 \times 25}{4 \times 25} = \frac{25}{100} = 25\%$, which is equivalent to doing $\frac{1}{4} \times 100 = 25\%$.

Similarly, if we were to compare the scores of a candidate in two tests, in one of which he obtained 60 out of 80 and in the other 75 out of 90, we can use percent values:

- In the first test: $\frac{60}{80} \times 100 = 75\%$
- In the second test: $\frac{75}{90} \times 100 = 83.33\%$

Thus, the candidate fared better in the second test.

There are two aspects of percent calculation:

- **Calculating a given percent of some number:**

  If we were to calculate: What is 20% of 120?

  This implies that we need to determine a number, which as a percent of 120 represents 20%.

  Following unitary method:

  If the value is 100, then the required number is 20

  => If the value is 120, the required number is $\frac{20}{100} \times 120 = 24$

  Thus, we have:

  $\boldsymbol{x\%\ \text{of}\ y = \frac{x}{100} \times y}$

- **Calculating the percent value of a number with respect to another:**

  If we were to calculate: 30 represents what percent of 150?

  This implies that if we have 30 out of 150, then what is the number that we have out of 100?

  Following unitary method:

  If the value is 150, then the required number is 30

  => If the value is 100, the required number is $\frac{30}{150} \times 100 = 20$

Thus, we have:

$$\textbf{x is what \% of y} = \frac{x}{y} \times 100$$

Some fractions and their percent equivalents are shown below:

Fraction	Percent	Fraction	Percent	Fraction	Percent	Fraction	Percent
$\frac{1}{1}$	100%	$\frac{1}{6}$	16.66%	$\frac{1}{11}$	9.09%	$\frac{1}{16}$	6.25%
$\frac{1}{2}$	50%	$\frac{1}{7}$	14.28%	$\frac{1}{12}$	8.33%	$\frac{1}{17}$	5.88%
$\frac{1}{3}$	33.33%	$\frac{1}{8}$	12.50%	$\frac{1}{13}$	7.69%	$\frac{1}{18}$	5.55%
$\frac{1}{4}$	25%	$\frac{1}{9}$	11.11%	$\frac{1}{14}$	7.14%	$\frac{1}{19}$	5.26%
$\frac{1}{5}$	20%	$\frac{1}{10}$	10%	$\frac{1}{15}$	6.66%	$\frac{1}{20}$	5.00%

Percent change: Percent change is calculated as the change between the initial and final values calculated as a percent of the initial value.

Thus, we have:

$$\textbf{Percentage change} = \frac{\textbf{Final value (F)} - \textbf{Initial value (I)}}{\textbf{Initial value (I)}} \times \textbf{100}$$

Given that the percent change be $n\%$, we have: $n\% = \left(\frac{F - I}{I}\right) \times 100$.

Rearranging the terms, we get:

$$\frac{F - I}{I} = \frac{n}{100} \Rightarrow \frac{F}{I} - 1 = \frac{n}{100}$$

$$\Rightarrow \frac{F}{I} = 1 + \frac{n}{100}$$

$$\Rightarrow \boldsymbol{F = I\left(1 + \frac{n}{100}\right)}$$

Thus, in order to get the final value, the initial value is to be multiplied with $\left(1 + \frac{n}{100}\right)$.

Some examples are shown below:

Percent change	Fraction multiplied	Percent change	Fraction multiplied
100% increase	$1+\frac{100}{100}=2$	100% decrease	$1-\frac{100}{100}=0$
50% increase	$1+\frac{50}{100}=\frac{3}{2}$	50% decrease	$1-\frac{50}{100}=\frac{1}{2}$
33.33% increase	$1+\frac{33.33}{100}=\frac{4}{3}$	33.33% decrease	$1-\frac{33.33}{100}=\frac{2}{3}$
25% increase	$1+\frac{25}{100}=\frac{5}{4}$	25% decrease	$1-\frac{25}{100}=\frac{3}{4}$
20% increase	$1+\frac{20}{100}=\frac{6}{5}$	20% decrease	$1-\frac{20}{100}=\frac{4}{5}$

Let us take a few examples:

(1) If the price of an article becomes $150 after increasing it by 20%, what was the initial price of the article?

Explanation:

Since the price is increased by 20%, we have:

$$F = I\left(1+\frac{20}{100}\right)$$

$$=> I = \frac{F}{1+\frac{20}{100}} = \frac{150}{1+\frac{1}{5}} = \frac{150}{\frac{6}{5}} = 150 \times \frac{5}{6} = \$125$$

Alternately, we can say that the final price of the article is $(100+20)\,\% = 120\%$ of the initial price.

Thus, we have:

$150 = 120\%$ of I

$$=> I = 150 \times \frac{100}{120} = \$125$$

(2) After two successive discounts of 20% each, the price of a television becomes $12800. What was the original price?

Explanation:

Let original price = $\$x$.

Thus, we have:

$$x\left(1 - \frac{20}{100}\right)\left(1 - \frac{20}{100}\right) = 12,800$$

$$=> x = 12,800 \times \frac{5}{4} \times \frac{5}{4} = \$20,000$$

Successive percent change: When back-to-back percent changes happen, then the overall percent change can be calculated using the following relation:

Let there be successive percent changes of $x\%$ and $y\%$. If the overall percent change be $z\%$, then:

$$\boldsymbol{z = \left(x + y + \frac{xy}{100}\right)\%}$$

Here x and y have to be taken with their sign, positive for increase and negative for decrease.

Let us take an example:

If there is a 20% decrease followed by a 30% increase in the price of a commodity, what is the overall percent change in price of the commodity?

Explanation:

Let the price of the commodity be \$100.

Thus, after 20% decrease in price, the price = $(100 - 20)\%$ or 80% of \$100 = \$80.

Thus, after a 30% increase in price, the final price

$$= (100 + 30)\% \text{ or } 130\% \text{ of } \$80 = \$\left(\frac{130}{100} \times 80\right) = \$104$$

Thus, overall percent change (from \$100 to \$104) = $\frac{104 - 100}{100} \times 100 = 4\%$

Alternate approach:

We can use the successive percent change rule shown above as follows:

We have: $x = -20, \ y = 30$

Thus, overall percent change

$$= -20 + 30 - \frac{20 \times 30}{100} = -20 + 30 - 6 = 4\%.$$

Percent change in product of two terms: The net percent change in the product of two terms, when one changes by $x\%$ and the other changes by $y\%$, is also given by the same formula:

$$\boldsymbol{\left(x + y + \frac{xy}{100}\right)\%}$$

Here x and y have to be taken with their sign, positive for increase and negative for decrease.

Let us take a few examples:

(1) If the length of a rectangle increases by 20% and the breadth decreases by 30%, what is the percent change in area of the rectangle?

Explanation:

Let the length and breadth of the rectangle be $100x$ and $100y$, respectively.

Thus, the area = Length $\times$ Breadth = $10,000xy$

New length = $(100 + 20)\,\%$ of $100x = 120x$

New breadth = $(100 - 30)\,\%$ of $100y = 70y$

New area = $120x \times 70y = 8,400xy$

Thus, percent change in area = $\dfrac{10,000xy - 8,400xy}{10,000xy} \times 100 = 16\%$ (decrease)

Alternate approach:

We can use the successive percent change rule shown above as follows:

We have: $x = 20,\ y = -30$

Thus, overall percent change

$= 20 - 30 - \dfrac{20 \times 30}{100} = -10 - 6 = -16\%.$

(2) The price of an item is increased by 20%. As a result, the number of items sold decreases by 10%. What is the percent change in the revenue generated as a result?

Explanation:

Revenue = Price $\times$ Number of items sold

Thus, percent change in revenue (a product of two commodities) is $20 - 10 - \dfrac{20 \times 10}{100} = 8\%$

Percent change in the sum of two commodities: Let there be two commodities whose values are A and B, which have percent changes of $x\%$ and $y\%$, the net percent change in the sum of the two is given by:

$$\left(\frac{Ax + By}{A + B}\right) \times 100\%$$

Here x and y have to be taken with their sign, positive for increase and negative for decrease.

Let us take an example:

Let the monthly salaries of a couple are \$1,000 for the husband and \$1,500 for the wife. If the husband's salary increases by 10% and the wife's salary increases by 20%, what is the overall percent increase in their combined salary?

Explanation:

Increased salary of the husband = (100 + 10) % of \$1,000 = \$1,100

Increased salary of the wife = (100 + 20) % of \$1,500 = \$1,800

Thus, new combined salary = \$(1,100 + 1,800) = \$2,900

Initial combined salary = \$(1,000 + 1,500) = \$2,500

Thus, percent increase = $\frac{2,900 - 2,500}{2,500} \times 100 = 16\%$

Alternate approach:

We can use the percent change rule shown above as follows:

We have: $A = 1,000,\ B = 1,500,\ x = 10,\ y = 20$

Thus, net percent change

$$= \left(\frac{Ax + By}{A + B}\right) \times 100\%$$

$$= \frac{1,000 \times 10 + 1,500 \times 20}{1,000 + 1,500} \times 100 = \frac{100\,(10 \times 10 + 15 \times 20)}{100\,(10 + 15)} \times 100$$

$$= \frac{400}{25} \times 100 = 16\%$$

Change of base: If the percent difference between two values, expressed in terms of one is to be calculated in terms of the other, we use the change of base concept.

- **A is x% more than B => B is $\left(\frac{100x}{100 + x}\right)$% less than A**
- **A is x% less than B => B is $\left(\frac{100x}{100 - x}\right)$% more than A**

Let us take an example:

If A is 20% more than B, by what percent is B less than A?

Explanation:

We can say that: If B is 100, A = (100 + 20) % of 100 = 120.

Thus, B is $\left(\frac{120 - 100}{120}\right) \times 100 = 16.66\%$ less than A.

Alternate approach:

We can use the change of base rule shown above as follows:

Since A is 20% more than B, we have: $x = 20$

Thus, the required percent value $= \left(\frac{100 \times 20}{100 + 20}\right)\% = \frac{2000}{120}\% = 16.66\%$

11.2 Profit, Loss, and Discount

Some important points regarding profit, loss and discount are:

- The price at which an article is purchased or manufactured is the Cost Price
- The price at which the article is sold is the Selling Price
- Profit or Loss is the difference between the Cost Price and Selling Price:
  - Profit = Selling Price − Cost Price
  - Loss = Cost Price − Selling Price
- Percent Profit or Percent Loss are calculated on the Cost Price
- Marked price or List price is the price on which discounts are applicable resulting in the final Selling Price
- Percent Discount is calculated on the Marked (List) Price

Some important relations:

- If an article having cost C is sold at $P\%$ profit, then the selling price S is
  $S = (100 + P)\%$ of C
  $=> S = \frac{100 + P}{100} \times C$
  $=> \boldsymbol{S = C \times \left(1 + \frac{P}{100}\right)}$
- If an article having cost C is sold at $L\%$ loss, then the selling price S is
  $S = (100 - L)\%$ of C
  $=> S = \frac{100 - L}{100} \times C$
  $=> \boldsymbol{S = C \times \left(1 - \frac{L}{100}\right)}$
- If an article having marked (list) price M is sold at $D\%$ discount, then the selling price S is
  $S = (100 - D)\%$ of M
  $=> S = \frac{100 - D}{100} \times M$
  $=> \boldsymbol{S = M \times \left(1 - \frac{D}{100}\right)}$
- If an article having cost C is sold at a price S, the percent profit or loss is
  $\boldsymbol{\left(\frac{S - C}{C}\right) \times 100\%}$
  If the above value comes positive, it is a profit, else it is a loss
- If an article having marked price M is sold at a price S, the percent discount is
  $\boldsymbol{\left(\frac{M - S}{M}\right) \times 100\%}$

Let us take a few examples:

(1) A shopkeeper has a stock of pens, each having the same cost price. If he sells 4 pens at the price of the cost of 5 pens and he sells all the pens he has with him, what is his percent profit?

Explanation:

We know that:

Cost price of 5 pens = Selling price of 4 pens = \$20 (assumed)

(Note: We assumed the price as \$20 since 20 is divisible by both 4 and 5)

Thus, the cost of 1 pen = $\$\frac{20}{5} = \$4$

Selling price of 1 pen = $\$\frac{20}{4} = \$5$

Thus, percent profit = $\frac{5-4}{4} \times 100 = 25\%$

Alternately, we can say that since the shopkeeper sells 4 pens for the cost of 5, he makes a profit of $(5 - 4) = 1$ pen for every 4 pens.

Thus, his percent profit = $\frac{1}{4} \times 100 = 25\%$

Point to be noted:

If Selling Price of n articles = Cost price of m articles

Percent profit or loss

$$= \left(\frac{m-n}{n}\right) \times 100\%$$

If the above is positive, then it is a profit, else it is a loss

(2) A shopkeeper has a stock of pens, each having the same cost price. If for every 6 pens that he sells at cost price, he gives away 1 pen for free, and he sells all the pens he has with him, what is his percent loss?

Explanation:

Since the shopkeeper gives away 1 pen for free for every 6 pens sold at cost price, we have:

Cost price of 6 pens = Selling price of 7 pens = \$42 (assumed)

(Note: We assumed the price as \$42 since 42 is divisible by both 6 and 7)

Thus, the cost of 1 pen = $\$\frac{42}{6} = \$7$

Selling price of 1 pen = $\$\frac{42}{7} = \$6$

Thus, percent loss = $\frac{7-6}{7} \times 100 = 14.28\%$

Point to be noted:

If y articles are given for free for every x articles sold at cost price, we can say that $(x + y)$ articles are sold at the cost price of x articles.

Thus, percent loss

$$= \left(\frac{y}{x+y}\right) \times 100\%$$

(3) A shopkeeper has a stock of pens, each having the same cost price. If for every 10 pens that he sells at a price 20% higher than the cost, he gives away 1 pen for free. If he sells all the pens he has with him, what is his percent profit?

Explanation:

Let the cost price of a pen be \$10.

Thus, selling price of a pen = $(100 + 20)\%$ of \$10 = \$12.

The man gives 1 pen for free for every 10 pens sold at a higher price.

Thus, selling price of $(10 + 1) = 11$ pens = $\$(10 \times 12) = \$120$

Cost price of 11 pens = $\$(11 \times 10) = \$110$

Thus, percent profit = $\frac{120 - 110}{110} \times 100 = 9.11\%$

Alternate approach:

Apparently, the shopkeeper sells each pen at a price 20% above the cost price.

Thus, we have:

The selling price of 1 pen = $(100 + 20)\%$ of the cost price = $\frac{6}{5}$ times the cost price

=> The selling price of 5 pens = Cost price of 6 pens ... (i)

The man gives 1 pen for free for every 10 pens sold at a higher price.

Thus, from (i):

The selling price of 10 pens = Cost price of 12 pens

Thus, the man sells 10 pens for the cost of 12 pens and gives 1 pen for free.

Thus, the selling price of $(\mathbf{10} + \mathbf{1}) = \mathbf{11}$ pens is the same as the cost of 12 pens = \$132 (assumed).

Thus, the selling price of each pen = $\$\frac{\mathbf{132}}{\mathbf{11}} = \$\mathbf{12}$

The cost price of each pen = $\$\frac{\mathbf{132}}{\mathbf{12}} = \$\mathbf{11}$

Thus, percent profit = $\frac{\mathbf{12} - \mathbf{11}}{\mathbf{11}} \times \mathbf{100} = \mathbf{9.11}\%$

(4) An article is sold at 20% profit. Had it been sold at \$60 less than the selling price, there would have been a loss of 5%. What is the cost price of the article?

Explanation:

Difference in Selling prices = \$60

Difference in % Profit = $\mathbf{20} - (-\mathbf{5}) = \mathbf{25}\%$ of the Cost price

Thus, we have:

25% of Cost price = \$60

=> Cost price = $\$\left(\mathbf{60} \times \frac{\mathbf{100}}{\mathbf{25}}\right) = \$\mathbf{240}$

(5) A man sold two tables for \$96 each, gaining 20% on one and losing 20% on the other. What is his overall profit or loss percent?

Explanation:

For the first table:

Selling price = \$96 = (100 + 20) % of Cost price

=> Cost price = $\$\left(96 \times \frac{100}{120}\right) = \$80$

For the second table:

Selling price = \$96 = (100 − 20) % of Cost price

=> Cost price = $ $\left(96 \times \frac{100}{80}\right)$ = $120

Thus, total cost price = $ (80 + 120) = $200

Total selling price = $ (96 × 2) = $192

Thus, overall percent loss = $\frac{200 - 192}{200} \times 100 = 4\%$

Point to be noted:

If two items are sold, **each at the same price**, one at a **profit of *k*%** and the other at a **loss of *k*%**, there is an overall **loss** given by: $\left(\frac{k^2}{100}\right)\%$.

(6) In which of the following cases, is the discounted price the greatest?

(A) Successive discounts of 10% and 20%
(B) Successive discounts of 20% and 10%
(C) Successive discounts of 25% and 5%
(D) Single discount of 30%

Explanation:

We know that the overall percent change for successive percent changes of x% and y% is given by $\left(x + y + \frac{xy}{100}\right)\%$

Thus, going by options:

(A) Net percent change = $-10 - 20 + \frac{(-10)(-20)}{100} = -28\%$

(B) Net percent change = $-20 - 10 + \frac{(-20)(-10)}{100} = -28\%$

(C) Net percent change = $-25 - 5 + \frac{(-25)(-5)}{100} = -28.75\%$

Since the minimum discount is 28% for both options A and B, the greatest price would also be for options A and B.

11.3 Ratio, Proportion and Variation

Ratio: A ratio is defined as the multiple or part a quantity of one kind bears with another quantity of the same kind.

For example:

Let us have two quantities $\boldsymbol{a}$ and $\boldsymbol{b}$, such that $\boldsymbol{a} = 30$ and $\boldsymbol{b} = 45$

Since $30 \times \frac{3}{2} = 45$, we have:

$\boldsymbol{a} \times \frac{3}{2} = \boldsymbol{b} \Rightarrow \frac{\boldsymbol{a}}{\boldsymbol{b}} = \frac{2}{3}$

$\Rightarrow \boldsymbol{a} : \boldsymbol{b} = 2 : 3$

Thus, the ratio of a and b is 3 : 2

Some important relations

- If $A : B = p : q$, we have:
  - $A = \left(\frac{p}{p+q}\right) \times (A + B)$
  - $B = \left(\frac{q}{p+q}\right) \times (A + B)$
- Ratio remains same when the two quantities of a ratio are multiplied or divided by the same number. Thus, we have:

  $\frac{\boldsymbol{a}}{\boldsymbol{b}} = \frac{\boldsymbol{a} \times \boldsymbol{k}}{\boldsymbol{b} \times \boldsymbol{k}}$

  $\Rightarrow \boldsymbol{a} : \boldsymbol{b} = (\boldsymbol{a} \times \boldsymbol{k}) : (\boldsymbol{b} \times \boldsymbol{k})$

- If the ratio of two quantities is less than '1', adding the same positive number to both quantities of the ratio increases the value. Thus, we have:
  - If $\frac{a}{b} < 1$, and k is a positive number:

    $\frac{\boldsymbol{a} + \boldsymbol{k}}{\boldsymbol{b} + \boldsymbol{k}} > \frac{\boldsymbol{a}}{\boldsymbol{b}}$

    For example: $\frac{a}{b} = \frac{2}{3} \Rightarrow \frac{a+1}{b+1} = \frac{3}{4} > \frac{2}{3}$
  - Similarly, if $\frac{a}{b} < 1$, and k is a positive number:

    $\frac{\boldsymbol{a} - \boldsymbol{k}}{\boldsymbol{b} - \boldsymbol{k}} < \frac{\boldsymbol{a}}{\boldsymbol{b}}$

    For example: $\frac{a}{b} = \frac{2}{3} \Rightarrow \frac{a-1}{b-1} = \frac{1}{2} < \frac{2}{3}$

- If the ratio of two quantities is greater than '1', adding the same positive number to both quantities of the ratio decreases the value. Thus, we have:

  - If $\frac{a}{b} > 1$, and k is a positive number:
    $\frac{\boldsymbol{a + k}}{\boldsymbol{b + k}} < \frac{\boldsymbol{a}}{\boldsymbol{b}}$
    For example: $\frac{a}{b} = \frac{3}{2} => \frac{a+1}{b+1} = \frac{4}{3} < \frac{3}{2}$

  - Similarly, if $\frac{a}{b} > 1$, and k is a positive number:

    $\frac{\boldsymbol{a - k}}{\boldsymbol{b - k}} > \frac{\boldsymbol{a}}{\boldsymbol{b}}$
    For example: $\frac{a}{b} = \frac{3}{2} => \frac{a-1}{b-1} = \frac{2}{1} > \frac{3}{2}$

- If $\frac{a}{b} = \frac{c}{d}$, we have:
  Let $\frac{a}{b} = \frac{c}{d} = k$
  $=> a = b \times k$ and $c = d \times k$
  $=> a + c = k\,(b + d)$
  $=> \frac{a+c}{b+d} = k$
  $=> \frac{\boldsymbol{a}}{\boldsymbol{b}} = \frac{\boldsymbol{c}}{\boldsymbol{d}} = \frac{\boldsymbol{a + c}}{\boldsymbol{b + d}}$

- If $\frac{\mathrm{a}}{\mathrm{b}} = \frac{\mathrm{c}}{\mathrm{d}}$, we have:

  Let $\frac{a}{b} = \frac{c}{d} = k$
  $=> a = b \times k$ and $c = d \times k$
  Thus, we have:

  - $a + b = bk + b = b\,(k + 1)$, $a - b = bk - b = b\,(k - 1)$
  - $c + d = dk + d = d\,(k + 1)$, $c - d = dk - d = d\,(k - 1)$

  $=> \frac{\boldsymbol{a + b}}{\boldsymbol{a - b}} = \frac{\boldsymbol{c + d}}{\boldsymbol{c - d}}$

- If we have: $\boldsymbol{k} \times \boldsymbol{a} = \boldsymbol{l} \times \boldsymbol{b} = \boldsymbol{m} \times \boldsymbol{c}$, we have:

  $\boldsymbol{a : b : c} = \frac{\mathbf{1}}{\boldsymbol{k}} : \frac{\mathbf{1}}{\boldsymbol{l}} : \frac{\mathbf{1}}{\boldsymbol{m}}$

Proportion: If the ratio of two quantities is equal to the ratio of other two quantities then the four quantities are said to be in proportion.

Thus, if $a : b = c : d$, then a, b, c, and d are in proportion.

Variation:

- If two quantities a & b are such that any increase in b produces a proportionate increase in a, and vice-versa, then a is 'directly proportional' to b.

  Thus, we have:

  $\boldsymbol{a \propto b}$

  $=> a = kb$, where k is a constant

  $=> \boldsymbol{\frac{a}{b}}$ = **constant**

- If two quantities a and b are such that any increase in b will bring about a proportionate decrease in a and vice-versa, then a is inversely proportional to b.

  Thus, we have:

  $\boldsymbol{a \propto \frac{1}{b}}$

  $=> a = \frac{k}{b}$, where k is a constant

  $=> a \times b$ = constant

- If A, B and C are three quantities, such that $A \propto B$ and $B \propto C$

  $=> \boldsymbol{A \propto C}$

- Joint variation: If A, B and C are three quantities, such that $A \propto B$ if C is constant and $A \propto C$ if B is constant, we have:

  $\boldsymbol{A \propto (B \times C)}$, where both B and C vary

Let us take a few examples:

(1) If $\frac{p}{3} = \frac{q}{8} = \frac{7p - 2q}{k}$, what is the value of k?

Explanation:

We know that:

$\frac{p}{3} = \frac{q}{8} = \frac{7p - 2q}{k}$

Let $\frac{p}{3} = \frac{q}{8} = \frac{7p - 2q}{k} = r$

$=> p = 3r$ and $q = 8r$

Thus, we have:

$7p = 21r$ and $2q = 16r => 7p - 2q = 5r$

Thus, we have:

$\frac{7p - 2q}{k} = r$

$=> \frac{5r}{k} = r => k = 5$

(2) An alloy has iron and copper mixed in the ratio 2 : 3. If 21 pounds of iron is mixed with 15 pounds of the above alloy, what would be the percent of copper in the final alloy?

Explanation:

Amount of copper present in 15 pounds of the original alloy = $\frac{3}{2+3} \times 15 = 9$ pounds

Since the quantity of copper remains unchanged, while the total alloy quantity becomes $(15 + 21) =$ 36 pounds, we have:

Percent of copper = $\frac{9}{36} \times 100 = 25\%$

(3) The cost of manufacturing an article is partly fixed and partly varies with the number of articles to be manufactured. If the cost of manufacturing 10 articles is \$100 and the cost of manufacturing 20 articles is \$180, what is the fixed cost involved?

Explanation:

Let the cost of manufacturing an article be given by C.

Let the fixed cost be f and the variable cost be v.

Thus, we have:

$v \propto n$, where n is the number of articles

$=> v = kn$, where k is a constant of proportionality

Thus, we have:

$C = f + v => C = f + kn$

For $n = 10: \ C = 100 = f + 10k \ldots$(i)

For $n = 20: C = 180 = f + 20k \ldots$(ii)

From (ii) – (i):

$80 = 10k => k = 8$

Thus, from (i):

$100 = f + 80 \Rightarrow f = 20$

Thus, the fixed cost is \$20.

Alternate approach:

The cost of manufacturing 10 articles is \$100 and that for 20 articles is \$180

Thus, the incremental cost for $(20 - 10) = 10$ articles = \$ $(180 - 100) = \$80$

Since the cost of manufacturing 10 articles is \$100, the balance amount, i.e. \$ $(100 - 80) = \$20$ is the fixed cost.

(4) A telecom operator charges a fixed amount of $\$f$ for the first $\frac{1}{4}$ minutes of a call. Thereafter, for every $\frac{1}{2}$ minutes, it charges $\$v$. If the total charge for a call be $\$c$, what is the expression for the duration of the call in terms of f, v and c?

Explanation:

Let the duration of the call be x minutes.

Charges for the first $\frac{1}{4}$ minutes = $\$f$

Duration of call remaining = $\left(x - \frac{1}{4}\right)$ minutes.

Call charge every $\frac{1}{2}$ minute = $\$v$

Thus, call charge for $\left(x - \frac{1}{4}\right)$ minutes =

$$\$\left\{\left(\frac{v}{\frac{1}{2}}\right) \times \left(x - \frac{1}{4}\right)\right\} = \$\left\{2v\left(x - \frac{1}{4}\right)\right\}$$

Thus, total call charge = $\$\left\{f + 2v\left(x - \frac{1}{4}\right)\right\}$

Thus, we have:

$$f + 2v\left(x - \frac{1}{4}\right) = c$$

$$\Rightarrow 2v\left(x - \frac{1}{4}\right) = c - f$$

$$\Rightarrow x - \frac{1}{4} = \frac{c - f}{2v}$$

$$\Rightarrow x = \frac{c - f}{2v} + \frac{1}{4}$$

11.4 Mixtures

Mixture problems involve forming a mixture from two or more things, and determining some quantity (percentage, price, etc.) of the final mixture.

Mixture problems can be seen in any of (or similar to) the following scenarios:

- Combining multiple varieties of commodities having different prices and determining the average price of the mixture
- Combining multiple alloys with different percent purity and determining the percent purity of the mixture
- Combining multiple solutions with different components and determining the ratio of the components in the mixture

To find the mean value of a mixture:

- When X and x quantities of A and B, costing C and cost c, respectively, are mixed, the average cost, M, of mixture is given by:

$$M = \frac{CX + cx}{X + x}$$

- When two solutions A and B, having volume V and v, containing $R\%$ concentration of ingredient X and $r\%$ of ingredient X, respectively, are mixed, the concentration, $M\%$, of ingredient X in the mixture is given by:

$$M = \frac{V\left(\frac{R}{100}\right) + v\left(\frac{r}{100}\right)}{V + v} \times 100 = \frac{VR + vr}{V + v}$$

Let us take a few examples:

(1) Three varieties of tea priced at \$10 per pound, \$15 per pound and \$24 per pound are mixed together in the ratio 2 : 1 : 5, respectively. What is the average price of the resulting mixture?

Explanation:

We know that the three varieties are mixed in the ratio 2 : 1 : 5

Let the quantities of the three varieties be 2 pounds, 1 pound and 5 pounds, respectively.

Thus, total prices of the three varieties of tea are:

$\$ (2 \times 10) = \$20$, $\$ (1 \times 15) = \$15$ and $\$ (5 \times 24) = \$120$

Thus, combined price of $(2 + 1 + 5) = 8$ pounds of tea $= \$ (20 + 15 + 120) = \$155$

Thus, average price of the mixture $= \$\frac{155}{8} = \$19\frac{3}{8}$

(2) Three varieties of tea priced at \$20 per pound, \$15 per pound and \$24 per pound are mixed together. If the average price of the resulting mixture is \$20 per pound, which of the following is a possible ratio in which the three varieties of tea were mixed?

(A) 1 : 4 : 5

(B) 2 : 3 : 8

(C) 3 : 8 : 12

(D) 5 : 8 : 10

Explanation:

Let the quantities of the three varieties of tea used are x pounds, y pounds and z pounds.

Thus, total prices of the three varieties of tea are:

$\$\,(x \times 20) = \$20x$, $\$\,(y \times 15) = \$15y$ and $\$\,(z \times 24) = \$24z$

Thus, combined price of $(x + y + z)$ pounds of tea = $\$\,(20x + 15y + 24z)$

Thus, average price of the mixture = $\$\left(\frac{20x + 15y + 24z}{x + y + z}\right)$

Thus, we have:

$$\frac{20x + 15y + 24z}{x + y + z} = 20$$

$$=> 20x + 15y + 24z = 20x + 20y + 20z$$

$$=> 5y = 4z => \frac{y}{z} = \frac{4}{5}$$

Since x is not there in the above equation, any value of x can be used. Observe that the average price of the mixture is the same as the price per pound of the variety having quantity x.

Thus, the presence or the absence of that particular variety will not affect the average price of the mixture.

Thus, we need to choose those options where $y : z = 4 : 5$

The correct answers are options A and D.

(3) Iron ore mined from Site A has 72% iron, while that from Site B has 56% iron. A certain quantity of both ores are taken and melted to form a substance having 65% iron. If the quantity of iron ore taken from Site A is 5 pounds more than that from Site B, what is the weight of the ore used from Site B?

(A) 12

(B) $16\frac{1}{2}$

(C) $17\frac{1}{2}$

(D) 20

Explanation:

Let the quantity of ore taken from Site B be x pounds and that taken from Site A be $(x + 5)$ pounds.

Thus, quantity of iron (in pounds) present in the mixture

$= 56\%$ of $x + 72\%$ of $(x + 5)$

$$= \frac{56x}{100} + \frac{72(x + 5)}{100} = \frac{128x + 360}{100}$$

Total quantity of the mixture $= x + (x + 5) = (2x + 5)$ pounds

Thus, the percent of iron in the mixture

$$= \frac{128x + 360}{100(2x + 5)} \times 100\%$$

Thus, we have:

$$\frac{128x + 360}{100(2x + 5)} \times 100 = 65$$

$=> 128x + 360 = 130x + 325$

$=> 2x = 35 => x = 17\frac{1}{2}$

Note: Percent of iron present in ores from Site A and Site B = 72% and 56%

If the two ores were mixed in equal ratio, percent of iron in the mixture $= \frac{72 + 56}{2} = 64\%$

Since the actual percent of iron in the mixture is 65%, i.e. slightly greater than the above value, there should be slightly higher quantity of ore from Site A.

Alternate approach:

Let the quantities of ores from Site A and Site B be a and b pounds respectively.

Thus, we have:

$$\frac{72a + 56b}{a + b} = 65$$

$=> 72a + 56b = 65a + 65b => 7a = 9b$

$=> a = \frac{9}{7}b$

But, we also know that: $a = b + 5$

$=> b + 5 = \frac{9}{7}b => \frac{2}{7}b = 5$

$=> b = 17\frac{1}{2}$

Method of alligation:

Alligation is the method by which we calculate the ratio in which two different ingredients of known values are mixed to produce a mixture of a given value. The value of the ingredient may refer to the price, composition, etc.

In the alligation method, the amount of working required to get to the answer is considerably reduced.

Let us look into the method for different scenarios:

- When P quantity of dearer ingredient A of cost C and p quantity of cheaper ingredient B of cost c (where $C > c$) are mixed to get an average cost of mixture M:

$$\frac{\textbf{Weight of cheaper ingredient}}{\textbf{Weight of dearer ingredient}} = \frac{\textbf{Cost of dearer} - \textbf{Cost of mixture}}{\textbf{Cost of mixture} - \textbf{Cost of cheaper}}$$

The rule can also be represented in the following way:

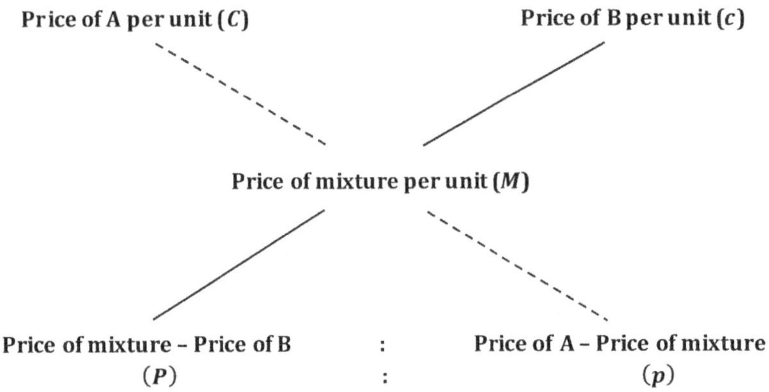

- When P quantity of a solution A having $R\%$ of constituent X and p quantity of a solution B having $r\%$ of constituent X (where $R > r$) are mixed to get an average concentration of constituent X as $M\%$:

$$\frac{\textbf{Weight of higher conc.solution}}{\textbf{Weight of lower conc.solution}} = \frac{\textbf{Conc. in higher} - \textbf{Conc. in mixture}}{\textbf{Conc. in mixture} - \textbf{Conc. in lower}}$$

The rule can also be represented in the following way:

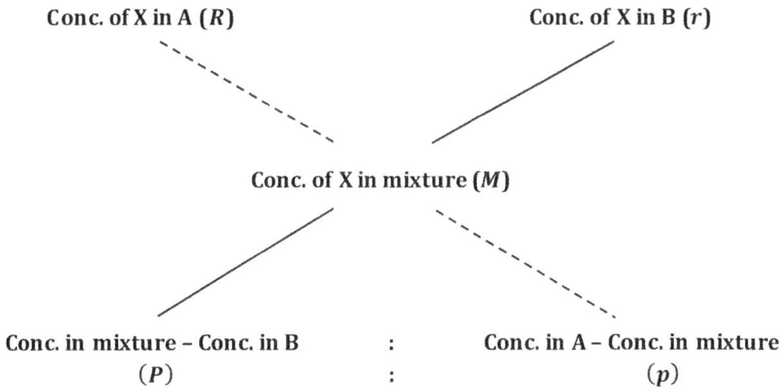

Let us take a few examples:

(1) A merchant mixes x kg of tea costing \$12 per kg and y kg of another variety costing \$20 per kg. If the price of the mixture is \$15 per kg, what is the ratio of x and y?

Explanation:

We have: x kg of tea priced at \$12 per kg and y kg of tea priced at \$20 per kg.

Thus, the average price of the mixture

$$= \$\left(\frac{12x + 20y}{x + y}\right)$$

Thus, we have:

$$\frac{12x + 20y}{x + y} = 15$$

$$=> 12x + 20y = 15x + 15y$$

$$=> 3x = 5y => \frac{x}{y} = \frac{5}{3}$$

Alternate approach:

Let us solve using the method of alligation:

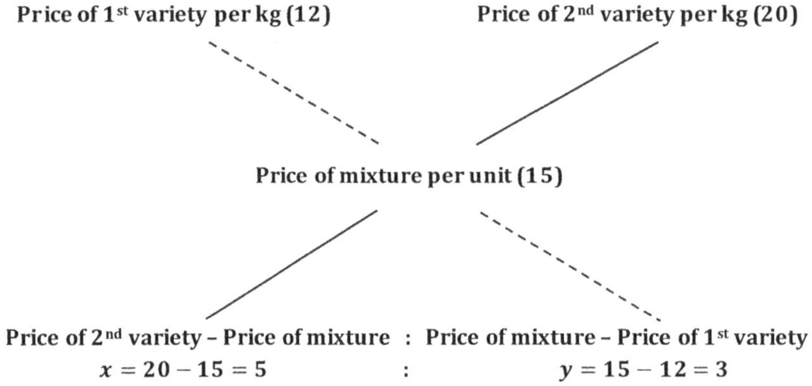

Thus, the required ratio of $x : y = 5 : 3$

(2) A solution A contains alcohol and spirit in the ratio 4 : 1 and the solution B contains alcohol and spirit in the ratio 2 : 3. In what proportion should the solutions to be taken from A and B to form a mixture having alcohol and spirit in the ratio of 1 : 1?

Explanation:

Let us consider the concentration of alcohol in the solutions.

Concentration of alcohol in solution A = $\left(\frac{4}{4+1}\right) \times 100 = 80\%$

Concentration of alcohol in solution B = $\left(\frac{2}{2+3}\right) \times 100 = 40\%$

Let the quantity taken from solution A and solution B be x and y, respectively.

Thus, average concentration of alcohol in the mixture

$= \frac{80x + 40y}{x + y}$

Concentration of alcohol in the final mixture = $\left(\frac{1}{1+1}\right) \times 100 = 50\%$

Thus, we have:

$\frac{80x + 40y}{x + y} = 50$

$=> 80x + 40y = 50x + 50y$

$=> 30x = 10y => \frac{x}{y} = \frac{1}{3}$

Alternate approach:

Let us solve using the method of alligation:

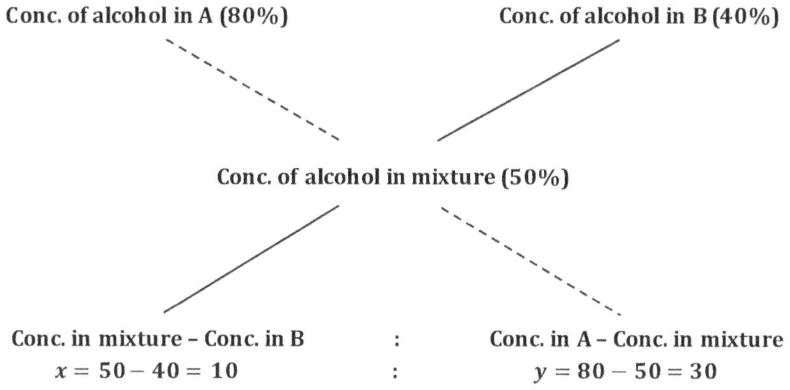

Thus, the required ratio of $x : y = 1 : 3$

Removal and Replacement

Say, a vessel contains a solution of different ingredients in some ratio. A part of the solution is removed and some quantity of one of the existing ingredients is added to the solution.

Let us take an example:

A container has 100 liters of spirit. 10 liters of the contents are taken out and replaced with petrol. This process is repeated one more time. What is the concentration of spirit finally present in the container?

Explanation:

The idea behind the problem is as follows:

If a vessel having capacity V has two constituents A and B in the ratio $a : b$, and a part of the contents of the vessel, say p% are removed, the remaining vessel would still have A and B in the same ratio $a : b$. Also, the contents that are removed, would also have A and B in the ratio $a : b$.

Since p% of the contents of the vessel are removed, the part remaining is $(100 - p)\%$ of the vessel. Thus, the remaining parts of the constituents are also $(100 - p)\%$ of their initial values.

Applying the above concept in the problem, we have:

Initial contents of the vessel = 100 liters spirit.

Since 10 liters are removed and replaced with petrol, we have 90 liters spirit and 10 liters petrol.

After this, again 10 liters of the above solution of 100 liters is removed.

Thus, $\frac{10}{100} \times 100 = 10\%$ of the solution is removed.

Thus, spirit remaining = (100 – 10) = 90% of 90 = $\frac{90}{100} \times 90 = 81$ liters.

Similarly, alcohol remaining = (100 – 10) = 90% of 10 = $\frac{90}{100} \times 10 = 9$ liters.

Since 10 liters alcohol is added again, the final quantity of alcohol = 10 + 9 = 19 liters.

Thus, ratio of spirit and alcohol finally = 81 : 19

11.5 Computation

Computational problems are typically those which cannot be categorized into any fixed category.

Such problems are usually solved using basic arithmetic skills applied along with logical reasoning skills.

There can be various types of computational problems. A few of the typical ones are mentioned below:

Alphanumeric problems: An alphanumeric problem is an arithmetic problem involving addition, subtraction, multiplication, and in some rare cases, division, where the digits of the numbers are replaced with letters.

Careful observation of the patterns is necessary to determine the numeric value of each letter.

Let us take an example:

In the multiplication shown below, $1ABCDE$ is a six-digit number. What is the value of $(A + B + C + D + E)$?

$$\begin{array}{r} 1\,A\,B\,C\,D\,E \\ \times \qquad\quad 3 \\ \hline \underline{A\,B\,C\,D\,E\,1} \end{array}$$

Explanation:

In the above multiplication, we observe:

- $E \times 3$ ends in 1 => E must be 7 => There is a carry of 2
- $D \times 3 + 2$ ends in E, i.e. 7 => D must be 5 => There is a carry of 1
- $C \times 3 + 1$ ends in D, i.e. 5 => C must be 8 => There is a carry of 2
- $B \times 3 + 2$ ends in C, i.e. 8 => B must be 2 => There is no carry
- $A \times 3$ ends in B, i.e. 2 => A must be 4 => There is a carry of 1
- $1 \times 3 + 1 = 4$ => $A = 4$, as obtained above

Thus, we have:

$A + B + C + D + E = 4 + 2 + 8 + 5 + 7 = 26$

Symbol based problems: In symbol based problems, a symbol is defined, based on which, the calculations are to be done.

Let us take an example:

If $r \# s = r^s + s^r$ and $p @ 0 = p$, where @ implies one of $+$, $-$, $\times$, $\div$, what is the value of $(2 @ 3) \# 5$?

Explanation:

We have: $p @ 0 = p$, where @ implies one of $+$, $-$, $\times$, $\div$

Thus, we may have:

- @ implies '+': $p @ 0 = p + 0 = p$
- @ implies '−': $p @ 0 = p - 0 = p$

Working with the above scenarios:

- $(2 @ 3) \# 5 = (2 + 3) \# 5 = 5 \# 5 = 5^5 + 5^5 = 2\left(5^5\right)$
- $(2 @ 3) \# 5 = (2 - 3) \# 5 = (-1) \# 5 = (-1)^5 + 5^{(-1)} = -1 + \frac{1}{5} = -\frac{4}{5}$

Thus, the possible solutions are: $2\left(5^5\right)$ OR $-\frac{4}{5}$

Calculation intensive: Certain problems may be simply calculation based, where the calculation may be simplified using some basic identities or rules.

Let us take an example:

What is the approximate value of $\left\{(0.11 + 0.07)^2 - (0.11 - 0.07)^2\right\}\%$ of $\left\{\frac{100}{(0.11)^2 - (0.07)^2}\right\}$?

(A) 2.4

(B) 3.6

(C) 4.3

(D) 8.1

Explanation:

Let's take the first part of the expression

$\left\{(0.11 + 0.07)^2 - (0.11 - 0.07)^2\right\}\%$

$= \left[\left(0.11^2 + 2 \times 0.11 \times 0.07 + 0.07^2\right) - \left(0.11^2 - 2 \times 0.11 \times 0.07 + 0.07^2\right)\right]\%$

$= \left(0.11^2 + 2 \times 0.11 \times 0.07 + 0.07^2 - 0.11^2 + 2 \times 0.11 \times 0.07 - 0.07^2\right)\%$

$= \frac{4 \times 0.11 \times 0.07}{100}$; the terms 0.11^2 & 0.07^2 got cancelled

Let's take the second part of the expression

$$\frac{100}{(0.11)^2 - (0.07)^2}$$

$$= \frac{100}{(0.11 + 0.07)\,(0.11 - 0.07)}$$

$$= \frac{100}{0.18 \times 0.04}$$

Thus, the required value (taking both the parts of the expression)

$$= \left(\frac{4 \times 0.11 \times 0.07}{100}\right) \times \left(\frac{100}{0.18 \times 0.04}\right)$$

$$= 4 \times \left(\frac{0.11 \times 0.07}{0.18 \times 0.04}\right)$$

$$= 4 \times \left(\frac{11 \times 7}{18 \times 4}\right)$$

$$= 4 \times \frac{77}{72}$$

Since $2 > \frac{77}{72} > 1$, the value should be slightly greater than 4 (but definitely less than 8).

The correct answer is option C.

11.6 Average

Average of a set of numbers is a number expressing the central or typical value of the set. It is calculated as the ratio of the sum of the set of numbers and the number of terms in the set.

Thus, for a set of n numbers $x_1, x_2, x_3, \ldots x_n$, the average (arithmetic mean) is calculated as

$$\overline{x} = \frac{\text{Sum of the } n \text{ numbers}}{n} = \frac{x_1 + x_2 + x_3 + \cdots + x_n}{n}$$

Thus, we have:

Sum of the n numbers $= \overline{x} \times n$

Some important points:

- If a constant number k is added to or subtracted from each term of a set of numbers, the average of the set would also increase by k or decrease by k, respectively.

  Thus, we have:

  If the average of $x_1, x_2, x_3, \ldots x_n$ is $\overline{x}$, then we have:

  - Average of $x_1 + k, x_2 + k, x_3 + k, \ldots x_n + k$ is $\overline{x} + k$
  - Average of $x_1 - k, x_2 - k, x_3 - k, \ldots x_n - k$ is $\overline{x} - k$

- If each term of a set of numbers is multiplied by or divided by a constant number k, the average of the set would also be multiplied by k or divided by k, respectively.

  Thus, we have:

  If the average of $x_1, x_2, x_3, \ldots x_n$ is $\overline{x}$, then we have:

  - Average of $x_1 \times k, x_2 \times k, x_3 \times k, \ldots x_n \times k$ is $\overline{x} \times k$
  - Average of $\frac{x_1}{k}, \frac{x_2}{k}, \frac{x_3}{k}, \ldots \frac{x_n}{k}$ is $\frac{\overline{x}}{k}$

Weighted average:

If the average value of a set of n observations is x_n, the average value of a set of m observations is x_m, and so on, the combined average of all $(n + m + \ldots)$ observations is given as

$$\overline{x} = \frac{x_n \times n + x_m \times m + \ldots}{n + m + \ldots}$$

Let us take an example:

In a class of 5 students, 3 students scored 20 and 2 students scored 15 in a test. What is the average score of the class?

Simply speaking, the sum of the scores is $20 + 20 + 20 + 15 + 15 = 90$

Thus, the average score = $\frac{90}{5} = 18$

Using the concept of weighted average, we have:

Average score = $\frac{20 \times 3 + 15 \times 2}{3 + 2} = \frac{90}{5} = 18$

Additional rules:

- If n numbers having an average x_n, are added to an existing set of m numbers having an average x_m, we have:

  Final average = Initial average + $\frac{n\,(x_n - x_m)}{m + n}$

  Let us take an example:

  Average of 4 numbers is 30. What is the new average if the numbers 66 and 64 are included in the set?

  Explanation:

  Initial total = $4 \times 30 = 120$

  Final total = $120 + 66 + 64 = 250$

  Thus, final average = $\frac{250}{6} = 41\frac{2}{3}$

  Alternate approach:

  The average of the two new numbers = $\frac{66 + 64}{2} = 65$

  Using the relation above, we have:

  Final average = $30 + \frac{2\,(65 - 30)}{4 + 2} = 30 + \frac{2 \times 35}{6} = 30 + 11\frac{2}{3} = 41\frac{2}{3}$

  Note: This approach offers a quick calculation for the difference between the initial and final average values.

- If n numbers having an average x_n, are removed from an existing set of m numbers having an average x_m, we have:

  Final average = Initial average − $\frac{n\,(x_n - x_m)}{m - n}$

Let us take an example:

Average of 6 numbers is 30. What is the new average if the numbers 24 and 26 are removed from the set?

Explanation:

Initial total = $6 \times 30 = 180$

Final total = $180 - 24 - 26 = 130$

Thus, final average = $\frac{130}{4} = 32\frac{1}{2}$

Alternate approach:

The average of the two numbers which are removed = $\frac{24 + 26}{2} = 25$

Using the relation above, we have:

Final average $= 30 - \frac{2\,(25 - 30)}{6 - 2} = 30 - \left(-\frac{10}{4}\right) = 32\frac{1}{2}$

- When a few incorrect numbers are taken while calculating the average, we have:

Actual average = Incorrect average + $\frac{\text{Sum of correct terms} - \text{Sum of incorrect terms}}{\text{Total number of terms}}$

Let us take an example:

The average of 10 numbers is 48. It was later found out that two numbers had been taken wrongly: 36 was taken as 54 and 46 was taken as 32. What is the actual average?

Explanation:

Actual average $= 48 + \frac{(36 + 46) - (54 + 32)}{10} = 48 + \left(-\frac{4}{10}\right) = 47\frac{3}{5}$

- When the average of a group changes as a result of one term being replaced by another, we have:

New term = Replaced term + (Increase in average $\times$ Number of terms)

Let us take an example:

The average of a group of 6 increases by 3 when 72 is replaced by another number. What is the value of the new number included in place of 72?

Explanation:

Let the new number be x and the initial average be y

Thus, initial total = $6y$ => Final total = $6y - 72 + x$

Thus, final average = $\frac{6y - 72 + x}{6}$

Thus, we have: $\frac{6y - 72 + x}{6} = 3 + y$ => $6y - 72 + x = 18 + 6y$

=> $x = 72 + 18 = 90$

Alternate approach:

Using the above relation:

Value of the new number = $72 + 3 \times 6 = 90$

11.7 Speed, Time and Distance

Speed refers to the rate of travel, or, in other words, the distance covered in unit time.

Thus, we have the basic rule:

$$\text{Speed } (S) = \frac{\text{Distance travelled } (D)}{\text{Time taken } (T)} \Rightarrow D = S \times T$$

Few important relations:

- If the ratio of speeds of two bodies, A and B, is $a : b$, the ratio of the times taken by A and B to cover the same distance d is $\frac{d}{a} : \frac{d}{b} = \frac{1}{a} : \frac{1}{b} = b : a$.

- If the ratio of speeds of A and B is $a : b$, the ratio of the distances travelled by them in the same time t is $t \times a : t \times b = a : b$.

- If the ratio of time of travel of A and B is $a : b$, the ratio of the distances travelled by them with the same speed s is $s \times a : s \times b = a : b$.

Let us take an example:

Walking at $\left(\frac{3}{4}\right)^{\text{th}}$ of his usual speed, a person reaches his destination 10 minutes late. What is the usual time he takes (had he travelled at his usual speed) to cover the same distance?

Explanation:

Let the usual time be x minutes.

In both situations, the distance covered is the same.

Thus, the ratio of time taken is the reciprocal of the ratio of speeds.

Thus, assuming his usual speed to be s, we have:

$$\frac{x}{x+10} = \frac{\frac{3}{4}s}{s}$$

$$\Rightarrow \frac{3}{4} = \frac{x}{x+10}$$

$$\Rightarrow 3x + 30 = 4x \Rightarrow x = 30$$

Average Speed: If a man covers different distances at different speeds, the average speed is given by:

$$\text{Average Speed} = \frac{\text{Total Distance travelled}}{\text{Total Time taken}}$$

We have the possible scenarios:

- If $d_1, d_2, \ldots d_n$ are distances travelled, taking time $t_1, t_2, \ldots t_n$ respectively, we have:

$$\text{Average speed} = \frac{d_1 + d_2 + \cdots + d_n}{t_1 + t_2 + \cdots + t_n}$$

- If a body travels at speeds $s_1, s_2, \ldots s_n$, for time $t_1, t_2, \ldots t_n$ respectively, we have:

$$\text{Average speed} = \frac{s_1t_1 + s_2t_2 + \cdots + s_nt_n}{t_1 + t_2 + \cdots + t_n}$$

  - If the time of travel is t in each case, we have:

$$\text{Average speed} = \frac{s_1t + s_2t + \cdots + s_nt}{t + t + \cdots + t} = \frac{s_1 + s_2 + \cdots + s_n}{n}$$

If $n = 2$, we have:

$$\text{Average speed} = \frac{s_1 + s_2}{2}$$

- If $d_1, d_2, \ldots d_n$ are distances travelled at speeds $s_1, s_2, \ldots s_n$ respectively, we have:

$$\text{Average speed} = \frac{d_1 + d_2 + \cdots + d_n}{\frac{d_1}{s_1} + \frac{d_2}{s_2} + \cdots + \frac{d_n}{s_n}}$$

  - If the same distance d is travelled at the above speeds, we have:

$$\text{Average speed} = \frac{d + d + \cdots + d}{\frac{d}{s_1} + \frac{d}{s_2} + \cdots + \frac{d}{s_n}} = \frac{n}{\frac{1}{s_1} + \frac{1}{s_2} + \cdots + \frac{1}{s_n}}$$

If $n = 2$, we have:

$$\text{Average speed} = \frac{2}{\frac{1}{s_1} + \frac{1}{s_2}} = \frac{2s_1s_2}{s_1 + s_2}$$

Let us take an example:

A man travels from City A to City B. If he covers the first half of the distance at 60 miles per hour and the remaining at 40 miles per hour, what is the average speed of the man?

How does the answer change if the man had covered the first half of the total time taken at 60 miles per hour and the remaining at 40 miles per hour?

Explanation:

Let the total distance be $2d$ miles.

Thus, he travelled d miles at 60 miles per hour and the remaining d miles at 40 miles per hour.

Thus, total time he took for the journey $= \frac{d}{60} + \frac{d}{40} = \frac{5d}{120} = \frac{d}{24}$ hours.

Thus, average speed = $\frac{2d}{\frac{d}{24}}$ = 48 miles per hour.

Alternate Approach 1:

We can assume any suitable value of the distance such that it is divisible by 60 and 40.

Let the distance for each case be 120 miles.

Thus, the man travelled 120 miles at 60 miles per hour and the remaining 120 miles at 40 miles per hour.

Thus, total time he took for the journey = $\frac{120}{60} + \frac{120}{40}$ = 5 hours.

Total distance = 120 + 120 = 240 miles.

Thus, average speed = $\frac{240}{5}$ = 48 miles per hour.

Alternate Approach 2:

Using the relation given above, we have:

Average speed = $\frac{2s_1s_2}{s_1 + s_2} = \frac{2 \times 60 \times 40}{60 + 40} = \frac{4800}{100}$ = 48 miles per hour.

Looking at the second part of the problem:

Let the time he travelled at 60 miles per hour and 40 miles per hour be t hours each.

Thus, total distance = $60t + 40t = 100t$ miles.

Total time = $t + t = 2t$ hours.

Thus, average speed = $\frac{100t}{2t}$ = 50 miles per hour.

Alternate approach:

Using the relation given above, we have:

Average speed = $\frac{s_1 + s_2}{2} = \frac{60 + 40}{2}$ = 50 miles per hour.

Stoppage time: Let the distance covered by a body be d. The average speed of the body excluding stoppage time be s_e, and the average speed including stoppage time be s_i. We need to determine the average stoppage time per hour.

Time taken to cover the distance d had there been no stoppages = $\frac{d}{s_e} = t_e$

Actual time taken to cover the distance d since there were stoppages $= \frac{d}{s_i} = t_i$

Thus, total stoppage time during the journey $= t_i - t_e$

$$= \frac{d}{s_i} - \frac{d}{s_e} = t_s$$

Thus, stoppage time per hour

$$= \frac{t_s}{t_i} \times 60 = \frac{\frac{d}{s_i} - \frac{d}{s_e}}{\frac{d}{s_i}} \times 60$$

$$= \frac{s_e - s_i}{s_e} \times 60$$

Let us take an example:

A train maintains an average speed of 60 miles per hour while traveling between two cities A and B. However, since the train stops at multiple stations between the two cities, its average speed comes down to 40 miles per hour. What is the average stoppage time per hour?

Explanation:

We can assume a suitable distance so that it is divisible by 60 and 40.

Let the distance be 120 miles.

Time taken, assuming the train does not stop between the cities $= \frac{120}{60} = 2$ hours.

Actual time taken by the train (including stoppages) $= \frac{120}{40} = 3$ hours.

Thus, total stoppage time $= 3 - 2 = 1$ hour.

Thus, the train had a stoppage time of 1 hour in a total of 3 hours.

Thus, stoppage time per hour $= \frac{1}{3}$ hours $= \frac{1}{3} \times 60 = 20$ minutes.

Relative Speed: Relative speed refers to the speed of a moving body with respect to another moving body.

- If two bodies A and B move in the same direction with speeds S_A and S_B $(S_A > S_B)$, their relative speed is given as $(S_A - S_B)$

  If B is d distance ahead of A, time taken by A to catch up with B $= \frac{d}{S_A - S_B}$.

- If two bodies A and B are moving in opposite directions with speeds S_A and S_B $(S_A > S_B)$, their relative speed is given as $(S_A + S_B)$

  If A and B are separated by distance d, time taken for them to meet $= \frac{d}{S_A + S_B}$.

A few important points:

- Time taken by a body of length x to pass a stationary object of negligible length, such as, a signal post or a pole or a stationary man equals the time taken by the body to cover a distance equaling its own length of x.

- Time taken by a body moving at speed S of length x to pass a moving object of negligible length, such as, a man running at speed s equals $\left(\frac{x}{S - s}\right)$.
- Time taken by a body of length x to pass a stationary object of length y equals the time taken by the body to cover a distance of $(x + y)$.

Effective or Resultant Speed: Resultant speed refers to the speed of a body which moves in a moving medium.

Let the speed of a body without the effect of any medium be S and the speed of the medium itself be M.

Thus, we have:

- If the body travels against the flow of the medium (also known as 'Upstream motion' in the context of a man swimming/rowing against a flowing river):

  Resultant speed (Upstream) = $S - M = U$

- If the body travels with the flow of the medium (also known as 'Downstream motion' in the context of a man swimming/rowing along a flowing river):

  Resultant speed (Downstream) = $S + M = D$

- Speed of the body without the effect of medium (also known as speed in 'Still water' in the context of a man swimming/rowing along a flowing river):

$$S = \frac{U + D}{2}$$

- Speed of the medium (or 'River' in the context of a man swimming/rowing along a flowing river):

$$M = \frac{D - U}{2}$$

The diagram depicting the situation of a man rowing in a river is shown below:

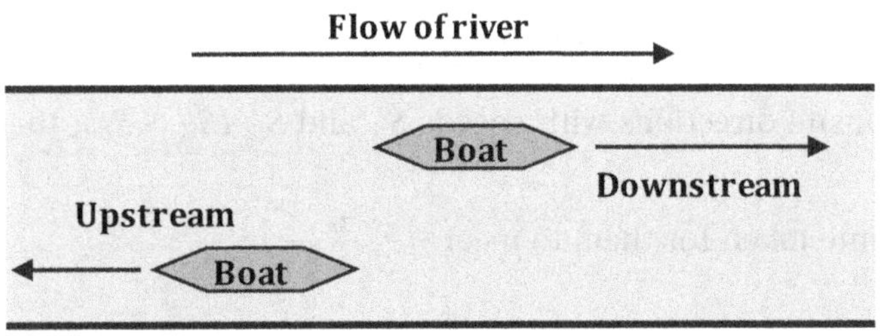

Circular track movement: If two objects P and Q move around a circular track of length L with speeds S_P and S_Q, we have the following relations:

- Time taken to meet anywhere on the track for the first time $(t) = \dfrac{L}{\text{Relative speed of P and Q}}$
- Time taken to meet at the starting point for the first time $(T) = \text{LCM}\left(\dfrac{L}{S_P}, \dfrac{L}{S_Q}\right)$

Let us take an example:

If P and Q move along a circular track of length 1,200 meters with speeds 4 meters per second and 6 meters per second, determine:

- Time taken for them to meet on the track for the first time
- Time taken for them to meet at the starting point for the first time

Explanation:

To determine the time taken for P and Q to meet on the track, we need to determine their relative speed, for which, we need to know their directions of motion.

Since their directions are not known, let us assume:

- **Same direction**: Relative speed = 6 – 4 = 2 meters per second.

  For them to meet again, Q has to cover one lap, i.e. 1,200 meters more than what P covers in the same time.

  Thus, distance = 1,200 meters => Time taken to meet $= \dfrac{1,200}{2} = 600$ seconds.

- **Opposite direction**: Relative speed = 6 + 4 = 10 meters per second.

  For them to meet again, P and Q together have to cover one lap, i.e. 1,200 meters.

  Thus, distance = 1,200 meters => Time taken to meet $= \dfrac{1,200}{10} = 120$ seconds.

To determine the time taken by P and Q to meet at the starting point for the first time, the directions of motion are not required.

Time taken by P to reach the starting point $= \dfrac{1,200}{4} = 300$ seconds.

Time taken by Q to reach the starting point $= \dfrac{1,200}{6} = 200$ seconds.

Thus, time taken by both to reach the starting point simultaneously

= LCM of 300 and 200 = 600 seconds.

Let us take a few additional examples:

(1) A man travels for 6 hours. He covers the first half of the distance at 20 miles per hour and the rest at 30 miles per hour. What is the total distance?

Explanation:

Let the distance be $2d$ miles.

Thus, total time taken = $\frac{d}{20} + \frac{d}{30} = \frac{d}{12}$ hours.

Thus, we have:

$\frac{d}{12} = 6 \Rightarrow d = 72$

Alternate approach:

Since the man covers equal distances at speeds of 20 miles per hour and 30 miles per hour, we have:

Average speed = $\frac{2 \times 20 \times 30}{20 + 30}$ = 24 miles per hour.

Total time of travel = 6 hours.

Thus, total distance covered = Average speed $\times$ Total distance = $24 \times 6 = 144$ miles.

Thus, one-way distance = $\frac{144}{2}$ = 72 miles.

(2) A person goes around the three sides of an equilateral triangle shaped field at speeds of 10, 12 and 20 miles per hour and reaches back to his starting point. What is his average speed during the journey?

Explanation:

Let us assume a suitable length of each side of the triangle so that it is divisible by 10, 12 and 20.

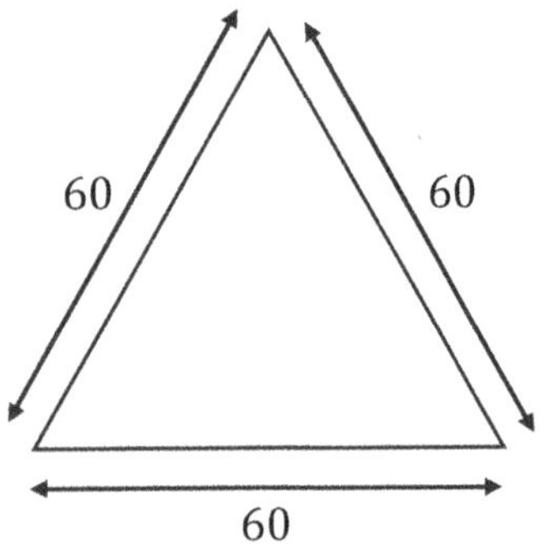

Let the distance be 60 miles.

Thus, time taken to cover the perimeter of the triangle

$= \frac{60}{10} + \frac{60}{12} + \frac{60}{20} = 14$ hours.

Total distance $= 3 \times 60 = 180$ miles.

Thus, average speed $= \frac{180}{14} = 12\frac{6}{7}$ miles per hour.

(3) A is twice as fast as B and B is thrice as fast as C. If C covers a distance in 54 minutes, how long will A take to cover the same distance?

Explanation:

We know that: B is thrice as fast as C.

Since C takes 54 minutes to cover a particular distance, time taken by B to cover the same distance

$= \frac{54}{3} = 18$ minutes

We also know that: A is twice as fast as B.

Since B takes 18 minutes to cover a particular distance, time taken by A to cover the same distance

$= \frac{18}{2} = 9$ minutes

(4) A man covers a certain distance between his house and office. If he travels at an average speed of 30 miles per hour, he is late by 10 minutes. However, if he travels at an average speed of 40 miles per hour, he reaches his office 5 minutes early. What is the distance between his house and office?

Explanation:

Let the distance be x miles.

Time taken to cover the distance at 30 miles per hour $= \frac{x}{30}$ hours.

Time taken to cover the distance at 40 miles per hour $= \frac{x}{40}$ hours.

Thus, time difference $= \frac{x}{30} - \frac{x}{40} = \frac{x}{120}$ hours.

Since the man was 10 minutes late in the first case and was 5 minutes early in the second, the time difference $= 5 - (-10) = 15$ minutes $= \frac{1}{4}$ hour.

Thus, we have:

$$\frac{x}{120} = \frac{1}{4} => x = 30$$

For the above type of problems, we may remember a simple rule:

$$\text{Distance} = \frac{\text{Product of the speeds}}{\text{Difference of the speeds}} \times \text{Difference in time}$$

Thus, applying the above rule, we have:

$$\text{Distance} = \frac{30 \times 40}{40 - 30} \times \left(\frac{5}{60} - \left(\frac{10}{60}\right)\right) = \frac{1,200}{10} \times \frac{15}{60} = 30$$

(5) A man covers a certain distance at his normal speed. Had he moved 3 miles per hour faster, he would have taken 40 minutes less to reach the destination. However, if he had moved 2 miles per hour slower, he would have taken 40 more minutes to reach the destination. What is the normal speed of the man?

Explanation:

$$\text{Distance} = \frac{\text{Product of the speeds}}{\text{Difference of the speeds}} \times \text{Difference in time}$$

Let the normal speed be S.

Case 1: New speed = $(S + 3)$.

$$\text{Distance} = \frac{S(S+3)}{(S+3) - S} \times \frac{40}{60} = \frac{2S\,(S+3)}{9} \ldots \text{(i)}$$

Case 2: New speed = $(S - 2)$.

$$\text{Distance} = \frac{S(S-2)}{S - (S-2)} \times \frac{40}{60} = \frac{S\,(S-3)}{3} \ldots \text{(ii)}$$

Thus, from (i) and (ii), we have:

$$\frac{2S\,(S+3)}{9} = \frac{S\,(S-3)}{3} => 2S + 6 = 3S - 9$$

$$=> S = 15$$

Alternate approach:

Let the distance be d miles and the normal speed be s miles per hour.

Thus, we have:

Case 1:

$$\frac{d}{s} - \frac{d}{s+3} = \frac{40}{60} => \frac{3d}{s(s+3)} = \frac{2}{3} \ldots \text{(i)}$$

Case 2:

$$\frac{d}{s-2} - \frac{d}{s} = \frac{40}{60} => \frac{2d}{s(s-3)} = \frac{2}{3} \ldots \text{(ii)}$$

From (i) and (ii), we have:

$$\frac{3d}{s(s+3)} = \frac{2d}{s(s-3)} => s = 15$$

(6) A man takes 8 hours to walk to a certain place and ride back. However, he could have taken 4 hours less if he had covered both ways by riding. How long would he have taken to walk both ways?

Explanation:

Walking time + Riding time = 8 hours

Riding time (both ways) = 8 – 4 = 4 hours.

=> Riding time (one way) = $\frac{4}{2}$ = 2 hours.

=> Walking time (one way) = 8 – 2 = 6 hours

Hence, walking time (both ways) = 6 + 6 = 12 hrs.

Alternate approach:

Two ways riding saves a time of 4 hours. It implies that one way riding takes 4 hours less than one-way walking. Thus, one-way walking takes 4 hours more than one way riding.

Thus, walking both ways will take 8 + 4 = 12 hours.

11.8 Time and Work

Let us look at the basic rules:

- If M_1 men do a work W_1 in D_1 days and M_2 equally efficient men do a work W_2 in D_2 days, we have:

  $$\frac{M_1 D_1}{W_1} = \frac{M_2 D_2}{W_2}$$

  Let us take an example:

  If 20 men can complete a work in 10 days, how long would 40 men take to complete a work twice as large?

  1 man would be able to complete the work in $20 \times 10 = 200$ days.

  Hence, 40 men would complete the work in $\frac{200}{40} = 5$ days.

  Thus, 40 men would complete double the work in $5 \times 2 = 10$ days.

  Alternately, substituting the values in the above relation:

  $$\frac{20 \times 10}{W} = \frac{40 \times D_2}{2W} => D_2 = 10$$

- If the number of hours (or days) required to complete a work is t, the fraction of the work completed in one hour (or one day) $= \frac{1}{t}$

  For example: If A can do a work in 20 days, part of the work done in 1 day $= \frac{1}{20}$

- If the fraction of work completed in one hour (or one day) is f, the time taken to complete the entire work $= \frac{1}{f}$

  For example: If A can complete $\frac{1}{4}$ of a work in 1 day, time taken by A to complete the entire work $= \frac{1}{\frac{1}{4}} = 4$ days

- If two men take x and y days to do the same work, the ratio of their efficiencies

  $$= \frac{1}{x} : \frac{1}{y} = y : x$$

Let us take a few examples:

(1) A and B can do a piece of work in 15 days and 20 days, respectively. In how many days will they finish the work, if both of them work together?

Explanation:

Fraction of work completed by A in 1 day = $\frac{1}{15}$

Fraction of work completed by B in 1 day = $\frac{1}{20}$

Thus, fraction of work completed by A and B together in 1 day = $\frac{1}{15} + \frac{1}{20} = \frac{7}{60}$

Thus, time taken by them to complete the entire work = $\frac{1}{\frac{7}{60}} = \frac{60}{7} = 8\frac{4}{7}$ days.

Alternate approach:

In such questions, it is best to assume the total work to be the LCM of the number of days each person takes to complete the work (rather than working with fractions).

Let the work = LCM of 15 & 20 = 60 units.

Work completed by A in 1 day = $\frac{60}{15}$ = 4 units.

Work completed by B in 1 day = $\frac{60}{20}$ = 3 units.

Thus, work completed by A and B together in 1 day = 4 + 3 = 7 units.

Hence, number of days that both of them working together will take = $\frac{60}{7} = 8\frac{4}{7}$ days.

(2) A can complete a work in 24 days. B is 60% more efficient than A. What is the number of days it would take A and B to complete the same piece of work working together?

Explanation:

Let B take x days to complete the work.

Ratio of efficiency of A and B = 100 : 160

Thus, ratio of time taken by A and B = $\frac{1}{100} : \frac{1}{160}$ = 160 : 100 = 8 : 5

Thus, we have:

$\frac{8}{5} = \frac{24}{x} \Rightarrow x = 15$

Let the total work = LCM of 15 and 24 = 120 units.

Thus, work done by A per day = $\frac{120}{24}$ = 5 units and work done by B per day = $\frac{120}{15}$ = 8 units.

Thus, total work done by A and B per day = 5 + 8 = 13 units.

Thus, time taken = $\frac{120}{13} = 9\frac{3}{13}$ days.

Alternate approach:

Let the required time be t days.

Ratio of efficiency of A and B = 100 : 160

=> Ratio of efficiency of A and A-B together = 100 : (100 + 160) = 100 : 260

Thus, ratio of time taken by A and A-B together = $\frac{1}{100} : \frac{1}{260}$ = 260 : 100

=> $\frac{260}{100} = \frac{24}{t}$

=> $t = 24 \times \frac{100}{260} = \frac{120}{13} = 9\frac{3}{13}$

(3) 2 men and 3 boys can do a piece of work in 10 days while 3 men and 2 boys can do the same work in 8 days. What is the ratio of efficiency of a man and a boy?

Explanation:

We know that:

$$\frac{M_1 D_1}{W_1} = \frac{M_2 D_2}{W_2}$$

The work in both cases is the same.

Also, we have:

M_1 = (2 men + 3 boys), M_2 = (3 men + 2 boys), D_1 = 10 days, D_2 = 8 days

=> (2 men + 3 boys) × 10 = (3 men + 2 boys) × 8

=> 20 men + 30 boys = 24 men + 16 boys

=> 4 men = 14 boys

=> Ratio of efficiencies of a man and a boy = $\frac{14}{4} = \frac{7}{2}$

(4) A and B finish a work in t days. A alone takes $(t + 12)$ days and B alone takes $(t + 3)$ days What is the value of t?

Explanation:

Part of work done by A in 1 day = $\left(\frac{1}{t+12}\right)$

Part of work done by B in 1 day = $\left(\frac{1}{t+3}\right)$

Part of work done by A and B together in 1 day = $\left(\frac{1}{t}\right)$

Thus, we have:

$$\frac{1}{t} = \frac{1}{t+12} + \frac{1}{t+3}$$

$$=> \frac{1}{t} = \frac{2t+15}{t^2+15t+36}$$

$$=> t^2 + 15t + 36 = 2t^2 + 15t$$

$$=> t^2 = 36$$

$$=> t = 6$$

Pipes and cisterns: A pipe connected with a tank or a cistern or a reservoir, that fills it, is known as an inlet. Similarly, a pipe connected with a tank or a cistern or a reservoir, emptying it, is known as an outlet.

Let us take a few examples:

(1) A pipe can fill a tank in 3 hours. Because of a leak in the tank, it took 4 hours to fill. How long will it take for the leak to drain out the entire tank?

Explanation:

Let the capacity of the tank = LCM of 3 & 4 = 12 liters.

Quantity filled by the filling pipe per hour = $\frac{12}{3}$ = 4 liters.

Quantity filled by both in 1 hour = $\frac{12}{4}$ = 3 liters.

Hence, the leak empties 4 – 3 = 1 liter per hour.

Thus, the time taken by the leak to empty the tank = $\frac{12}{1}$ = 12 hours.

(2) A leak can empty a completely filled cistern in 4 hours. If a pipe, which admits 12 liters of water per hour, is opened, the cistern is emptied in 6 hours. What is the capacity of the cistern?

Explanation:

Let the capacity of the cistern be x liters.

Since the leak can empty the cistern in 4 hours, the rate of emptying = $\frac{x}{4}$ liters per hour.

After the filling pipe, which fills at the rate of 12 liters per hour is opened, the effective rate of emptying = $\left(\frac{x}{4} - 12\right)$ liters per hour.

Thus, time, in hours, taken to empty the cistern

$$= \frac{x}{\frac{x}{4} - 12} = \frac{4x}{x - 48}$$

Thus, we have:

$$\frac{4x}{x - 48} = 6$$

$$=> \ 4x = 6x - 288$$

$$=> \ x = 144$$

(3) Pipes P and Q are inlet pipes that can fill a tank in 3 hours and 5 hours, respectively. Pipe R is an outlet pipe. Starting with an empty cistern, if pipes P and Q are opened for 2 hours and 3 hours, respectively, and pipe R is opened for 3 hours, the entire cistern gets filled. What is the time taken by pipe R to empty the completely filled cistern?

Explanation:

Let the capacity of the tank be the LCM of the time taken by pipes P and Q

= LCM of 3 and 5 = 15 liters.

Thus, filling rate of pipe P per hour = $\frac{15}{3}$ = 5 liters per hour

Filling rate of pipe Q per hour = $\frac{15}{5}$ = 3 liters per hour

Let the emptying rate of pipe R = x liters per hour

We know that to fill the cistern, the pipes P, Q and R are opened for 2 hours, 3 hours and 3 hours, respectively.

Thus, we have:

$5 \times 2 + 3 \times 3 - x \times 3 = 15$

$=> \ x = \frac{4}{3}$

Thus, the pipe R empties at the rate of $\frac{4}{3}$ liter per hour.

Thus, time taken by pipe R to empty the cistern = $\frac{15}{\frac{4}{3}} = \frac{45}{4} = 11\frac{1}{4}$ hours.

11.9 Simple & Compound Interest

If a person borrows some amount of money for a certain time (T), then

- Principal (P) is the amount of money borrowed
- Interest (I) is the additional money to be paid
- Amount (A) is the sum of principal and interest
- Rate (R%) is the interest that the borrower has to pay for every \$100 per year

Simple Interest: If the interest on a certain period is calculated only on the principal borrowed, it is Simple Interest.

Compound Interest: If the interest is calculated on the amount after each year (period) instead of only on the principal, it is Compound Interest.

The table below gives an example of how Simple Interest and Compound Interest operate. The principal at the beginning of 1st year is \$1,000 and the rate of interest is 10% per annum. The details for 3 years are shown:

	Simple Interest				Compound Interest			
Year	Principal	Interest for the year	Cumulative interest	Amount	Principal	Interest for the year	Cumulative interest	Amount
1	1,000	100	100	1,100	1,000	100	100	1,100
2	1,000	100	200	1,200	1,000	110	210	1,210
3	1,000	100	300	1,300	1,000	121	331	1,331
	A few important points							
	The interest for every year (not cumulative interest) is a constant				The interest for every year increases by R% of the interest in the previous year			
	The amount every year increases by a constant value				The amount every year increases by R% of the interest in the previous year			
	$I = \frac{P \times R \times T}{100}$; $A = P + I = P\left(1 + \frac{R \times T}{100}\right)$				$A = P\left(1 + \frac{R}{100}\right)^T$; $I = A - P = P\left\{\left(1 + \frac{R}{100}\right)^T - 1\right\}$			

Concept of compounding: In compound interest, the interest need not always be collected annually. Interest may be collected semi-annually, quarterly, etc. In such cases, the formula for the amount under compound interest is modified to:

$$A = P\left(1 + \frac{\frac{R}{n}}{100}\right)^{(T \times n)}$$

Here, n represents the number of compounding, for example: $n = 2$ implies compounding is done semi-annually, while $n = 4$ implies compounding is done quarterly.

Difference between Compound Interest and Simple Interest after 2 years:

Let us consider Principal P and Rate of interest $R\%$. The difference can be easily observed from the table below:

Year	Simple Interest for the year	Compound Interest for the year
1	$R\%$ of $P = x$	$R\%$ of $P = x$
2	$R\%$ of $P = x$	$(R\%$ of $P) + (R\%$ of $x) = x + R\%$ of x
Total	$\mathbf{2x}$	$\mathbf{2x + R\%}$ **of** $\mathbf{x}$

Compound Interest and Simple Interest is the same for a certain sum of money at the same rate for the first year.

Difference after 2 years with interest calculated yearly:

Compound Interest – Simple Interest

$= \text{R\% of } x = R\% \text{ of } (R\% \text{ of } P)$

$$= P\left(\frac{R}{100}\right)^2$$

Let us take a few examples:

(1) In how many years will a principal of \$400 yield an interest of \$100 given that the rate of interest and the number of years are numerically equal?

Explanation:

Let Rate of interest (%) = Time = k.

Since the principal (P) is \$400, we have:

$$\text{Simple Interest} = \frac{P \times R \times T}{100}$$

$$\Rightarrow\ 100 = \frac{400 \times k \times k}{100}$$

$$\Rightarrow\ k^2 = 25 \Rightarrow k = 5$$

Thus, the required time is 5 years.

(2) In how many years at 10% simple interest per annum will the amount become 5 times the principal?

Explanation:

Since the amount becomes 5 times the principal, the interest accumulated must be 4 times the principal.

Let the principal be P and the time taken be T.

Thus, we have:

$$\frac{P \times 10 \times T}{100} = 4P$$

=> $T = 40$

(3) A sum of money invested under simple interest triples in 6 years. In how many years would the sum of money become 27 times its value?

What would be the answer if the sum of money was invested under compound interest?

Explanation:

Let the sum of money (Principal) be P.

Thus, the amount after 6 years = $3P$.

Thus, interest accumulated in 6 years = $3P - P = 2P$

Thus, for the interest to be $2P$, time taken is 6 years.

If the sum of money has to be 27 times, i.e. the amount is $27P$, interest accumulated $= 27P - P = 26P$

Thus, for the interest to be $26P$, time taken

$$= \frac{6}{2} \times 26 = 78 \text{ years.}$$

Thus, time taken is 78 years.

Let us now see the scenario under compound interest:

We know that the value of the sum becomes 3 times in 6 years.

Thus, we have:

P becomes $3P$ in 6 years

=> $3P$ becomes $3 \times 3P = 9P$ in another 6 years, i.e. $6 + 6 = 12$ years

=> $9P$ becomes $3 \times 9P = 27P$ in another 6 years, i.e. $12 + 6 = 18$ years.

Thus, time taken is 18 years.

(4) A sum of money invested under simple interest amounts to \$1,500 after 5 years and \$1,600 after 7 years. What is the rate of interest?

Explanation:

Let the sum of money invested (Principal) be \$$P$.

Let the rate of interest be R%.

Thus, amount after 5 years:

$$P + \frac{5PR}{100} = 1,500 \ldots \text{(i)}$$

Amount after 7 years:

$$P + \frac{7PR}{100} = 1,600 \ldots \text{(ii)}$$

From (ii) - (i):

$$P\left(\frac{2R}{100}\right) = 100 => \frac{PR}{100} = 50 \ldots \text{(iii)}$$

Substituting the above in (i):

$$P + 5 \times 50 = 1,500 => P = 1,250$$

Thus, from (iii):

$$\frac{1,250 \times R}{100} = 50 => R = 4$$

Alternate approach:

The interest accumulated in $7 - 5 = 2$ years = \$1,600 − \$1,500 = \$100

Thus, interest accumulated each year = $\$\frac{100}{2} = \$50$

Thus, interest accumulated in 5 years = \$ (5 × 50) = \$250

Thus, Principal = \$1,500 − \$250 = \$1,250

Thus, interest per year on \$1,250 is \$50

$$=> \text{Rate of interest} = \frac{50}{1,250} \times 100 = 4\%$$

(5) A sum of money invested under compound interest gives \$240 interest in the second year and \$288 interest in the third year. What is the value of the sum invested?

Explanation:

We know that in compound interest, the interest for every year increases by $R\%$ of the interest in the previous year (where $R\%$ is the rate of interest).

Thus, percent increase in the interest from the second year to the third year

$$= \frac{288 - 240}{240} \times 100 = \frac{48}{240} \times 100 = 20\%$$

Thus, the percent increase in the interest from the first year to the second year is also 20%.

Thus, we have:

$(100 + 20)\,\% = 120\%$ of the Interest in the first year = Interest in the second year

=> 120% of the Interest in the first year = \$240

$$=> \text{Interest in the first year} = \$\left(\frac{240}{120} \times 100\right) = \$200$$

Since the first year interest is simply $R\%$ of the invested sum, we have:

20% of Invested sum = \$200

$$=> \text{Invested sum of money} = \$\left(\frac{200}{20} \times 100\right) = \$1{,}000$$

11.10 Sets

A Set is a collection of well-defined things. The objects of a set are called its elements or members.

Nomenclature:

- **Universal Set:** All sets are assumed to be contained in a set called the universal set.
- **Null Set:** An empty set having no elements in it is called a null set.
- **Disjoint Set:** Two sets are disjoint if they have no elements in common.
  For example: $A = \{1, 2, 3\}$ and $B = \{4, 5, 6\}$ are two disjoint sets.
- **Cardinal Number of a Set:** The Cardinal number of a finite set A is the number of elements of the set, denoted by $n(A)$. For example: For $A = \{1, 2, 3\}, n(A) = 3$

Venn Diagrams: A Venn diagram is a pictorial representation of sets represented by enclosed areas in a plane. The universal set is represented by a rectangle, and the other sets are represented by areas lying within it.

Set Operations:

- **Union:** The Union of two sets A and B, i.e. $A \cup B$, is a set that contains all the elements contained in set A or set B.

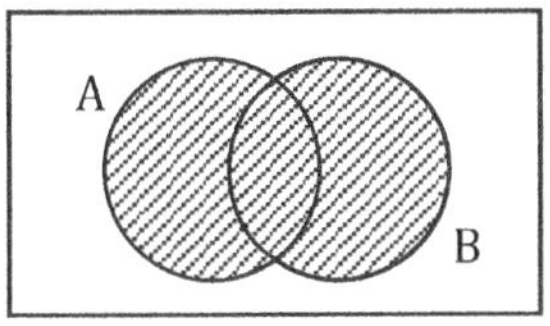

For example: If $A = \{2, 3\}$, $B = \{1, 3, 5\}$

$=> A \cup B = \{1, 2, 3, 5\}$

- **Intersection:** Intersection of two sets A and B, i.e. $A \cap B$, that contains the elements common to both sets A and B.

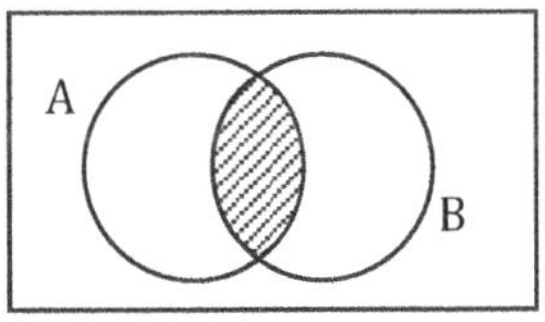

For example: For $A = \{2, 3\}$, $B = \{1, 3, 5\}, A \cap B = \{3\}$

Let us take an example:

In a survey, it was found that 65 people keep dogs as pets, 70 people keep cats, 40 people keep birds, 20 people keep both dogs and cats, 10 people keep both cats and birds, 5 people keep both birds and dogs and 3 people keep all three. How many people participated in the survey if it is known that each person has at least one pet?

Explanation:

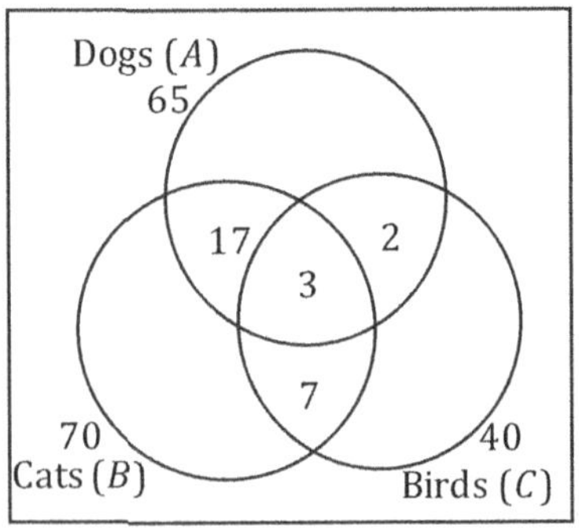

$A = 65, \ B = 40, \ C = 70$

$A \cap B = 20, \ B \cap C = 10, \ C \cap A = 5$

$A \cap B \cap C = 3$

The total number of people surveyed

$= A \cup B \cup C$

$= A + B + C - (A \cap B + B \cap C + C \cap A) + A \cap B \cap C$

$= 65 + 40 + 70 - (20 + 10 + 5) + 3 = 143$

11.11 Statistics

Statistics is the study of the collection, analysis and interpretation of data. Some of the most commonly used statistical measures are the Mean, Median, Range, and Standard Deviation.

Let us discuss them.

Arithmetic mean:

- For a set of n numbers $x_1,\ x_2,\ x_3,\ \dots\ x_n$, the arithmetic mean is calculated as:

$$\overline{x} = \frac{\text{Sum of the } n \text{ numbers}}{n} = \frac{x_1 + x_2 + x_3 + \ \cdots + x_n}{n}$$

- If $x_1,\ x_2,\ x_3,\ \dots\ x_n$ be n observations and $f_1,\ f_2,\ f_3,\ \dots\ f_n$, be their corresponding frequencies, their arithmetic mean is calculated as:

$$\overline{x} = \frac{x_1 \times f_1 + x_2 \times f_2 + \dots x_n \times f_n}{f_1 + f_2 + \cdots + f_n}$$

For ease of calculations, the mean can be calculated easily using the method of 'deviation about the mean', as shown below:

Let the assumed mean be x_a

The observations are modified by subtracting x_a from each of the values as follows:

$x_1 \equiv x_1 - x_a$

$x_2 \equiv x_2 - x_a$

...

...

$x_n \equiv x_n - x_a$

The means are now calculated as shown:

- For a set of n numbers $x_1,\ x_2,\ x_3,\ \dots\ x_n$, the arithmetic mean is calculated as:

$$\overline{x} = x_a + \frac{(x_1 - x_a) + (x_2 - x_a) + (x_3 - x_a) + \ \cdots + (x_n - x_a)}{n}$$

- If $x_1,\ x_2,\ x_3,\ \dots\ x_n$ be n observations and $f_1, f_2,\ f_3,\ \dots\ f_n$, be their corresponding frequencies, their arithmetic mean is calculated as:

$$\overline{x} = x_a + \frac{(x_1 - x_a) \times f_1 + (x_2 - x_a) \times f_2 + \dots (x_n - x_a) \times f_n}{f_1 + f_2 + \cdots + f_n}$$

Let us take a few examples:

(1) What is the mean of 123, 134, 128, 139, 141, and 126?

Explanation:

$$\text{Mean} = \frac{123 + 134 + 128 + 139 + 141 + 127}{6} = 132$$

Alternate approach:

Let the assumed mean = 134.

Thus, the deviations are:

- $123 - 134 = -11$
- $134 - 134 = 0$
- $128 - 134 = -6$
- $139 - 134 = 5$
- $141 - 134 = 7$
- $127 - 134 = -7$

$$\Rightarrow \text{Mean} = 134 + \frac{-11 + 0 - 6 + 5 + 7 - 7}{6} = 134 - \frac{12}{6} = 132$$

(2) What is the mean of the observations shown below?

Observations	**Frequencies**
11	14
7	8
15	6
13	12

Explanation:

We have:

Observations	**Frequencies**	**Observation × Frequency**
11	14	$11 \times 14 = 154$
7	8	$7 \times 8 = 56$
15	6	$15 \times 6 = 90$
13	12	$13 \times 12 = 156$
	Total = 40	**Total = 154 + 56 + 90 + 156 = 456**

$$\Rightarrow \text{Mean} = \frac{456}{40} = 11.4$$

Alternate approach:

Let the assumed mean be 11.

Thus, we have:

Observations	Modified Observation = Observation − Assumed mean	Freq.	Mod. Obs. × Freq.
11	11 − 11 = 0	14	0 × 14 = 0
7	7 − 11 = − 4	8	(− 4) × 8 = − 32
15	15 − 11 = 4	6	4 × 6 = 24
13	13 − 11 = 2	12	2 × 12 = 24
		Total = 40	**Total = 16**

=> Mean = $11 + \frac{16}{40} = 11.4$

Median:

Median refers to the middle value of all observations arranged in ascending or descending order.

Thus, we have:

- **If n is odd:** Median = Value of the $\left(\frac{n+1}{2}\right)^{th}$ observation.
- **If n is even:** Median = Average of $\left(\frac{n}{2}\right)^{th}$ and $\left(\frac{n}{2}+1\right)^{th}$ observations.

Let us take a few examples:

(1) What is the median of 123, 134, 128, 139, 141 and 126 ?

Explanation:

After arranging data in ascending order, we have:

123, 126, 128, 134, 139, 141

Since there are 6 (even) terms, we have:

$$\text{Median} = \frac{\left(\frac{6}{2}\right)^{th} \text{ term} + \left(\frac{6}{2}+1\right)^{th} \text{ term}}{2} = \frac{3^{rd} \text{ term} + 4^{th} \text{ term}}{2}$$

$$123, 126, \underset{3^{rd}}{\underline{128}}, \underset{4^{th}}{\underline{134}}, 139, 141$$

$$= \frac{128 + 134}{2} = 131$$

(2) What is the median of the observations shown below?

Observations	Frequencies
11	12
7	8
15	6
13	14

Explanation:

After arranging the observations in ascending order, we have:

Observations	Frequencies	Cumulative Frequency
7	8	8
11	12	12 + 8 = 20
13	14	14 + 20 = 34
15	6	6 + 34 = 40
	Total = 40	

Since there are 40 observations, we have:

$$\text{Median} = \frac{\left(\frac{40}{2}\right)^{\text{th}} \text{term} + \left(\frac{40}{2} + 1\right)^{\text{th}} \text{term}}{2} = \frac{20^{\text{th}} \text{ term} + 21^{\text{st}} \text{ term}}{2}$$

Observations	Frequencies	Cumulative Frequency	
7	8	8	
11 <—	12	12 + 8 = 20 <—	**20th term**
13 <—	14	14 + 20 = 34 <—	**21st term**
15	6	6 + 34 = 40	
	Total = 40		

$$\text{Median} = \frac{11 + 13}{2} = 12$$

Range:

Range refers to the difference between the values of maximum and minimum observations.

For example, the range of the set of observations 11, 13, 23, 21, 13, 17, 19 is 23 − 11 = 12

Standard Deviation:

Standard deviation shows the spread of the observations. If the observations are far away from the mean, the standard deviation would be high; similarly, if the observations are close to the mean, the standard deviation would be low.

For a set of n numbers $x_1, x_2, x_3, \ldots x_n$, the Standard Deviation (s) is calculated as:

If $\overline{x}$ is the mean of the observations:

$$s = \sqrt{\frac{(x_1 - \overline{x})^2 + (x_2 - \overline{x})^2 + \cdots + (x_n - \overline{x})^2}{n}}$$

Let us take an example:

(1) What is the standard deviation of 12, 15, 16 and 17 ?

$$\text{Mean} = \overline{x} = \frac{12 + 15 + 16 + 17}{4} = 15$$

$$=> s = \sqrt{\frac{(12 - 15)^2 + (15 - 15)^2 + (16 - 15)^2 + (17 - 15)^2}{4}}$$

$$= \sqrt{\frac{9 + 0 + 1 + 4}{4}} = \sqrt{3.5} = 1.87$$

A few important rules:

For a set of n numbers: $x_1, x_2, x_3, \ldots x_n$

- If a constant number k is added to (or subtracted from) each term of the above set of observations, we have:
    - The Mean increases (or decreases) by k
    - The Median increases (or decreases) by k
    - The Range remains unchanged
    - The Standard Deviation remains unchanged

- If a constant number k is multiplied (divided) to each term of the above set of observations, we have:
    - The Mean is multiplied (divided) by k
    - The Median is multiplied (divided) by k
    - The Range is multiplied (divided) by k
    - The Standard Deviation is multiplied (divided) by k

- If a new term, whose value is equal to the mean, is introduced in the above set of observations, we have:
    - The Mean remains unchanged
    - Nothing can be concluded about Median; it depends on the distribution
    - The Range remains unchanged
    - The Standard Deviation decreases

- If the terms of a set of observations arranged in ascending or descending order have a constant difference between consecutive terms, i.e., they form an Arithmetic Sequence, we have: Mean = Median

Different ways of representing data:

Let us discuss the data given in a tabular form:

The following table gives the production of different electronic goods (in thousands) by a company 'X' in the years 2014 and 2015:

A table:

Item	2013	2014
TVs	120	180
Laptops	240	270
Hard Disks	100	130
DVDs	140	120
Mobiles	200	300

The same data can be represented in different forms as shown below:

A pie chart:

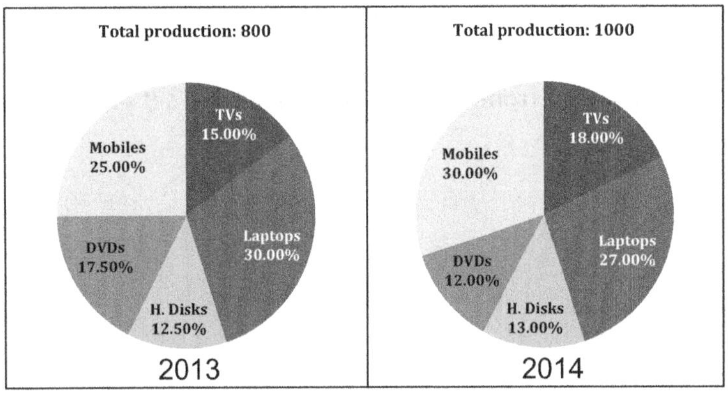

The same data can be represented in the form of a bar graph as shown below:

A bar graph:

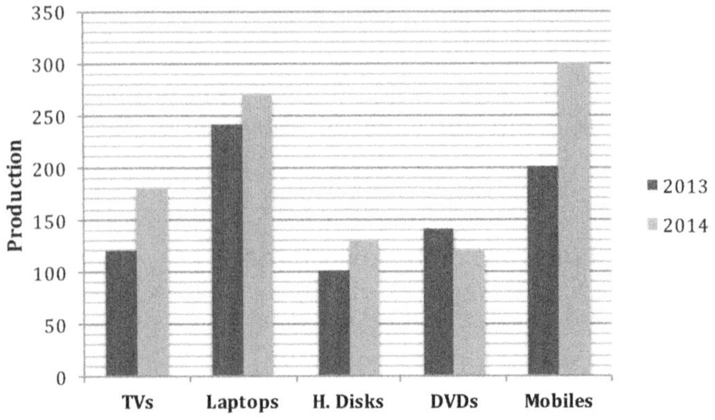

A stacked bar graph:

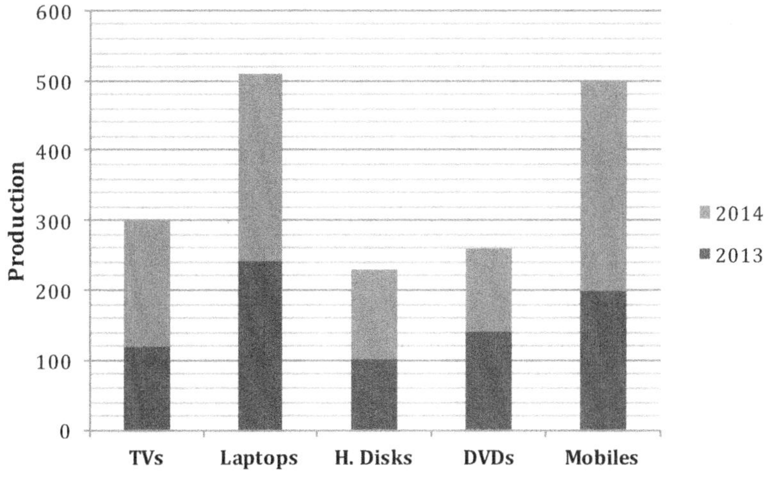

A horizontal bar graph:

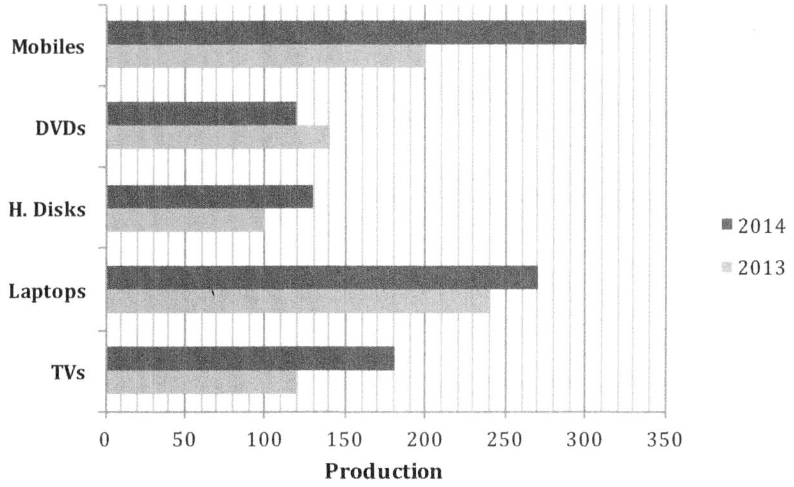

A line graph:

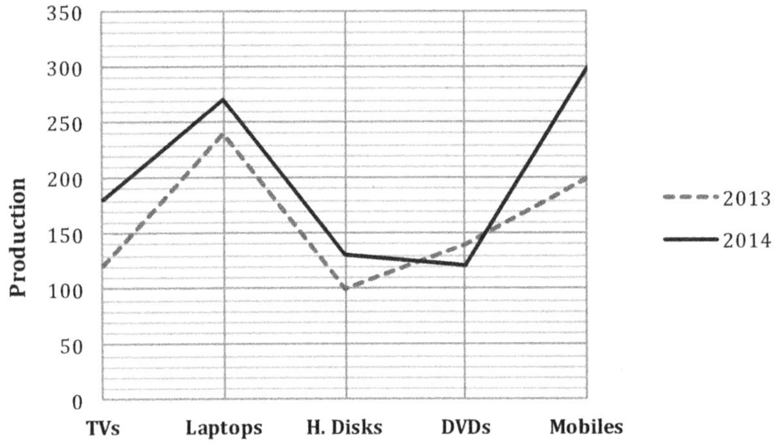

A scatter plot:

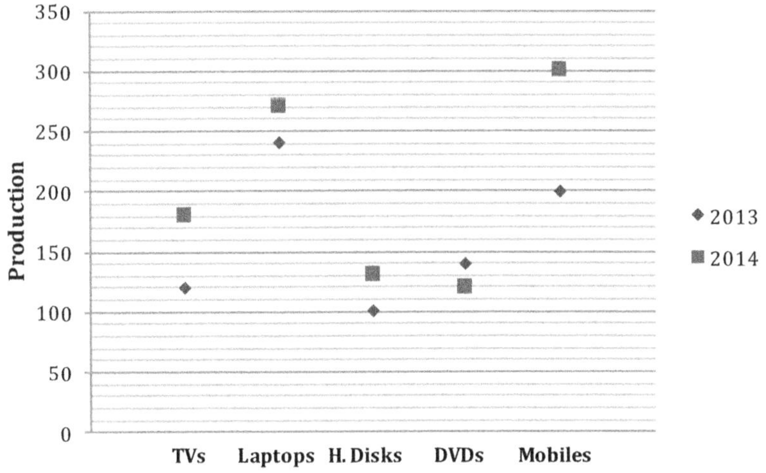

Few questions based on the above graphs:

- **Table:**

What is the percentage change in the production of TVs from 2013 to 2014?

Explanation:

Percentage change

$$= \frac{\text{Final value} - \text{Initial value}}{\text{Initial value}} \times 100 = \frac{180 - 120}{120} \times 100$$

$$= \frac{60}{120} \times 100 = 50\%$$

- **Table:**

What is the average production value of all electronic items (in thousands) in 2013?

Explanation:

Here, we simply need to take the average of all the production values in the year 2013.

$$\text{Average} = \frac{120 + 240 + 100 + 140 + 200}{5} = \frac{800}{5} = 160.$$

- **Bar graph:**

What is the net difference in total production (in thousands) between 2013 and 2014?

Explanation:

We can simply add up the values of 2013 and 2014 and subtract them.

Alternately, we can find the difference between each food item in each year and add up the differences (with sign):

Item	**2013**	**2014**	**Difference**
TVs	120	180	+60
Laptops	240	270	+30
Hard Disks	100	130	+30
DVDs	140	120	− 20
Mobiles	200	300	+100
Net difference			**200**

- **Line graph:**

What is the percentage change in production of mobiles from 2013 to 2014 ?

Explanation:

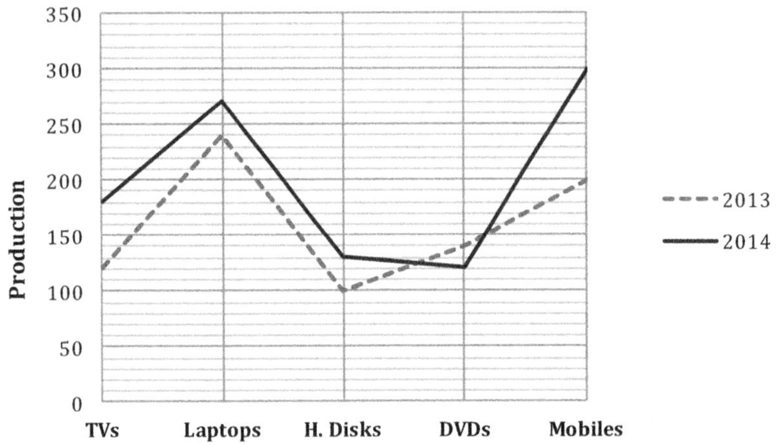

Percent increase in Mobiles

$$= \frac{\text{Production in 2014} - \text{Production in 2013}}{\text{Production in 2013}} \times 100\% = \frac{300 - 200}{200} \times 100\% = 50\%$$

Thus, the answer is 50%.

- **Pie chart:**

What is the angle subtended at the center by Laptops in 2013?

Explanation:

We need to convert the % values given in the pie chart to degrees.

We know that 100% corresponds to 360°.

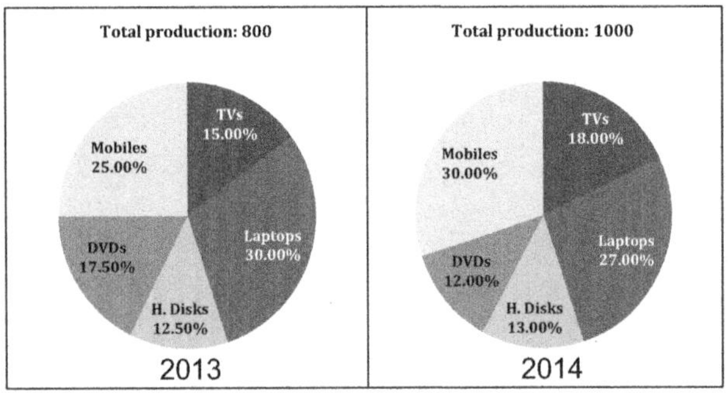

In 2013, Laptops are represented by 30% in the pie chart.

Thus, the corresponding degree measure

$\frac{30}{100} \times 360^\circ = 108^\circ$

- **Pie chart:**

What is the percentage change in production of TVs from 2013 to 2014?

Explanation:

We need to understand that in the first pie chart, share of TVs is 15% of 800, while in the second pie chart, it is 18% of 1,000.

Thus, number of TVs in 2013 = $\frac{15}{100} \times 800 = 120$

The number of TVs in 2014 = $\frac{18}{100} \times 1,000 = 180$

Thus, required percent change

$= \frac{180 - 120}{120} \times 100 = 50\%$

- **Scatter plot:**

If, in 2014, the ratio of the selling price of a Mobile to the selling price of a Laptop is 1 : 2, what is the ratio of the sales proceeds of Mobiles to the sales proceeds of Laptops in 2014?

Explanation:

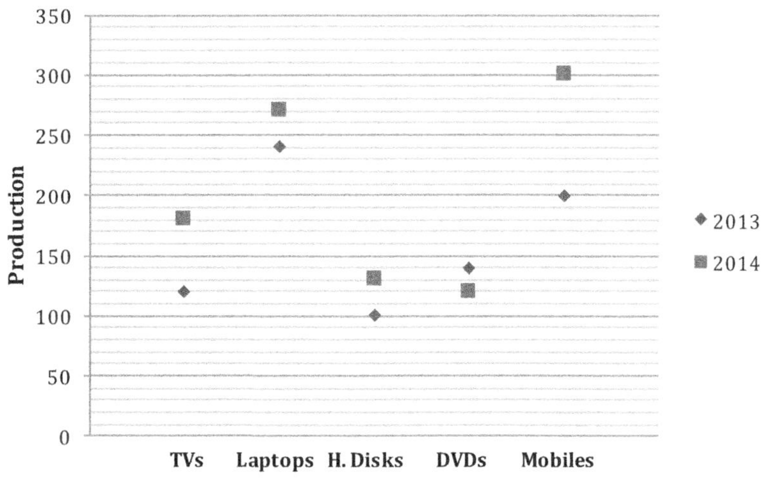

Production of Mobiles in 2014 = 300

Production of Laptops in 2014 = 270

Thus, ratio of production of Mobiles to that of Laptops

$$= \frac{300}{270} = \frac{10}{9}$$

Ratio of the selling price of a Mobile to the selling price of a Laptop $\approx \frac{1}{2}$

Hence, ratio of the sales proceeds of Mobiles to the sales proceeds of Laptops

$$= \frac{10}{9} \times \frac{1}{2} = \frac{5}{9}$$

In the next chapter, you will find 100 exam-like questions. Best of luck!

Chapter 12

Problem Solving & Data Analysis: Practice Questions

12.1 Computation

1. A solution leaks out of a jar at the rate of p liters for every q hours. If the solution costs 10 dollars per liter, what is the cost, in dollars, of the amount of the solution that will leak out in r hours?

   (A) $\frac{10q}{pr}$

   (B) $\frac{10p}{qr}$

   (C) $\frac{10pr}{q}$

   (D) $\frac{10qr}{p}$

2. **SPR:**

   A man refueled his car one day. However, the very next day, the price of fuel fell by \$2.5 per gallon. The man calculated that he could have saved \$75 had he fueled one day later. What quantity (in gallons) of fuel had the man purchased?

3. Each new instance of good behavior in a school provides increasing award points to students on his or her 'behavior reflection sheet.' A student gets one point for the first such instance, and the incremental value of award points for each new instance is 3 points. In which numbered instance will a student achieve 16 points, a qualification to get an outstanding behavior reward?

   (A) 4

   (B) 5

   (C) 6

   (D) 7

4. A man distributed his entire wealth between his two sons, John and Jack, such that John got three times as much wealth as Jack. If John got \$5,000 more than Jack, how much wealth did the father have?

   (A) \$2,500

   (B) \$7,500

   (C) \$10,000

   (D) \$12,500

5. A boy took an examination that had +3 marks for a correct response and −1 for a wrong one. If there were a total of 60 questions and he attempted all of them to get a score of 100, how many questions did he get wrong?

   (A) 5

   (B) 20

   (C) 40

   (D) 45

6. In an award ceremony, books were distributed to a school. Two-thirds of the books were distributed to students from lower-middle standard. Of the remaining, one-third were distributed to students from high school. The remaining books were kept in the school library. What percentage of the total books were stocked in the school library?

(A) 6.67%

(B) 20%

(C) 26.7%

(D) 40%

7. A glacier surged at the rate of $\frac{11}{5}$ feet per hour. How many days does it take the glacier to surge $\frac{1}{2}$ mile? (1 mile = 5,280 feet)

(A) 8

(B) 24

(C) 36

(D) 50

8. **SPR:**

The operation '#' represents one of addition, subtraction, or multiplication of integers. If 2#0 = 2, what is the value of 1#0?

9. The population of a country increases at the rate of 30,000 people every month. The population in 2012 was 360 million. In which year would the population of the country be 378 million?

(A) 2060

(B) 2061

(C) 2062

(D) 2063

12.2 Ratio & Proportion

10. John and Mary have few marbles. The number of marbles with them is 75 times the ratio of the number of marbles with John to that with Mary. If Mary has 30 marbles, how many marbles does John have?

(A) 20

(B) 30

(C) 45

(D) 50

11. John has twice as much money as Peter. If John gives $50 to Peter, the ratio of the amount of money between them is reversed. How much money is with them?

(A) 50

(B) 100

(C) 150

(D) 200

12. **SPR:**

Some TVs at a showroom are LCD, and the rest are LED. If the ratio of the number of LED TVs to the number of LCD TVs at the showroom is 7 : 6, and there are a total of 260 TVs at the showroom, how many of the TVs are LED?

13. **SPR:**

In University P, there are 105 more male teachers than female teachers. If the university were to hire 14 more female teachers, the ratio of the number of male teachers to the number of female teachers would then be 16 : 9. How many female teachers are at the university?

14. A shopkeeper could sell only $\left(\frac{4}{5}\right)^{\text{th}}$ of the stock at the rate of \$3 per item. If 100 items remained unsold, what amount did he receive from the sale?

    (A) \$240

    (B) \$1,200

    (C) \$1,250

    (D) \$1,300

15. At a garment shop, the ratio of the number of shirts to the number of trousers is 4 to 5, and the ratio of the number of jackets to the number of shirts is 3 to 8. If the ratio of the number of sweaters to the number of trousers is 6 to 5, what is the ratio of the number of jackets to the number of sweaters?

    (A) 9 to 25

    (B) 1 to 3

    (C) 1 to 4

    (D) 3 to 5

12.3 Average

16. **SPR:**

An auditorium has 35 rows with 30 seats in each row. During a program, there was an average (arithmetic mean) of 13 unoccupied seats per row for the front 12 rows and an average of 10 occupied seats per row for the back 23 rows. How many of the seats were occupied during the program?

17. **SPR:**

If the average (arithmetic mean) of w, x, y and z is n, what is the value of $(n - w) + (n - x) + (n - y) + (n - z)$?

12.4 Sets & Venn Diagrams

18. In an election between two candidates, Joseph and Judith, 60% of the total number of voters did not vote for Joseph, and 50% of the total number of voters did not vote for Judith. If it is known that each voter can vote for only one candidate, what percentage of voters did not cast their votes?

(A) 5

(B) 10

(C) 20

(D) 30

19. In an entrance examination, three-fifths of all candidates were male. One-third of all candidates failed to clear the entrance examination. One-fourth of all male candidates failed to clear the examination. Of the candidates who cleared the examination, what fraction is female?

(A) 32.5%

(B) 54.2%

(C) 67.5%

(D) 75.0%

20. **SPR:**

Of the school students who participated in a survey, 11 had taken a course in math, 23 had taken a course in science, and 12 had taken neither of the two courses. If 6 had taken both courses, how many students participated in the survey?

21. **SPR:**

Of the people who attended a training program, 70 percent were males. Some of the attendees were freelancers. If 300 people attended the program and 90 of the male attendees were not freelancers, what percent of the attendees were male freelancers?

22.

	Number of students interested in arts	Number of students not interested in arts	Total
Number of boys			36
Number of girls			
Total	52		

The table above shows the number of boys and girls interested in arts and not interested in arts in a particular class of a school. If the number of girls in the class is twice the number of students who are not interested in the arts, what is the total number of students in the class?

12.5 Profit & Loss

23. **SPR:**

A certain dealer replenishes his stock whenever it drops to 15 items. Every time he raises his stock to 40 items. He needs to replenish his stock once every two weeks. If the cost of each item is \$5, what is the dealer's total expense (in dollars) on replenishing his stock in a month?

24. **SPR:**

A man purchased bulbs at 4 for every \$6. However, he observed that 2 out of every 5 bulbs were defective. What was the price (in dollars) per usable bulb that the man paid?

25. **SPR:**

The selling price of an article is equal to the cost of the article plus the markup. If the markup on the television set is 25 percent of the cost, the markup is what percent of the selling price?

26. **SPR:**

A merchant sold an appliance for \$50 more than its cost. If the merchant's gross profit on the appliance was 20% of the selling price, what was the selling price of the appliance?

27. **SPR:**

A merchant sells shirts under an offer: "Buy 3, Get 5." What discount, in percent, does he offer?

28. A furniture shop sells a table for \$150 per piece and a chair for \$85 per piece. Last fortnight it sold five more tables than chairs. If the total revenue from the sale of only tables and chairs last fortnight was \$1,925, what was the total number of tables and chairs that the shop sold?

(A) 12
(B) 14
(C) 15
(D) 16

12.6 Percents

29. **SPR:**

What is the value of x, if 10% of 20% of $x = 4$.

30. **SPR:**

Joe has a bag containing two varieties of beans: a cheaper variety, costing \$6 each, and a costlier variety. The price of each costlier variety of beans is three times that of the cheaper variety of beans. If the bag has a total of one-third cheaper variety of beans, the price of the costlier beans forms what percent of the price of the bag?

31. A machine can be repaired for \$1,200 and will last for one year, while a new machine will cost \$2,800 and will last for two years. The average cost per year of the new machine is what percent greater than the cost of repairing the current machine?

(A) 7%

(B) 10%

(C) 16.67%

(D) 18.83%

32. A particular sales tax rate is \$0.82 per \$50. What is the rate, as a percent, which is three times as much as the rate mentioned?

(A) 492%

(B) 49.2%

(C) 4.92%

(D) 1.23%

33. Telescope X increases the visual range of a particular location from 90 kilometers to 270 kilometers, while Telescope Y increases the visual range of another location from 45 kilometers to 180 kilometers. By what percent is the percent increase in visual range using Telescope Y greater than that obtained using Telescope X?

(A) 33

(B) 50

(C) 75

(D) 100

12.7 Interest & Compounding

34. A sum of money invested under simple interest amounts to \$1,200 in 3 years and \$1,500 in 5 years. What is the rate at which the sum of money was invested?

(A) 10%

(B) 15%

(C) 20%

(D) 25%

35. In the beginning of the year, 35 percent of Company X's 120 customers were retailers, and after the 24-month period, 25 percent of its 240 customers were retailers. What was the simple annual percent growth rate in the number of retailers?

(A) 14.28%

(B) 21.43%

(C) 24.0%

(D) 30.0%

12.8 Speed/Time/Distance

36. A man has to drive 120 miles. He estimates that his average speed for the entire journey would be between 24 and 36 miles per hour. If he needs to reach his destination by 5 p.m. on the same day, which of the following gives the correct time within which he needs to start off on his journey?

(A) 11:30 am to 2 pm

(B) 12 noon to 1:30 pm

(C) 12 noon to 1:40 pm

(D) 12:20 pm to 2:30 pm

37. A car traveled 900 miles. It would have taken 2 hours less if its average speed had been greater by 10 miles per hour. What was the average speed, in miles per hour, of the car?

(A) 35

(B) 40

(C) 45

(D) 50

38. A car can run 440 miles per full tank of diesel on the highway and 360 miles per full tank in the city. If the car would run 5 fewer miles per gallon in the city than on the highway, how many miles per gallon would the car run in the city?

(A) 20.50

(B) 22.50

(C) 25.50

(D) 27.50

39. A bike running at a particular speed takes 5 seconds longer to travel one mile than it takes to run 50 miles per hour. At what speed, in miles per hour, is the bike traveling?

(A) 45

(B) 46.75

(C) 47.5

(D) 48

12.9 Probability

40. A box contains exactly 22 balls, of which 11 are white and the rest are black. If one ball is picked from this box at random, what is the probability that the ball will be black?

(A) $\frac{1}{3}$

(B) $\frac{1}{2}$

(C) $\frac{2}{3}$

(D) $\frac{3}{4}$

41. **SPR:**

A pair of die is rolled. What is the probability of getting the sum of numbers greater than 9 on their upper faces?

42. **SPR:**

A coin is tossed 3 times. What is the probability of getting 'Tails' on the first two tosses and a 'Head' on the third toss?

43. **SPR:**

A biased coin is tossed five times. It lands only on Head or Tail. The probability that it lands on Head = $\frac{2}{3}$. What is the probability of getting 'Head/Tails/Head/Tails/Head' in the order in five tosses?

44. **SPR:**

What is the probability that a team of 5 students chosen from a class comprising 10 boys and 12 girls would have 2 boys and 3 girls?

45. **SPR:**

The above figure shows a square dartboard. The outer-most square is of side 5 centimeters, the middle square is of side 3 centimeters, and the inner-most square is of side 2 centimeters. If a dart randomly hits the dartboard, what is the probability that it hits any of the black parts of the dartboard?

12.10 Time & Work

46. Working alone at their constant rates, each of the three machines, X, Y, and Z, produces 1,000 idential machine parts. X takes 3 hours, Y takes 2 hours, and Z takes 6 hours. How many hours will it take them, working together at their respective rates, to produce 3,000 machine parts?

(A) 1.75
(B) 2.50
(C) 2.75
(D) 3.00

47. Four adults consume food costing \$60 on a 3-day fishing trip. If a child consumes half the amount of food consumed by an adult, what would be the cost of food consumed by 6 adults and 3 children during a 4-day fishing trip?

(A) \$180
(B) \$150
(C) \$125
(D) \$90

48. An assembly machine produces 2,000 units of product per hour. Working 15 hours each day, another machine, thrice as efficient, will produce how many units of the same product in 10 days?

(A) 100,000
(B) 200,000
(C) 600,000
(D) 900,000

49. A water pump began filling an empty swimming pool with water and ran at a constant rate until the pool was full. At some point, the pool was half full, and $2\frac{1}{3}$ hours later, it was $\frac{5}{6}$ full. How many hours would it take the pump to fill the empty pool completely?

(A) 4
(B) $5\frac{1}{3}$
(C) 7
(D) $7\frac{1}{5}$

50. Working alone, machines A, B, and C can complete the work in 6, 9, and 12 hours, respectively. If each of them works alone for an hour, what part of the work will remain unfinished?

(A) $\frac{1}{3}$
(B) $\frac{13}{36}$
(C) $\frac{23}{36}$
(D) $\frac{2}{3}$

51. Working individually, machines A, B, and C can complete the work in 6, 9, and 12 hours, respectively. If each of them works individually for 1 hour in order A, B, and C and repeats the same till the work is finished, in how many hours will the work be finished?

(A) 3
(B) 6
(C) 8
(D) 9

12.11 Arithmetic Mean

52. The average (arithmetic mean) of 11 numbers in a list is 50. If two numbers are added to the list: one of them is greater than the mean by 6 and the other is less than the mean by 6, what is the value of the average (arithmetic mean) of the 13 numbers?

(A) 48

(B) 49

(C) 50

(D) 52

53. **SPR:**

The average (arithmetic mean) of 11 numbers in a list is 50. If one of the numbers is removed and the average (arithmetic mean) of the 10 numbers in the list remains the same, what is the value of the removed number?

54. A set consists of 11 distinct numbers. If x is a number in the list and is three times the average (arithmetic mean) of the 10 numbers other than x, then x is what fraction of the sum of the 11 numbers in the list?

55. If the average (arithmetic mean) of p, q, and 15 is equal to the average of p, q, 15, and 35, what is the average of p and q?

(A) 40

(B) 45

(C) 50

(D) 65

12.12 Data Analysis

Table

56.

	Finances of Company X		**Finances of Company Y**	
Attribute	**2010**	**2011**	**2010**	**2011**
Total Income	250	450	400	650
Net Profit	125	150	170	250
Cash Profit	80	100	128	215

The above table shows the finances of Company X and Y (all values are in million dollars) for the years 2010 and 2011. For either company, the percent increase in Net Profits from 2010 to 2011 was the same as that from 2011 to 2012, while the Total Income of each company doubled from 2011 to 2012.

Among the following options, for which of the following companies and in which year, the Cash Profit to Total Income ratio is the lowest?

(A) Company X in 2010

(B) Company X in 2011

(C) Company Y in 2010

(D) Company Y in 2011

57.

	Finances of Company X		Finances of Company Y	
Attribute	**2010**	**2011**	**2010**	**2011**
Total Income	250	450	400	650
Net Profit	125	150	170	250
Cash Profit	80	100	128	215

The above table shows the finances of Company X and Y (all values are in million dollars) for the years 2010 and 2011. For either company, the percent increase in Net Profits from 2010 to 2011 was the same as that from 2011 to 2012, while the Total Income of each company doubled from 2011 to 2012.

Which of the following differences is greater than $750 million?

(A) Difference between Total Income and Net Profits in 2011 for Company X

(B) Difference between Total Income and Net Profits in 2012 for Company X

(C) Difference between Total Income and Net Profits in 2011 for Company Y

(D) Difference between Total Income and Net Profits in 2012 for Company Y

58.

Sector	Net Income in the second quarter, 2016 (billion $)	Percent change from first quarter, 2016
Telecommunications	4.80	−20%
Power	7.20	+20%
Agriculture	5.00	−10%
Services	9.00	+200%
Information Technologies	2.50	−50%

The table above represents the combined net income of all German companies in five sectors for the second quarter of 2016 and the percent change from the first quarter.

Which sector had the lowest net income during the first quarter of 2016?

(A) Power

(B) Agriculture

(C) Services

(D) Information Technologies

59.

Sector	Net Income in the second quarter, 2016 (billion $)	Percent change from first quarter, 2016
Telecommunications	4.80	−20%
Power	7.20	+20%
Agriculture	5.00	−10%
Services	9.00	+200%
Information Technologies	2.50	−50%

The table above represents the combined net income of all German companies in five sectors for the second quarter of 2016 and the percent change from the first quarter.

What is the average (arithmetic mean) of net income (billion ($)) of all the five sectors in the first quarter in 2016?

(A) $3.912
(B) $4.567
(C) $5.112
(D) $6.504

Bar Diagram

60.

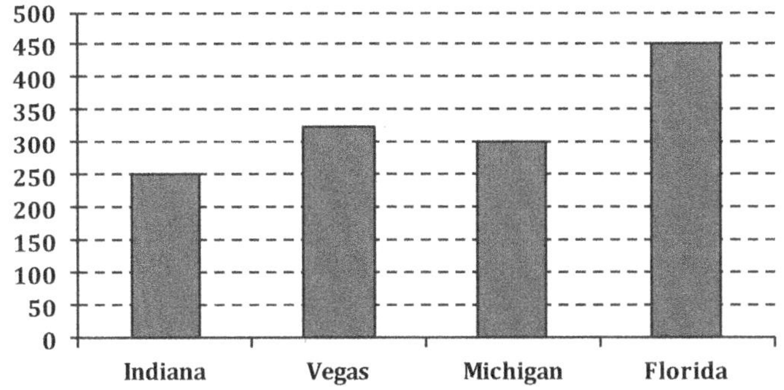

The barchart above shows the number of high-speed internet customers (all values are in thousands) for a particular company in four states in 2011. If the number of customers in Indiana and Michigan together was 20 percent of the total number of customers, the number of customers in Vegas represents what percent of the total number of customers?

(A) 16.4%
(B) 15.3%
(C) 12.4%
(D) 11.8%

Line Graph

61.

Year-wise ratios of imports to exports of a particular company X are shown in the graph above, with ratios presented on the X-axis and years presented on the Y-axis. If the exports were \$60 million in 2001 and the exports increased by \$2 million every year, what was the percent increase in imports in 2005 over 2001?

(A) 60.0%

(B) 75.0%

(C) 83.3%

(D) 126.7%

62. The graph below shows the distribution of 59 people on the basis of the scheduled time before and after they reached a meeting. The black curved line is the best estimation curve.

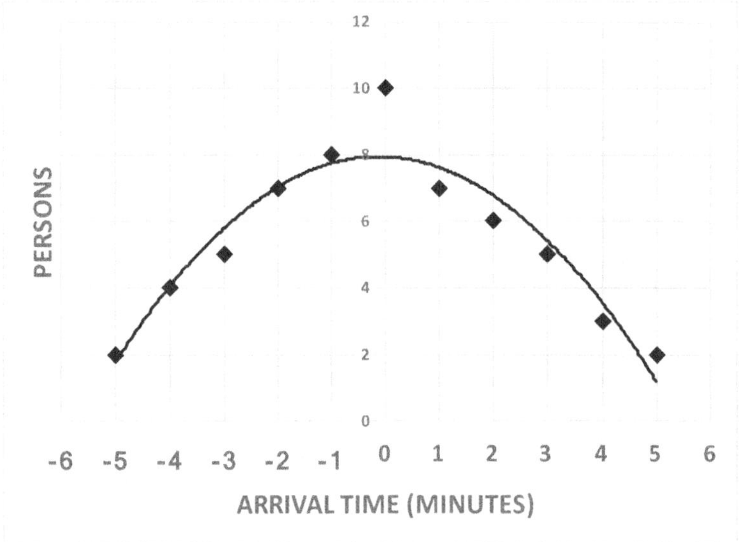

What was John's efficiency (output per hour) for the full 8-hour shift?

(A) 3.00

(B) 3.50

(C) 4.00

(D) 4.50

63. The distance from Allen's home to his workplace can be broken down into four distinct segments, as mentioned below:

- Segment 1: A 3-mile drive with a maximum permissible speed of 12 miles per hour
- Segment 2: A 12-mile drive with a maximum permissible speed of 20 miles per hour
- Segment 3: A 2-mile stretch with a maximum permissible speed of 6 miles per hour
- Segment 4: A 4-mile stretch with a maximum permissible speed of 8 miles per hour

Which of the following expressions, when evaluated, gives the average speed, m, in miles per hour, of Allen if he drove at the maximum permissible speeds for each segment of his journey?

(A) $m \div \frac{60}{101} = \frac{1}{21}$

(B) $m \times 101 = 21$

(C) $m \times \frac{101}{60} = 21$

(D) $m \times 21 = 101$

Chapter 13

Problem Solving & Data Analysis Answer-key

(1) D

(2) 30

(3) C

(4) C

(5) B

(6) D

(7) D

(8) 1

(9) C

(10) A

(11) C

(12) 140

(13) 103

(14) B

(15) C

(16) 434

(17) 0

(18) B

(19) A

(20) 40

(21) 40

(22) 68

(23) 250

(24) \$2.5 or 2.50

(25) 20

(26) 250

(27) 40

(28) C

(29) 200

(30) 85.71

(31) C

(32) C

(33) B

(34) C

(35) B

(36) C

(37) C

(38) B

(39) B

(40) B

(41) $\frac{1}{6}$

(42) 1/8

(43) 0.033

(44) 0.38

(45) $\frac{4}{5}$

(46) D

(47) B

(48) D

(49) C

(50) B

(51) C

(52) C

(53) 50

(54) B

(55) B

(56) B

(57) D

(58) C

(59) C

(60) D

(61) D

(62) C

(63) C

Chapter 14

Problem Solving & Data Analysis: Solutions to Practice Questions

14.1 Computation

1. Amount of solution leaked out in q hours = p liters.

   Thus, the amount of solution leaked out in 1 hour = $\frac{p}{q}$ liters.

   Thus, the amount of solution leaked out in r hours = $\frac{pr}{q}$ liters.

   Cost of 1 liter of the solution = \$10.

   Thus, cost of the solution leaked out in r hours = $\$\left(10 \times \frac{pr}{q}\right) = \$\left(\frac{10pr}{q}\right)$.

   The correct answer is option D.

2. The man's savings of \$75 corresponds to the price reduction on the quantity of fuel he purchased.

   Let he purchased x gallons.

   Thus: $2.5 \times x = 75 \Rightarrow x = \frac{75}{2.5} = 30$ gallons.

   Thus, the man had purchased 30 gallons of fuel.

   The correct answer is 30.

3. A student gets 1 point for the first instance of good behavior.

   After this, a student gets $1 + 3 = 4$ points for the second instance, $4 + 3 = 7$ points for the third instance and so on. The values have been tabulated below:

Instances	Points awarded for the instance
1	1
2	4
3	7
4	10
5	13
6	16

   Hence, on the 6th instance, a student would get 16 points to get outstanding behavior reward.

   The correct answer is option C.

4. Let us assume that Jack received $\$x$.

Thus, John received $\$3x$.

Thus, we have: $3x = 5,000 + x \Rightarrow x = 2,500$.

Thus, total wealth of the father $= x + 3x = 4x = \$(4 \times 2,500) = \$10,000$.

The correct answer is option C.

5. Let us assume that the boy got x responses wrong.

Thus, he must have got $(60 - x)$ responses correct (since there are in all 60 questions and he attempted all of them).

Thus, his marks from correct responses $= 3 \times (60 - x) = 180 - 3x$.

His marks from the wrong responses $= (-1) \times x = -x$.

Thus, his total marks $= (180 - 3x) + (-x) = 180 - 4x$.

Hence, we have: $180 - 4x = 100 \Rightarrow 4x = 80 \Rightarrow x = 20$.

Thus, he had got 20 responses wrong.

The correct answer is option B.

6. Let the number of books present initially be B.

Number of books distributed among students of lower-middle school $= \frac{2B}{5}$.

Number of books left $= B - \frac{2B}{5} = \frac{3B}{5}$.

Number of books distributed among students of high school $= \frac{3B}{5} \times \frac{1}{3} = \frac{B}{5}$.

Number of books left $= \frac{3B}{5} - \frac{B}{5} = \frac{2B}{5}$.

These books were stacked in the school library.

Thus, percentage of books stocked in the school library $= \frac{\frac{2B}{5}}{B} \times 100 = \frac{2B}{5B} \times 100 = 40\%$.

The correct answer is option D.

Alternate approach:

We can see that the fractions used in the problem are $\frac{2}{5}$ & $\frac{1}{3}$.

So, let us take the number of books as the LCM of 5, & 3 i.e. 15.

Thus, the number of books distributed among students of lower-middle school = $\frac{2}{5} \times 15 = 6$.

Number of books left = $15 - 6 = 9$.

Number of books distributed among students of high school = $9 \times \frac{1}{3} = 3$.

Number of books left = $9 - 3 = 6$.

These books were stacked in the school library.

Thus, percentage of books stocked in the school library = $\frac{6}{15} \times 100 = \frac{2}{5} \times 100 = 40\%$.

[Note that the calculations involved are much easier now than in the earlier approach]

7. Distance surged by the glacier in 1 hour

$$= \frac{11}{5} \text{ feet}$$

$$= \frac{\frac{11}{5}}{5280} \text{ miles} = \frac{11}{5 \times 5280} \text{ miles.}$$

Hours needed to surge $\frac{1}{2}$ miles

$$= \frac{\frac{1}{2}}{\frac{11}{5 \times 5280}}$$

$$= \frac{1 \times 5 \times 5280}{2 \times 11}$$

$$= 5 \times 240$$

Days needed to surge $\frac{1}{2}$ miles

$$= \frac{5 \times 240}{24} = 50$$

The correct answer is option D.

8. We know that the operation '#' represents one of addition, subtraction, or multiplication of integers.

Thus, we have:

- If '#' represents addition: $2\#0 = 2 + 0 = 2 => \text{LHS} = \text{RHS}$
- If '#' represents subtraction: $2\#0 = 2 - 0 = 2 => \text{LHS} = \text{RHS}$
- If '#' represents multiplication: $2\#0 = 2 \times 0 \neq 2 => \text{LHS} \neq \text{RHS}$

Thus, '#' represents either 'addition' or 'subtraction.'

Thus, we have:

$1\#0 = 1 + 0 = 1$

OR

$1\#0 = 1 - 0 = 1$

Thus, in either case, we have:

$1\#0 = 1$

The correct answer is '1.'

9. Total increase in population = 378 − 300 = 18 million

Increase in population per month = 30,000

Thus, increase in population per year = $30,000 \times 12 = 360,000 = 0.36$ million

Thus, number of years required for the increase = $\frac{18}{0.36} = 50$ years

Thus, the population would be 378 million in the year $(2012 + 50) = 2,062$

The correct answer is option C.

14.2 Ratio & Proportion

10. The key is to translate the problem into an equation.

If the number of marbles with John and Mary be j and m respectively, we have:

$j + m = 75 \times \left(\frac{j}{m}\right) \ldots \text{(i)}$

Now, we know that $m = 30$.

Substituting this value in (i): $j + 30 = 75 \times \left(\frac{j}{30}\right)$

$=> j + 30 = \frac{5j}{2} => \frac{5j}{2} - j = 30 => \frac{3j}{2} = 30 => j = 20.$

Thus, the number of marbles with John is 20.

The correct answer is option A.

11. Let Peter have $\$x$.

Thus, John has $\$2x$.

After John gives \$50 to Peter, the amount of money with John and Peter becomes $\$(2x - 50)$ and $\$(x + 50)$, respectively.

Since the ratio gets reversed, Peter now has twice the amount of money as John has.

Thus: $\frac{x + 50}{2x - 50} = 2 => x + 50 = 4x - 100 => 3x = 150.$

Thus, they together have \$150 (since the total amount $= x + 2x = 3x$).

[Note: There is no need to calculate the amount of money present with them individually since we have been asked to find the total amount. Since their amounts are x and $2x$, their total is $3x$, which we have already obtained.]

The correct answer is option C.

12. Let the number of LCD TVs and LED TVs be x and y, respectively.

Thus, we have:

$x + y = 260 \ldots$ (i)

The ratio of the number of LED TVs to the number of LCD TVs at the showroom is 7 to 6.

Thus, we have:

$\frac{y}{x} = \frac{7}{6}$

Using (i), we have:

$y = \left(\frac{7}{7 + 6}\right) \times 260$

$= 140$

The correct answer is '140.'

13. Let the number of male teachers and female teachers be m and f, respectively.

Since there are 105 more male teachers than female teachers, we have: $m = 105 + f \ldots \text{(i)}$

We know that, if 14 females were hired, the ratio of the number of male teachers to the number of female teachers would be 16 : 9.

Thus, we have:

$$\frac{m}{f+14} = \frac{16}{9} \ldots \text{(ii)}$$

Thus, from (i) and (ii), we have:

$$\frac{105+f}{f+14} = \frac{16}{9}$$

$=> 945 + 9f = 16f + 224$

$=> 7f = 721$

$=> f = 103$

The correct answer is '103.'

14. Since $\left(\frac{4}{5}\right)^{\text{th}}$ of the stock was sold, the remaining $\left(1 - \frac{4}{5}\right) = \left(\frac{1}{5}\right)^{\text{th}}$ of the stock was not sold.

Thus, $\left(\frac{1}{5}\right)^{\text{th}}$ of the total stock = 100.

Thus, $\left(\frac{4}{5}\right)^{\text{th}}$ of the total stock = $4 \times 100 = 400$.

Thus, 400 items were sold each at \$3.00.

Thus, total amount received = $\$ (400 \times 3) = \$1,200$.

The correct answer is option B.

15. We have:

$$\frac{\text{Number of shirts}}{\text{Number of trousers}} = \frac{4}{5} \ldots \text{(i)}$$

$$\frac{\text{Number of jackets}}{\text{Number of shirts}} = \frac{3}{8} \ldots \text{(ii)}$$

$$\frac{\text{Number of sweaters}}{\text{Number of trousers}} = \frac{6}{5}$$

Taking reciprocal on both sides:

$$=> \frac{\text{Number of trousers}}{\text{Number of sweaters}} = \frac{5}{6} \ldots \text{(iii)}$$

Thus, from the above three equations, we have:

$$\frac{\text{Number of jackets}}{\text{Number of sweaters}}$$

$$= \left(\frac{\text{Number of jackets}}{\text{Number of shirts}}\right) \times \left(\frac{\text{Number of shirts}}{\text{Number of trousers}}\right) \times \left(\frac{\text{Number of trousers}}{\text{Number of sweaters}}\right)$$

$$= \frac{3}{8} \times \frac{4}{5} \times \frac{5}{6}$$

$$= \frac{1}{4}$$

The correct answer is option C.

14.3 Average

16. Number of seats in each row = 30

Average number of seats unoccupied per row for the front 12 rows = 13

Thus, the average number of seats occupied per row for the front 12 rows

= 30 - 13

= 17

Thus, total number of seats occupied in the front 12 rows

= 17 × 12

= 204

Average number of seats occupied per row for the back 23 rows = 10

Thus, total number of seats occupied in the back 23 rows

= 10 × 23

= 230

Thus, total number of seats occupied

$= 204 + 230$

$= 434$

The correct answer is '434.'

17. Since the average of w, x, y and z is n, we have:

$$\frac{w + x + y + z}{4} = n$$

$=> w + x + y + z = 4n$

$=> (n - w) + (n - x) + (n - y) + (n - z) = 4n - (w + x + y + z)$

$= 4n - 4n$

$= 0$

The correct answer is '0.'

Note that the sum of deviations of all the elements in a set w.r.t. its mean is 0.

14.4 Sets & Venn Diagrams

18. Since there are only two candidates, we can say that if x% voters did not vote for a particular candidate, then $(100 - x)$% of the voters must have voted for that candidate.

Since 60% voters did not vote for Joseph, it implies that $(100 - 60) = 40$% voters voted for Joseph.

Also, since 50% voters did not vote for Judith, it implies that $(100 - 50) = 50$% voters voted for Judith.

Since no voter voted for both candidates, we have:

Thus, total percentage of voters who voted $= (40 + 50) = 90$%.

Thus, $(100 - 90) = 10$% of the voters did not cast their vote.

The correct answer is option B.

19. We see that the fractions used in the problem are $\frac{3}{5}, \frac{1}{3}$, & $\frac{1}{4}$.

This means that we would need to multiply the number of candidates with the above fractions.

Thus, it is best to assume a number for the total number of candidates which is divisible by 5, 3, 4 (i.e. the denominators of the fractions).

LCM of 5, 3, 4 = 60. So, let us assume that there are 60 candidates.

The following table gives the breakup of the candidates:

	Cleared	Not Cleared	Total
Male	$36 - 9 = 27$	$\frac{1}{4} \times 36 = 9$	$\frac{3}{5} \times 60 = 36$
Female	$40 - 27 = 13$	$20 - 9 = 11$	$60 - 36 = 24$
Total	$60 - 20 = 40$	$\frac{1}{3} \times 60 = 20$	60

The working is shown below:

Total male candidates = $\frac{3}{5} \times 60 = 36$.

Total female candidates = $60 - 36 = 24$.

Total candidates who did not clear the examination = $\frac{1}{3} \times 60 = 20$.

Male candidates who did not clear the examination = $\frac{1}{4} \times 36 = 9$.

Male candidates who cleared the examination = $36 - 9 = 27$.

Total candidates who cleared the examination = $60 - 20 = 40$.

Female candidates who cleared the examination = $40 - 27 = 13$.

Female candidates who did not clear the examination = $20 - 9 = 11$.

So, among the candidates who cleared the examination, females constitute = $\frac{13}{40} \times 100 = 32.5\%$.

The correct answer is option A.

20. The above information can be represented in the form of a Venn-diagram as shown below:

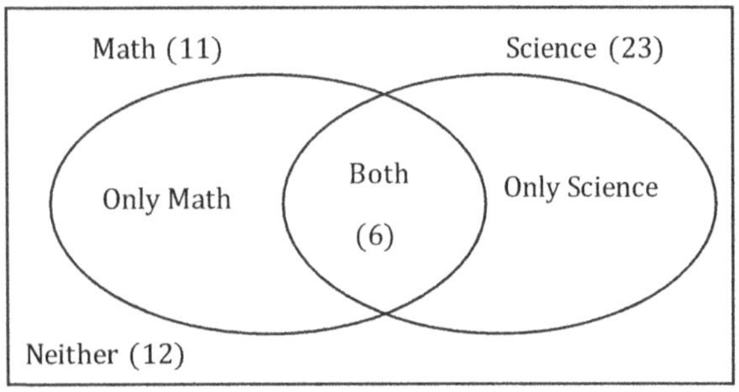

Total number of students

= (# Math) + (# Science) - (# Both) + (# Neither)

= 11 + 23 - 6 + 12

= 40

The correct answer is '40.'

21. Number of attendees = 300

Thus, the number of male attendees

= 70% of 300

= 210

Number of male attendees who were not freelancers = 90.

Thus, the number of males who were not freelancers

= 210 - 90 = 120

Thus, the required percent

$= \frac{120}{300} \times 100\%$

= 40%

The correct answer is '40.'

22.

	Number of students interested in arts	Number of students not interested in arts	Total
Number of boys			36
Number of girls			$2x$
Total	52	Let the number of students not interested in arts be x	

Thus, we have the total number of students:

$52 + x = 36 + 2x$

$=> x = 16$

=> The total number of students

$= 52 + x = 52 + 16$

$= 68$

The correct answer is '68.'

14.5 Profit & Loss

23. The number of items the dealer needs in order to replenish his stock $= 40 - 15 = 25$.

The cost of each item is \$5.

Hence, his total cost for one such replenishment $= \$5 \times 25 = \$125$.

He needs to replenish once every two weeks i.e. twice each month.

Thus, his cost of replenishing his stock in a month $\$125 \times 2 = \$250$.

The correct answer is '250.'

24. To solve the problem, it is helpful to realize that the number of bulbs can be any. So, we pick one that is suitable for doing the calculations. Here, the convenient number of bulbs in the lot is the one that is divisible by both 5 and 4.

Thus, let the number of bulbs be 20 (LCM of 4 and 5).

The cost of 20 bulbs $= \$\frac{6}{4} \times 20 = \$30$.

Number of defectives $= \frac{2}{5} \times 20 = 8$.

So, number of usable bulbs $= 20 - 8 = 12$.

So, price paid per usable bulb $= \$\frac{30}{12} = \$2.50$.

The correct answer is \$2.5 or 2.50.

Alternate approach:

Cost of a bulb $= \$\left(\frac{6}{4}\right) = \$1.50$.

Since 2 out of 5 bulbs are defective, it means that he had a loss of $\frac{2}{5} \times 100\% = 40\%$ or he paid \$1.50 for 60% of bulbs.

=> Realizable price per bulb $= \frac{1.50}{60\%} = \frac{1.50}{0.60} = \$2.50$.

25. Since the problem asks for a percent value, we can assume a suitable value for the cost price.

Let the cost price be \$100.

Thus, the markup

= 25% of \$100

= \$25

Thus, the selling price

= Cost price + Mark up

= \$125

Thus, the required percent

$= \frac{25}{125} \times 100\%$

= 20%

The correct answer is '20.'

26. Let the selling price of the appliance be $\$x$

Profit made on the appliance

= 20% of $\$x$

$= \$\left(\frac{x}{5}\right)$

Since the selling price is \$50 more than the cost price, we have:

$\frac{x}{5} = 50$

$=> x = 250$

The correct answer is '250.'

27. Let the selling price of a shirt is \$1

Selling price of 5 shirts $= 5 \times 1 = 5$

Offer price of 5 shirts $= 3 \times 1 = 3$

Discount = 5 − 3 = 2

Percent discount = $\frac{\text{Discount}}{\text{Selling price}} \times 100\% = \frac{2}{5} \times 100\% = 40\%$

The correct answer is '40.'

28. Let the number of chairs sold last week be h.

Since the number of tables sold was 5 more than the chairs, the number of tables sold = $(h + 5)$.

Selling price of each table = $150.

Selling price of each chair = $85.

Thus, total sales revenue = $ $(85h + 150(h + 5))$.

Thus, we have:

$85h + 150(h + 5) = 1,925$

$=> 235h = 1,925 - 750$

$=> h = \frac{1,175}{235} = 5$

Thus, the number of chairs sold = 5.

Thus, the number of tables sold = 5 + 5 = 10.

Thus, the total number of tables and chairs sold = 5 + 10 = 15.

The correct answer is option C.

Alternate approach:

We know that the number of tables sold is 5 more than the number of chairs sold.

The price of 5 tables = $ (150 × 5) = $750.

Removing this from the total, i.e. $1,925, we are left with $ (1925 − 750) = $1175.

This amount was obtained by selling equal numbers of tables and chairs.

Total price of one table and one chair = $ (150 + 85) = $235.

Thus, number of items sold for $1,175

$$= \frac{1,175}{235} \times 2 = 10$$

Thus, total number of tables and chairs sold = 5 + 10 = 15.

14.6 Percents

29. We have: 10% of 20% of x = 4

$=> \frac{10}{100} \times \left(\frac{20}{100} \times x\right) = 4$

$=> x = 4 \times \frac{100}{20} \times \frac{100}{10} = \frac{10{,}000}{50} = 200.$

The correct answer is '200.'

30. Given that the bag has a total of one-third cheaper variety of beans, one-third of the beans cost \$6 each. Let the number of beans of the cheaper variety and that of the costlier variety be x and $2x$, respectively.

Since the price of each bean of the costlier variety is three times that of each bean of the cheaper variety, we have:

Price of each bean of the costlier variety

$= \$(3 \times 6)$

$=\$18$

Total price of cheaper variety beans = $ (6x)

Total price of costlier variety beans = $ (2x × 18) = $ (36x)

Thus, total price of all beans in the bag = $ (6x + 36x) = $ (42x)

Thus, the required percent

$$= \left(\frac{36x}{42x}\right) \times 100\%$$

$$= \frac{600}{7}\%$$

$= 85.71\%$

The correct answer is '85.71.'

31. The cost of repairing the current machine = \$1,200

The cost of new machine = \$2,800

Since the new machine lasts for two years, the average cost per year = $\$\frac{2,800}{2}$ = \$1,400.

Thus, the required percentage = $\frac{1,400 - 1,200}{1,200} \times 100 = 16.67\%$.

The correct answer is option C.

32. Tax paid on \$50 = \$0.82

Thus, a tax which thrice as much as the above, would be $\$(0.82 \times 3) = \$2.46$ on \$50.

Thus, the new tax on \$100 = 2 × Tax paid on \$50

=> $\$(2.46 \times 2) = \$4.92$

Thus, this tax, expressed as a percentage = $\frac{4.92}{100} \times 100 = 4.92\%$.

The correct answer is option C.

33. For X, visual range increased from 90 kilometers to 270 kilometers.

Total increase in visual range of X = 270 – 90 = 180 kilometers

Thus, Percentage increase in visual range using X = $\frac{180}{90} \times 100 = 200\%$.

For Y, visual range increased from 45 kilometers to 180 kilometers.

Total increase in visual range of X = 180 – 45 = 135 kilometers

Thus, Percentage increase in visual range using Y = $\frac{135}{45} \times 100 = 300\%$.

Apparently it seems that the required answer is simply: 300% – 200% = 100%.

However, it is not so since we are here asked to find percent change not absolute change

Here absolute change is 300% – 200% = 100%, which we are going to compare with percent change in X which is 200%.

The required percent = $\frac{100}{200} \times 100 = 50\%$.

What this shows us is that the use of telescope Y is 50% more effective than that of telescope X.

The correct answer is option B.

14.7 Interest & Compounding

34. Amount after three years = \$1,200

Amount after five years = \$1,500

Thus, interest accumulated in 2 years = $\$ (1,500 - 1,200) = \$300$

Thus, interest accumulated per year = $\$ \left(\frac{300}{2}\right)$ = \$150 (since under simple interest, interest accumulated every year is constant)

Thus, interest accumulated in the first three years = $\$ (150 \times 3) = \$450$

Thus, principal amount invested = $\$ (1,200 - 450) = \$750$

Thus, on \$750 invested, interest accumulated is \$150 every year.

Thus, rate of interest = $\frac{150}{750} \times 100 = 20\%$

The correct answer is option C.

35. Number of retailers at the beginning of the year = 35% of 120 = $\frac{35}{100} \times 120 = 42$.

Number of retailers after the 24-month period = 25% of 240 = $\frac{25}{100} \times 240 = 60$.

Thus, increase in the number of retailers = $60 - 42 = 18$.

Let the simple annual percent growth rate in the number of retailers be $r\%$.

Thus, in two years (the 24-month period), increase in the number of retailers at $r\%$ rate

$= \left(\frac{42 \times r \times 2}{100}\right)$

(The value is calculated on 42 since 42 is the value at the start of the year)

Thus, we have:

$\frac{42 \times r \times 2}{100} = 18$

$=> r = \frac{18 \times 100}{2 \times 42} = 21.43\%$

The correct answer is option B.

14.8 Speed/Time/Distance

36. The man will take the longest if he travels at the lowest speed, i.e. 24 miles per hour.

Thus, time taken at 24 miles per hour $= \frac{120}{24} = 5$ hours.

The man will take the shortest time if he travels at the highest speed, i.e. 36 miles per hour.

Thus, time taken at 36 miles per hour $= \frac{120}{36} = 3$ hours 20 minutes.

Thus, in order to reach at 5 pm, the earliest by which the man needs to start is 5 hours before 5 pm, i.e. 12 noon; and the latest by which the man needs to start is 3 hours 20 minutes before 5 pm, i.e. 1:40 pm.

The correct answer is option C.

37. Say the average speed of the car is S.

T_1 = Time required to travel 900 miles with speed $S = \frac{900}{S}$ hours

S is increased by 10 miles per hour; so new speed $= (S + 10)$ mph

T_2 = Time required to travel 900 miles with speed $(S + 10) = \frac{900}{S+10}$ hours

Given that the difference between T_1 and T_2 is 2 hour

$$=> \frac{900}{S} - \frac{900}{S+10} = 2$$

$$=> 900\left(\frac{1}{S} - \frac{1}{S+5}\right) = 2$$

$$=> \frac{S+5-S}{S(S+5)} = \frac{1}{450}$$

$$=> 5 \times 450 = S^2 + 5S$$

$$=> S^2 + 5S - 2250 = 0$$

Roots of above quadratic equations are '45' and '−50'.

Since speed cannot be negative, −50 is ignored, so $S = 45$ mph.

The correct answer is option C.

38. The car runs 5 miles less per gallon in the city compared to on the highway.

Let's consider that if the car runs x miles per gallon on the highway, then it runs $(x - 5)$ miles per gallon in the city.

Capacity of the full tank of diesel on highway = $\frac{440}{x}$ gallons

Capacity of the full tank of diesel in the city = $\frac{360}{x-5}$ gallons

Since the above fractions must be equal, we have:

$\frac{440}{x} = \frac{360}{x-5}$

$=> \frac{11}{x} = \frac{9}{x-5}$

$=> 11x - 55 = 9x$

$=> 2x = 55$

$=> x = 27.5$

As car runs $(x - 5)$ miles per gallon in the city, required answer is $27.5 - 5 = 22.5$

The correct answer is option B.

39. Time required to travel 1 mile with speed 50 miles per hour

$= \frac{1}{50}$ hours $= \frac{1}{50} \times 60 \times 60 = \; = 72$ seconds

Now new time required = $72 + 5 = 77$ seconds $= \frac{77}{60 \times 60}$ hours.

Speed $= \frac{\text{Distance}}{\text{Time}}$

$= \frac{1 \text{ mile}}{77 \text{ seconds}} = \frac{1}{\frac{77}{60 \times 60}} = \frac{3600}{77} = 46.75$ miles per hour

The correct answer is option B.

14.9 Probability

40. There are total 22 balls, out of which 11 are black.

Probability that the ball will be black $= \frac{\text{Number of ways of drawing a black ball}}{\text{Number of ways of drawing any color ball}}$

$=> \frac{11}{22} = \frac{1}{2}$

The correct answer is option B.

41. Understand the following table.

Sum	Ways	Number of ways	Freq. distribution & Probabilities
2	{1,1}	1	$p = \frac{1}{36}$
3	{2,1}, {1,2}	2	$p = \frac{1}{18}$
4	{3,1}, {2,2}, {1,3}	3	$p = \frac{1}{12}$
5	{4,1}, {3,2}, {2,3}, {1,4}	4	$p = \frac{1}{9}$
6	{5,1}, {4,2}, {3,3}, {2,4}, {1,5}	5	$p = \frac{5}{36}$
7	{6,1}, {5,2}, {4,3}, {3,4}, {2,5}, {1,6}	6	$p = \frac{1}{6}$
8	{6,2}, {5,3}, {4,4}, {3,5}, {2,6}	5	$p = \frac{5}{36}$
9	{6,3}, {5,4}, {4,5}, {3,6}	4	$p = \frac{1}{9}$
10	{6,4}, {5,5}, {4,6}	3	$p = \frac{1}{12}$
11	{6,5}, {5,6}	2	$p = \frac{1}{18}$
12	{6,6}	1	$p = \frac{1}{36}$
	Total number of ways	**36**	

Memorise...

1. **If Sum $\leq$ 7, then Number of ways = Sum $-$ 1**
2. **If Sum $\geq$ 8, then Number of ways = 13 $-$ Sum**

We have to calculate the probability of getting the sum > 9. It would be

$$=> P(SUM > 9) = \frac{n\{\text{\# of ways of getting a sum of 10, 11 or 12}\}}{36} = \frac{3+2+1}{36} = \frac{1}{6}$$

The correct answer is '$\frac{1}{6}$.'

42. We know that the probability of getting a 'Head' or a 'Tail' on any toss for a coin = 1/2.

We want Tail, Tail, and Head (in this order) on three tosses, so the probability of getting TTH = $(1/2) \times (1/2) \times (1/2) = 1/8$.

The correct answer is '1/8.'

This question is different from a question asking for getting two Tails and a Head. In this case, one can get two Tails in I & II, or in I & III, or in II & III positions.

The answer would have been $3 \times 1/8 = 3/8$.

43. We know that the probability of getting a 'Head' or a 'Tail' for this coin is not the same as this is a biased coin.

We are given that the probability of getting a Head = $\frac{2}{3}$

Thus, the probability of getting a Tail = $1 - \frac{2}{3} = \frac{1}{3}$

We want Head/Tails/Head/Tails/Head (in this order) on 5 tosses;

so, the probability of getting HTHTH = $\frac{2}{3} \times \frac{1}{3} \times \frac{2}{3} \times \frac{1}{3} \times \frac{2}{3} = \frac{8}{3^5} = 0.033$.

The correct answer is '0.033.'

44. The number of ways of choosing 5 students out of 10 + 12 = 22 students

$= C_5^{22} = \frac{22.21.20.19.18}{1.2.3.4.5} = 26,334.$

The number of ways of choosing 2 boys out of 10 boys = $C_2^{10} = \frac{10.9}{1.2} = 45$.

Similarly, the number of ways of choosing 3 girls out of 12 girls

$= C_3^{12} = \frac{12.11.10}{1.2.3} = 220.$

Since this is a case of 'AND', we must multiply the number of ways, thus the answer is $45 \times 220 = 9900$.

Thus, the probability that the team of 5 would have 2 boys and 3 girls

$= \frac{\text{Teams with 2 boys and 3 girls}}{\text{Teams of 5 students}} = \frac{9900}{26334} = 0.38$

The correct answer is '0.38.'

45. Probability of hitting the black parts = $\frac{\text{Area of the black parts}}{\text{Area of the outermost square}}$

Area of the outermost square = $5 \times 5 = 25$ cm^2

Area of the middle square = $3 \times 3 = 9$ cm^2

Area of the black strip = $25 - 9 = 16$ cm^2

Area of the inner-most square (one of the black parts) = $2 \times 2 = 4$ cm^2

Total area of black parts = $16 + 4 = 20$ cm^2

Thus, the probability of hitting the black parts = $\frac{20}{25} = \frac{4}{5}$

The correct answer is '$\frac{4}{5}$.'

14.10 Time & Work

46. Number of parts produced by Machine X in 1 hour = $\frac{1,000}{3}$

Number of parts produced by Machine Y in 1 hour = $\frac{1,000}{2}$

Number of parts produced by Machine Z in 1 hour = $\frac{1,000}{6}$

Thus, if all three machines work together, the total number of parts produced in 1 hour

$= \frac{1,000}{3} + \frac{1,000}{2} + \frac{1,000}{6}$

$= 1,000\left(\frac{1}{3} + \frac{1}{2} + \frac{1}{6}\right)$

$= 1,000 \times 1 = 1,000$ parts

Thus, time taken to produce 3,000 parts = $\frac{3,000}{1,000}$ = 3 hours.

The correct answer is option D.

47. Cost of food consumed by 4 adults in 3 days = \$60

=> Cost of food consumed by 1 adult in 3 days = $\$\left(\frac{60}{4}\right) = \$15$

=> Cost of food consumed by 1 adult in 1 day = $\$\left(\frac{15}{3}\right) = \$5$

We know that one child consumes half the amount of food consumed by an adult.

Thus, 3 children consume food equivalent to $\frac{3}{2}$ adults.

Thus, 6 adults and 3 children are equivalent to $\left(6+\frac{3}{2}\right)=\frac{15}{2}$ adults.

Thus, cost of food consumed by $\frac{15}{2}$ adults in 1 day = $\$\left(5\times\frac{15}{2}\right)=\$\left(\frac{75}{2}\right)$

=> Cost of food consumed by $\frac{15}{2}$ adults in 4 days = $\$\left(\frac{75}{2}\times 4\right)=\$150$

The correct answer is option B.

48. Number of units produced per hour by the first machine = 2,000.

Thus, number of units produced per hour by the second machine = 6,000

(Since it is thrice as efficient as the first machine)

Since the second machine works 15 hours a day, number of units produced in a day

$= 6,000 \times 15 = 90,000$

Thus, number of units produced in 10 days = $90000 \times 10 = 900,000$

The correct answer is option D.

49. According to the given data, the pump filled $\left(\frac{5}{6}-\frac{1}{2}\right)=\frac{1}{3}$ of the pool in $2\frac{1}{3}=\frac{7}{3}$ hours

Thus, time taken by the pump to fill the entire pool = $\frac{\left(\frac{7}{3}\right)}{\left(\frac{1}{3}\right)}=\frac{7}{3}\times\frac{3}{1}=7$ hours

Thus, the time taken by the pump to fill the empty pool completely = 7 hours

The correct answer is option C.

50. The part of work completed by machine A in 1 hour = $\frac{1}{6}$

The part of work completed by machine B in 1 hour = $\frac{1}{9}$

The part of work completed by machine C in 1 hour = $\frac{1}{12}$

The part of work completed by the three machines in 3 hours = $\frac{1}{6}+\frac{1}{9}+\frac{1}{12}=\frac{13}{36}$

The part of work remains to be done = $1-\frac{13}{36}=\frac{23}{36}$

The correct answer is option B.

51. The part of work completed by machine A in 1 hour = $\frac{1}{6}$

The part of work completed by machine B in 1 hour = $\frac{1}{9}$

The part of work completed by machine C in 1 hour = $\frac{1}{12}$

The part of work completed by the three machines (in order ABC) in 3 hours = $\frac{1}{6} + \frac{1}{9} + \frac{1}{12} = \frac{13}{36}$

Since $\frac{13}{36} < 1$, part of work remains to be done

The part of work completed by the three machines (in order ABC-ABC) in $3 \times 2 = 6$ hours $= 2 \times \frac{13}{36} = \frac{13}{18}$

Since $\frac{13}{18} < 1$, some part of work still remains to be done

The part of work completed by the three machines (in order ABC-ABC-ABC) in $3 \times 3 = 9$ hours $= 3 \times \frac{13}{36} = \frac{13}{12}$

Since $\frac{13}{12} > 1$, the work would finish before the complete 9 hours.

We can conclude that the work would finish in more than 6 but less than 9 hours. There is only one option that qualifies, i.e, 8 hours, Option C - the correct answer.

For the sake of understanding, we'll solve it completely.

Let's try if the work is complete in order ABC-ABC-AB~~C~~; when Machine B operates

The part of work completed by the three machines in 6 hours + Part of work completed by Machine A in 1 hour + Part of work completed by Machine B in 1 hour $= \frac{13}{18} + \frac{1}{6} + \frac{1}{9} = \frac{13 + 3 + 2}{18} = \frac{18}{18} = 1$ (Complete work)

The work would be completes in order ABC-ABC-AB, i.e. in 8 hours.

The correct answer is option C.

14.11 Arithmetic Mean

52. Note that since the first number is greater than the mean by the same amount as the second number is lesser than the mean, the mean is not going to change. The changes brought by them will get neutralized.

Let's see how.

The sum of the 11 numbers = $11 \times 50 = 550$

The value of the two numbers are: $50 + 6 = 56$ and $50 - 6 = 44$

Thus, the average (arithmetic mean) of the 13 numbers

$$= \frac{550 + 56 + 44}{13} = \frac{550 + 50 + 6 + 50 - 6}{13} = \frac{(550 + 50 + 50) + (6 - 6)}{13} = \frac{650 + 0}{13} = 50$$

The correct answer is option C.

53. Note that if, after the removal of a number, the average (arithmetic mean) of the remaining numbers does not change, the value of the removed number is the same as the average (arithmetic mean). Thus, the answer is 50.

Let's see how.

The sum of the 11 numbers $= 11 \times 50 = 550$

Say the value of the removed number $= x$

Then the average (arithmetic mean) of the remaining 10 numbers $= \frac{550 - x}{10}$

It is given that the average (arithmetic mean) remains the same.

Thus, $\frac{550 - x}{10} = 50$

$\Rightarrow x = 50$

The correct answer is '50.'

54. Let the sum of the 10 numbers other than x be s.

Thus, we have:

$x = 3 \times \left(\frac{s}{10}\right)$

$\Rightarrow s = \frac{10x}{3}$

Thus, the sum of all the 11 numbers in the list $= s + x = \frac{10x}{3} + x = \frac{13x}{3}$

Thus, the required fraction $= \frac{x}{\frac{13x}{3}} = \frac{3}{13}$.

The correct answer is option B.

55. We have:

$$\frac{p + q + 15}{3} = \frac{p + q + 15 + 35}{4}$$

$=> 4p + 4q + 60 = 3p + 3q + 150$

$=> p + q = 90$

$=> \frac{p+q}{2} = \frac{90}{2} = 45$

Thus, the average of p and $q = \frac{p+q}{2} = 45$.

The correct answer is option B.

14.12 Data Analysis

Table

56. Working with the options:

(A) Company X in 2010 - Required ratio $= \frac{80}{250} = 0.32$

(B) Company X in 2011 - Required ratio $= \frac{100}{450} = 0.22$ (Lowest)

(C) Company Y in 2010 - Required ratio $= \frac{128}{400} = 0.32$

(D) Company Y in 2011 - Required ratio $= \frac{215}{650} = 0.331$

The correct answers is option B.

57. Working with the options:

(A) Required difference = 450 - 150 = \$300 million $\ngtr$ \$750 million - Does not satisfy

(B) Since the percent increase in Net Profits from 2010 to 2011 was the same as that from 2011 to 2012, we have:

$\frac{\text{Net Profit in 2011}}{\text{Net Profit in 2010}} = \frac{\text{Net Profit in 2012}}{\text{Net Profit in 2011}}$

$=> \frac{150}{125} = \frac{\text{Net Profit in 2012}}{150}$

$=> \text{Net Profit in 2012} = \frac{150 \times 150}{125} = \$180 \text{ million}$

Since the Total Income in 2012 is double the Total Income in 2011, we have:

Total income in 2012 = 2 × 450 = \$900 million

Required difference = 900 - 180 = \$720 million $\ngtr$ \$750 million - Does not satisfy

(C) Required difference = 650 - 250 = \$400 million $\ngtr$ \$750 million - Does not satisfy

(D) Since the percent increase in Net Profits from 2010 to 2011 was the same as that from 2011 to 2012, we have:

$$\frac{\text{Net Profit in 2011}}{\text{Net Profit in 2010}} = \frac{\text{Net Profit in 2012}}{\text{Net Profit in 2011}}$$

$$=> \frac{250}{170} = \frac{\text{Net Profit in 2012}}{250}$$

$$=> \text{Net Profit in 2012} = \frac{250 \times 250}{170} = 367.6 \approx \$368 \text{ million}$$

Since the Total Income in 2012 is double the Total Income in 2011, we have:

Total income in 2012 = 2×650 = \$1, 300 million

Required difference = 1,300 - 368 = \$932 million > \$750 million - Satisfies

The correct answers is option D.

58. Let the percent change from first quarter to the second quarter be p%

Thus, we have:

$$\text{Income in second quarter} = (\text{Income in first quarter}) \times \left(1 \pm \frac{p}{100}\right)$$

$$=> \text{Income in first quarter} = \frac{\text{Income in second quarter}}{\left(1 \pm \frac{p}{100}\right)}$$

In the relation above, (+) is used if the percent change is positive and (–) is used if the percent change is negative.

Thus, based on the above relation, we have:

Sector	**Net Income (billion \$) in second quarter**	**Percent change from first quarter**	**Net Income (billion \$) in first quarter**
Power	7.20	+20%	$\frac{7.20}{\left(1+\frac{20}{100}\right)} = \frac{7.20}{1.2} = 6.00$
Agriculture	5.00	−10%	$\frac{5.00}{\left(1-\frac{10}{100}\right)} = \frac{5.00}{0.90} = 5.56$
Services	9.00	+200%	$\frac{9.00}{\left(1+\frac{200}{100}\right)} = \frac{9.00}{3} = 3.00$ (**Smallest**)
Information Tech.	2.50	−50%	$\frac{2.50}{\left(1-\frac{50}{100}\right)} = \frac{2.50}{0.5} = 5.00$

The correct answer is option C.

59. We have already computed the net incomes of the four sectors in the first quarter in the previous question. Let us calculate the same for Telecommunications sector too and then calculate the average.

The net incomes of Telecommunications in the first quarter $= \dfrac{4.80}{\left(1 - \dfrac{20}{100}\right)} = \dfrac{4.80}{0.80} = 6.00$

Thus the average of the five sectors $= \dfrac{6 + 6 + 5.56 + 3 + 5}{5} = \$5.112$ billion.

The correct answer is option C.

Bar Diagram

60. Let us bring out the bar-chart.

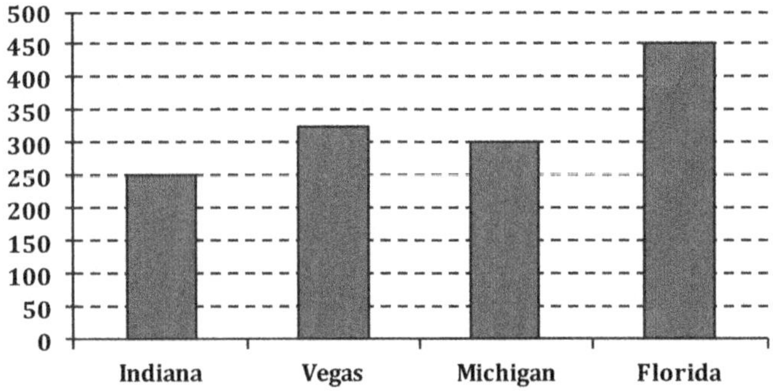

Number of customers in Indiana and Michigan together

= (250 + 300) thousand

= 550 thousand

Thus, 550 thousand represents 20% of the total number of customers of the company

=> 20% of total customers of the company = 550 thousand

=> Total customers of the company

$= 550 \times \dfrac{100}{20}$

= 2,750 thousand

From the bar graph, we find that the number of customers in Vegas = 325 thousand (the bar is in middle of 300 and 350)

Thus, the required percent

$= \frac{\text{Customers in Vegas}}{\text{Total customers}} \times 100$

$= \frac{325}{2750} \times 100\%$

$= 11.8\%$

The correct answers is option D.

Line Graph

61. Let us bring out the figure.

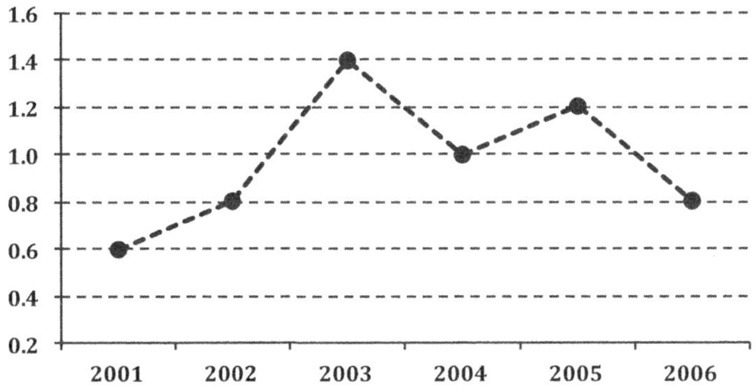

Exports in 2001 = \$60 million

From the graph, we find that $\frac{\text{Imports in 2001}}{\text{Exports in 2001}} = 0.6$

=> Imports in 2001 = $0.6 \times$ Exports in 2001

=> Imports in 2001 = 0.6×60 = \$36 million

We know that exports increased by \$2 million every year.

Thus, Exports in 2005 = \$ $(60 + 4 \times 2)$ million = \$68 million

From the graph, we find that $\frac{\text{Imports in 2005}}{\text{Exports in 2005}} = 1.2$

=> Imports in 2005 = $1.2 \times$ Exports in 2005

=> Imports in 2005 = 1.2×68 = \$81.6 million

Thus, percent increase in imports from 2001 to 2005

$= \frac{\text{Imports in 2005} - \text{Imports in 2001}}{\text{Imports in 2001}} \times 100\%$

$= \frac{81.6 - 36}{36} \times 100\%$

$= 126.7\%$

The correct answers is option D.

62. From the line graph, we find that in 8 hours, John made 32 units. One may also take it as 33, as the graph does not have minor grid lines. This would not make any impact since this is not a student-produced response question and the given choices are far apart.

Thus, efficiency $= \frac{\text{Output}}{\text{Time}} = \frac{32}{8} = 4$ units per hour

Do get confused that John made $32 - 5 = 27$ units; during the first 2 hours he made 5 units.

The correct answers is option C.

63. We have:

- Segment 1: A 3-mile drive with the speed of 20 miles per hour => Time $= \frac{3}{12} \times 60 = 15$ minutes
- Segment 2: A 12-mile drive with the speed of 20 miles per hour => Time $= \frac{12}{20} \times 60 = 36$ minutes
- Segment 3: A 2-mile stretch with the speed of 6 miles per hour => Time $= \frac{2}{6} \times 60 = 20$ minutes
- Segment 4: A 4-mile stretch with the speed of 8 miles per hour => Time $= \frac{4}{8} \times 60 = 30$ minutes

Total distance $= 3 + 12 + 2 + 4 = 21$ miles

Total time $= 101$ minutes $= \frac{101}{60}$ hour

Averaged speed $\times$ Total time $=$ Total distance

Thus, we have: $m \times \frac{101}{60} = 21$

The correct answer is option C.

Chapter 15

Concepts of Geometry & Trigonometry

15.1 Geometry

15.1.1 Lines

A **line** is assumed to extend indefinitely in both directions. There is one and only one line between two distinct points.

A **line segment** is the part of a line between two points called end points. A line segment is denoted by its end points. It is denoted by $\overline{XY}$.

When a line segment is extended indefinitely in one direction, it is called a **ray**. A ray has one endpoint.

Pairs of lines: A pair of straight lines may be either:

- **Parallel:** Parallel lines are lines in the same plane that never intersect, even if extended infinitely.

- **Intersecting:** A pair of lines in the same plane, if not parallel, must intersect at some point.

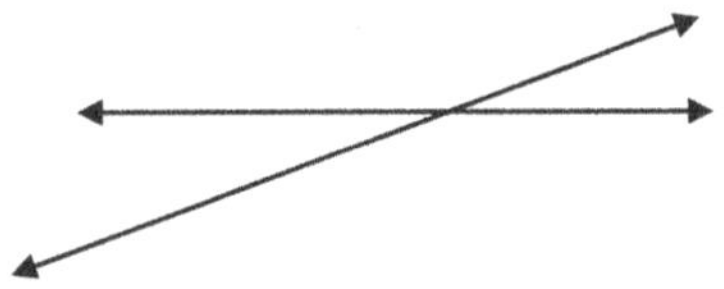

- **Perpendicular:** If the above lines intersect at right angles, then such lines are perpendicular to one another.

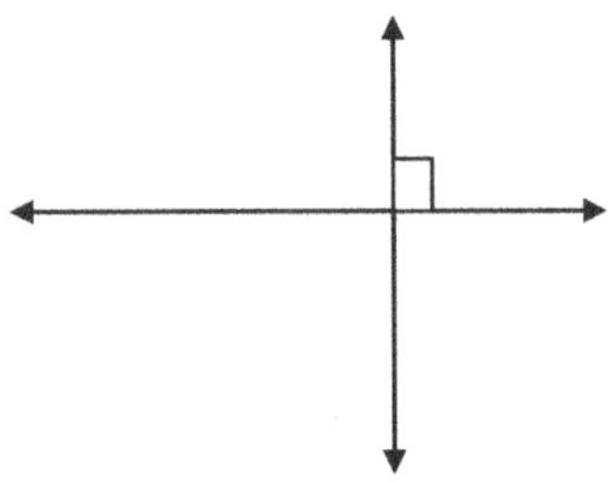

15.1.2 Angles

If two straight lines meet at a point they form an angle. The point is called the vertex of the angle and the lines are called the sides or rays of the angle.

Let us look at the important points associated with lines and angles:

1. An angle less than 90° is **acute**.	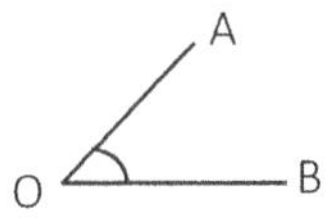
2. An angle equal to 90° is **right angle**.	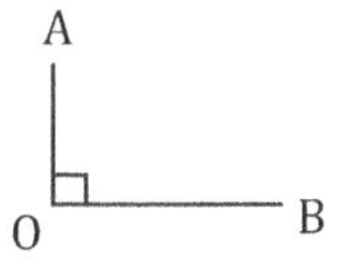
3. An angle greater than 90° but less than 180° is **obtuse**.	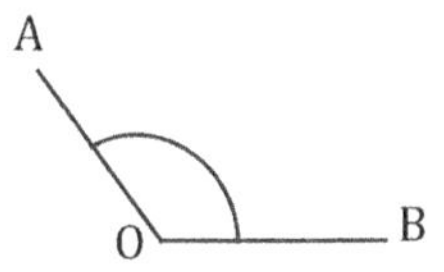
4. An angle of 180° is a **straight angle**.	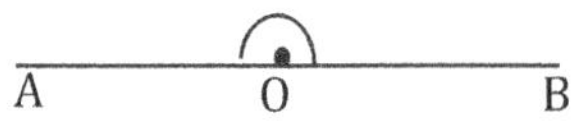
5. An angle greater than 180° but less than 360° is **reflex.**	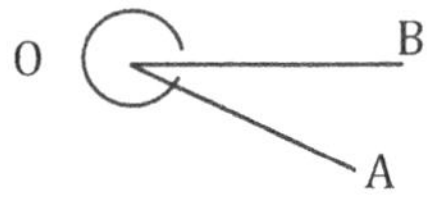
6. Two angles are **adjacent** if they have the same vertex with a common side and one angle is not inside the other. • ∠BOC and ∠COD are adjacent • ∠COD and ∠EOD are not adjacent	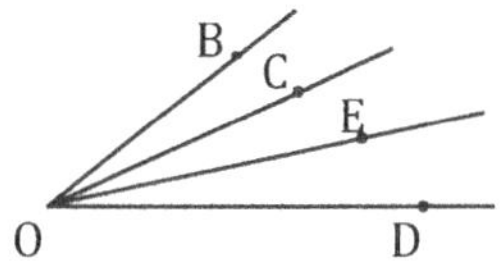
7. If the sum of two adjacent angles is 180°, then they are **supplementary**. • ∠AOC and ∠DOC are supplementary.	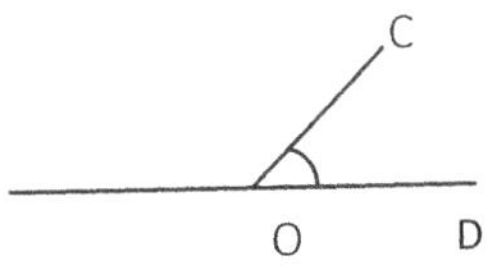
8. If the sum of two adjacent angles is 90°, then they are **complementary**. • ∠BOC and ∠DOC are complementary.	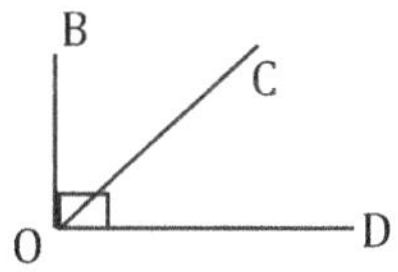

9. **Vertically Opposite Angles:** When two lines intersect, two pairs of vertically opposite angles are formed. Vertically opposite angles are equal. • Here, ∠C and ∠A are equal. • Similarly, ∠D and ∠B are equal.	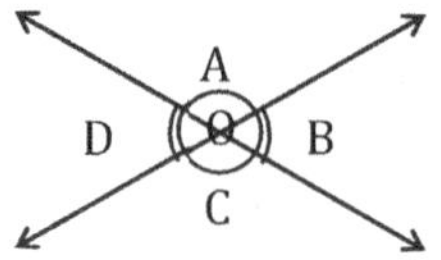
10. **Alternate and Corresponding Angles:** In the diagram below, AB and UV are parallel lines with DE as a transversal. (Note: A transversal refers to any straight line that intersects a pair of straight line.) This creates pairs of corresponding and alternate angle as shown below. • Alternate angles are equal ∠ Q = ∠ S and ∠ P = ∠ R • Interior angles on same side are supplementary ∠ P + ∠ S = ∠Q + ∠ R = ∠180° • Corresponding angles are equal ∠N =∠S; ∠M =∠R; ∠P =∠T; ∠Q =∠U	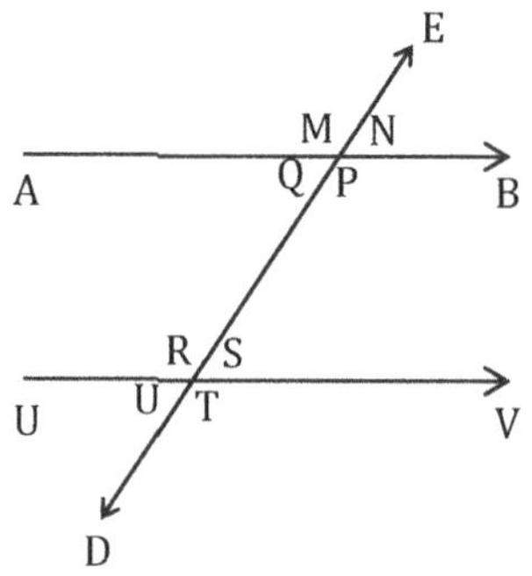

15.1.3 Triangles

A triangle is a figure bounded by three line segments in a plane. In the diagram, the line segments are AB, BC and CA, also called its sides; and A, B and C are called the vertices of the triangle. A triangle can also be called a 3-sided polygon.

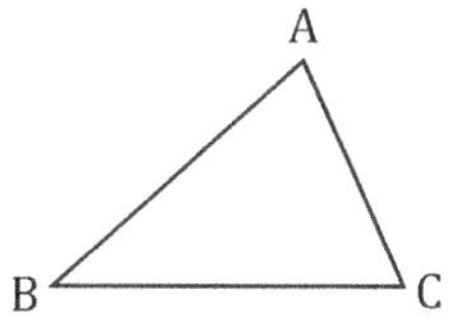

Let us look at the important points associated with triangles:

1. Sum of any two sides in a triangle is greater than the third side. Also, any side is less than half of the perimeter of the triangle. Note: What happens, if the sum of two sides is NOT greater than the third side? Let us observe using the diagram below. Let AB be one side of the triangle. Let the third vertex be C. To determine the position of C, we draw arc with radius AC and BC with centres at A and B respectively. The point of intersubsection of the arc is the position of C. If AC + BC $\ngtr$ AB, then the arc never intersects as shown below. Thus, there is no triangle at all! 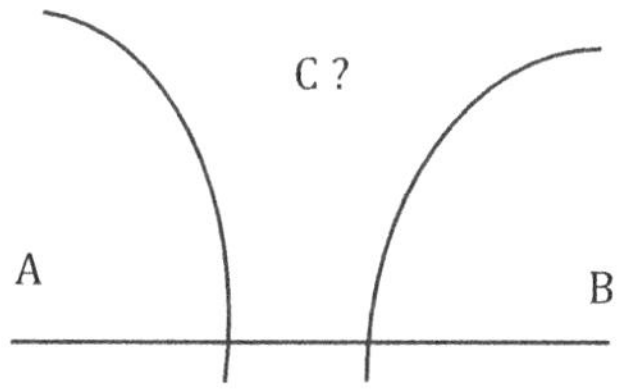 	Thus, in triangle ABC, we simultaneously have: $AB + BC > AC \ldots$ (i) $AB + AC > BC \ldots$ (ii) $AC + BC > AB \ldots$ (iii) Let us choose any of the three inequalities above, say (i): adding AC to both sides: $AB + BC + AC > AC + AC$ $\Rightarrow 2AC <$ Perimeter $\Rightarrow AC < \frac{1}{2} \times$ Perimeter Similarly, the inequalities for the other two sides follow.
2. The greatest side in a triangle has the greatest angle opposite to it and vice-versa. Similarly, the smallest side in a triangle has the smallest angle opposite to it and vice-versa.	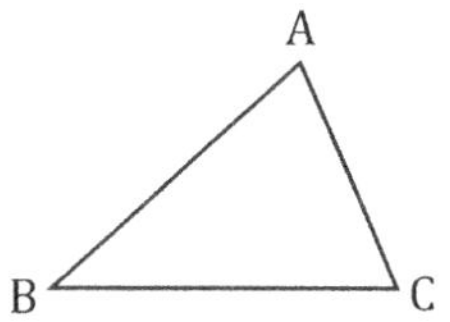 $AB > BC > AC \Rightarrow \angle C > \angle A > \angle B$

3. Sum of the interior angles is 180°. Sum of its exterior angles is 360°. In the figure alongside, the interior angles are i_a, i_b and i_c and the exterior angles are e_a, e_b and e_c. Note: Have you ever wondered: Why is the sum of the interior angles of a triangle equal to 180°? (Find out - At the end of this chapter)	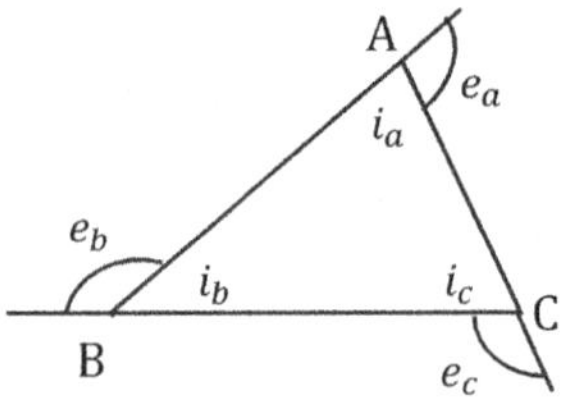 Thus, we have: $i_a + i_b + i_c = 180^\circ$ $e_a + e_b + e_c$ $= 180^\circ - i_a + 180^\circ - i_b + 180^\circ - i_c$ $= 180^\circ \times 3 - (i_a + i_b + i_c)$ $= 540^\circ - 180^\circ$ $= 360^\circ$
4. Any exterior angle in a triangle is equal to sum of the opposite interior angles.	This follows from the above rule #3. Let us see how: Since $i_a + i_b + i_c = 180^\circ$, we have: $i_a + i_b = 180^\circ - i_c$ $=> i_a + i_b = e_c$
5. A triangle with no two sides equal is called a **scalene** triangle.	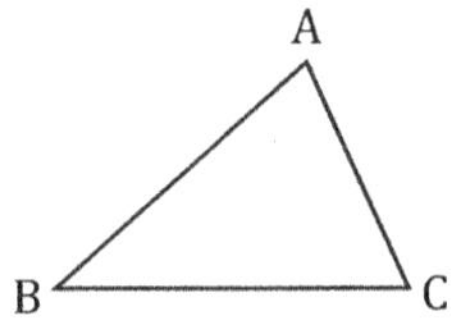 Here, AB $\neq$ BC $\neq$ CA, and $\angle C \neq \angle A \neq \angle B$

<table>
<tr>
<td>
6. If EXACTLY two sides of a triangle are equal, it is called an isosceles triangle.

The angles opposite to the equal sides are also equal.

Note: In an isosceles triangle, a perpendicular drawn from the vertex to the 'UNEQUAL' side bisects the side.
</td>
<td>
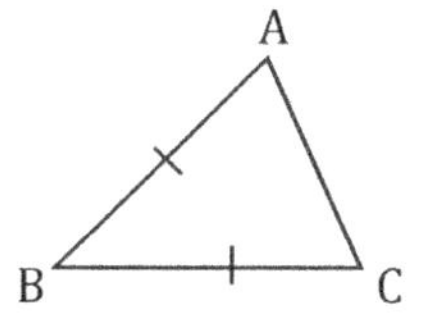

Here, AB = BC $\neq$ AC
=> $\angle$C =$\angle$A $\neq$ $\angle$B
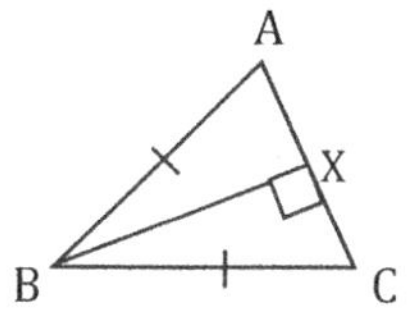

Here, BX $\perp$ AC => AX = CX, and $\angle$ABX =$\angle$CBX
</td>
</tr>
<tr>
<td>
7. If all the three sides of a triangle are equal, it is called an equilateral triangle.

Each angle is equal to 60°.

• Height of an equilateral triangle $= \frac{\sqrt{3}}{2} \times$ (Length of its side)
• Area of an equilateral triangle $= \frac{\sqrt{3}}{4} \times$ (Length of its side)^2
• Perimeter of an equilateral triangle = 3 × Length of its side
</td>
<td>
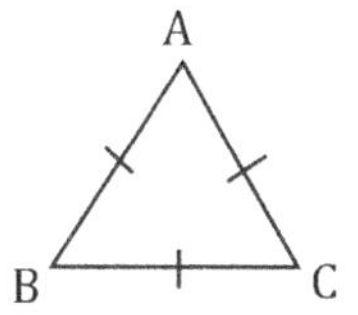

Here, AB = BC = CA

$\angle$C = $\angle$A = $\angle$B = 60°
</td>
</tr>
</table>

<table>
<tr>
<td>

8. If ONE angle of a triangle is 90°, it is called a **right-angled** triangle.

If ABC is right-angled at A, from Pythagoras' theorem we have:

$AB^2 + AC^2 = BC^2$

A few of the integer set of values of the sides AB, AC and BC satisfying Pythagoras' theorem are:

- **3, 4, 5**: $3^2 + 4^2 = 5^2$
- **6, 8, 10**: $6^2 + 8^2 = 10^2$
- **5, 12, 13**: $5^2 + 12^2 = 13^2$
- **7, 24, 25**: $7^2 + 24^2 = 25^2$
- **8, 15, 17**: $8^2 + 15^2 = 17^2$

</td>
<td>

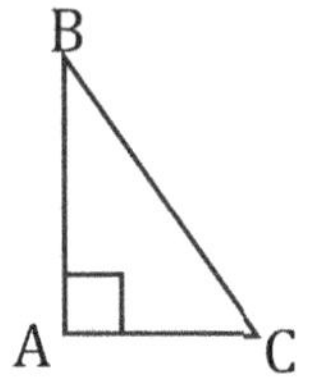

Another way of looking at Pythagoras' theorem:

The area of the square constructed on BC as its side equals the sum of areas of the squares constructed on AB and AC as the sides.

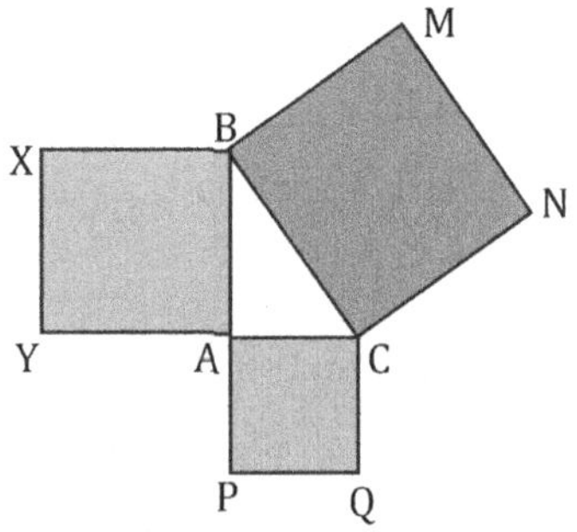

Thus, we have:

Area BMNC
= Area ACQP + Area ABXY

</td>
</tr>
<tr>
<td>

9. If ALL the angles of a triangle are less than 90°, it is called an **acute-angled** triangle.

Since ∠A, ∠B and ∠C are all acute, we have:

- $AB^2 + AC^2 > BC^2$
- $AB^2 + BC^2 > AC^2$
- $AC^2 + BC^2 > AB^2$

</td>
<td>

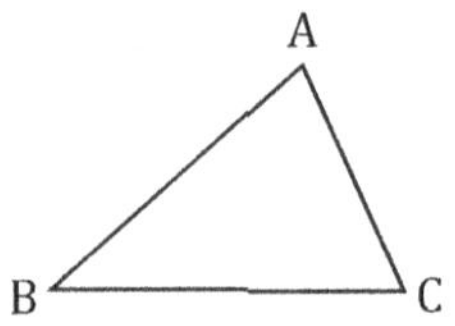

Here, ∠C < 90°, ∠A < 90° and ∠B < 90°

</td>
</tr>
<tr>
<td>

10. If ONE angle of a triangle is more than 90^0, it is called an **obtuse-angled** triangle.

Since C is obtuse, we have:

- $BC^2 + AC^2 < AB^2$

</td>
<td>

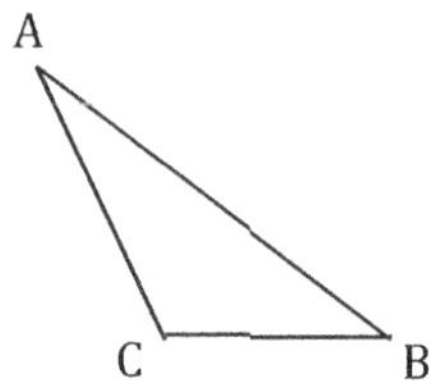

Here, ∠C > 90°, ∠A < 90° and ∠B < 90°

</td>
</tr>
</table>

11. **Area** of a triangle: • Area of $\triangle ABC = \frac{1}{2} \times$ Base$\times$ Height $= \frac{1}{2} \times BC \times AD$ $= \frac{1}{2} \times AC \times BE$ $= \frac{1}{2} \times AB \times CF$ Thus, from the above, we have: $BC \times AD = AC \times BE = AB \times CF$ Thus, we have: • $AD : BE : CF = \frac{1}{BC} : \frac{1}{AC} : \frac{1}{AB}$ Thus, the ratio of the altitudes equals the ratio of the reciprocals of the corresponding sides of a triangle.	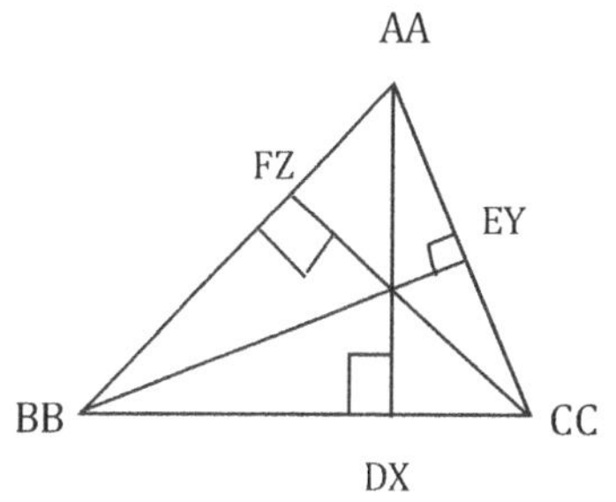
12. For triangles lying between the same set of parallel lines (l and m), the ratio of the areas of the triangles equals the ratio of the corresponding bases of the triangles. Since the triangles lie between the same set of parallel lines, their heights are the same, and hence, the ratio of their area equals the ratio of their bases.	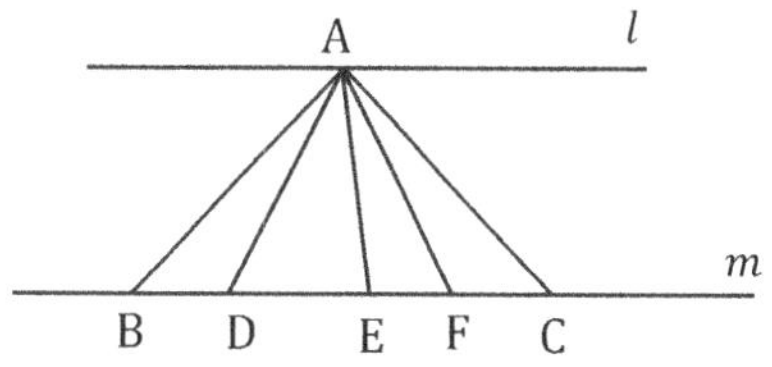 Here, lines from the vertex A are drawn to the base BC intersecting at D, E and F. Thus, triangles ABD, ADE, AEF and AFC lie between the same parallel lines (hence, they have the same height) => Area of $\triangle ABD$: Area of $\triangle ADE$: Area of $\triangle AEF$: Area of $\triangle AFC$ = BD : DE : EF : FC

Let us now look into relations between different triangles. There are two important areas to focus on: Congruency and Similarity. Let us discuss them one by one:

A. Congruent Triangles: Two triangles are congruent if they are identical to each other in all aspects. Thus, the corresponding sides and angles of the triangles are equal.

There are four ways to prove congruency:

(a) **SAS** (Side-Angle-Side)

Triangle ABC ≅ Triangle XYZ if AB = XY; ∠A = ∠X; AC = XZ (Here the angle is the included angle)

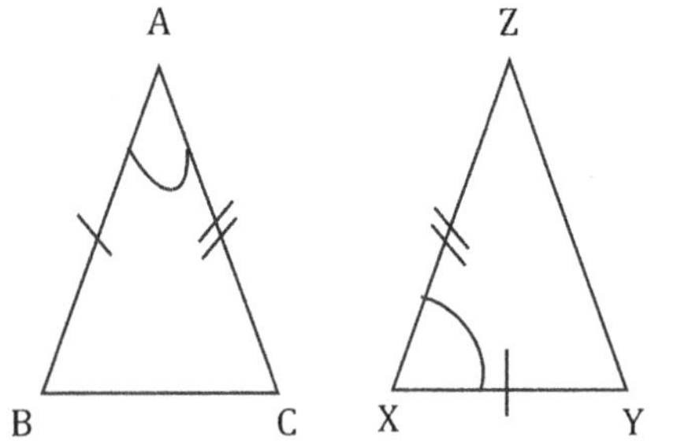

Note: Triangles MNP and DEF are NOT congruent:
Since: MN = DF; ∠N = ∠E; MP = DE
(∠N, ∠E are not the included angles)

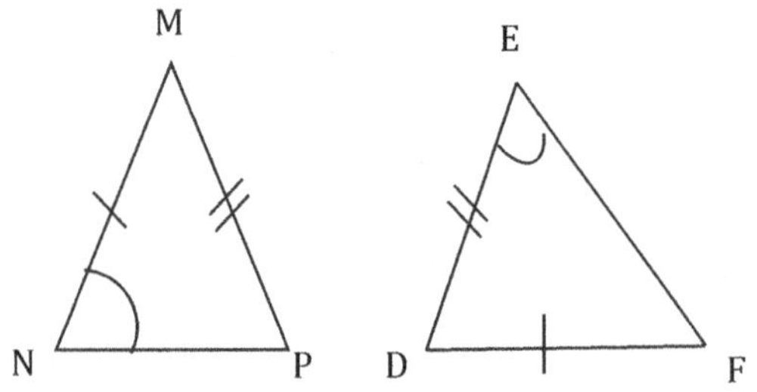

(b) **AAS** (Angle-Angle-Side)

Triangle ABC ≅ Triangle XYZ if ∠B = ∠Y; ∠C = ∠Z; AC = XZ.
Here, however, the side NEED NOT be the included side
(If two angles are equal, the third angle will also be equal)

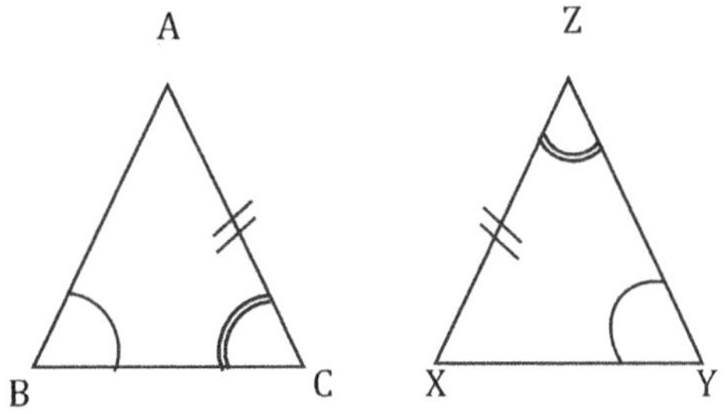

(c) **SSS** (Side-Side-Side)

Triangle ABC ≅ Triangle XYZ if AB = XY; AC = XZ; BC = YZ

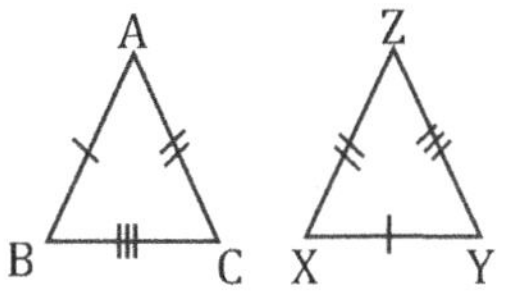

(d) **RHS** (Right angle-Hypotenuse-Side)

Triangle ABC ≅Triangle XYZ if $\angle B = \angle Y = 90°$;
AC = XZ (hypotenuse); AB = XY.

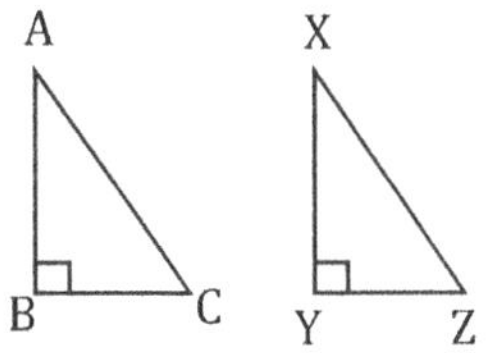

B. Similar Triangles: Similar triangles are similar in shape, need not have same size but are proportional i.e. the ratio of their sides is constant.

Two or more triangles can be called similar if:

(a) **AAA** (Angle-Angle-Angle)

Triangle ABC ~ Triangle XYZ if $\angle B = \angle Y$; $\angle C = \angle Z$
Note: Since $\angle A + \angle B + \angle C = \angle X + \angle Y + \angle Z = 180°$
$\Rightarrow \angle A = \angle X$

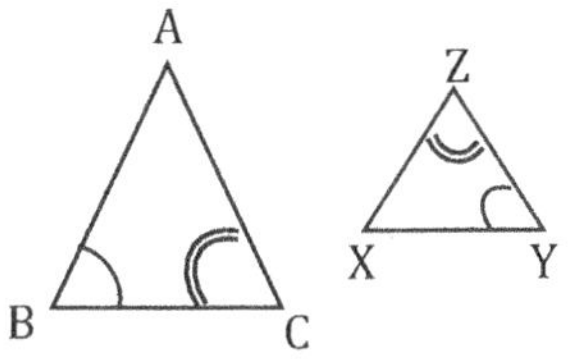

(b) **SAS** (Side ratio-Angle-Side ratio)

Triangle ABC ~ Triangle XYZ if $\angle A = \angle X$; $\frac{AB}{XY} = \frac{AC}{XZ}$
(Here the angle is the included angle)

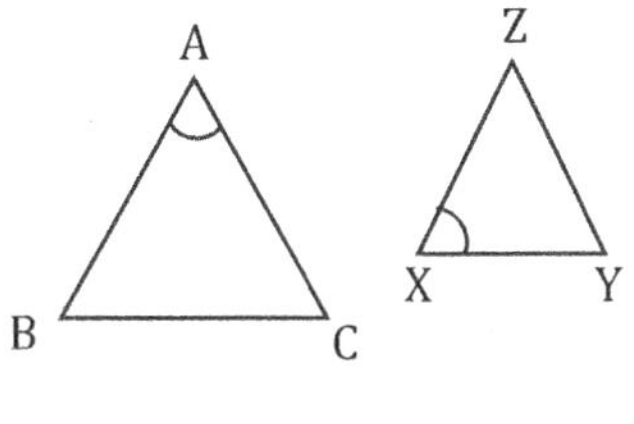

Let us look at three important results for similar triangles:

If two triangles are similar:

- Ratio of their corresponding sides equals the ratio of any corresponding length measure of the triangles; example: ratio of their altitudes or ratio of their medians, etc.
- Ratio of the areas of the triangles equals the SQUARE of the ratio of their corresponding sides:

  For any two similar triangles ABC and XYZ, we have:

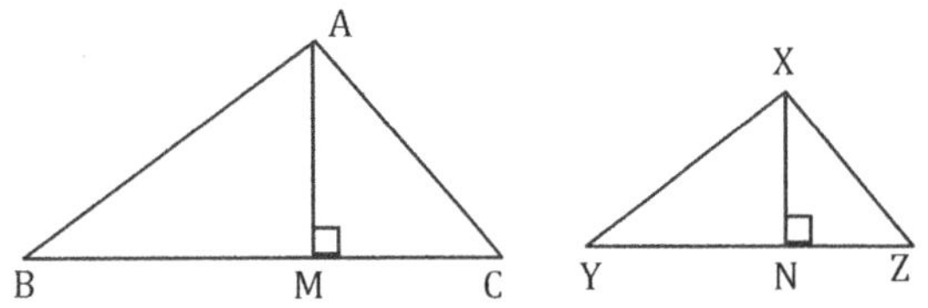

  Let the ratio of sides of the two triangles be k.

  Thus, we have:

$$\frac{AB}{XY} = \frac{BC}{YZ} = \frac{CA}{ZX} = \frac{AM}{XN} = k$$

$$\frac{(\text{Area of } \triangle \text{ ABC})}{(\text{Area of } \triangle \text{ XYZ})} = \left(\frac{\frac{1}{2} \times BC \times AM}{\frac{1}{2} \times YZ \times XN}\right) = \left(\frac{BC \times AM}{YZ \times XN}\right) = \left(\frac{BC}{YZ}\right) \times \left(\frac{AM}{XN}\right) = k \times k = k^2$$

- Ratio of their perimeters of the triangles equals the ratio of their corresponding sides.

A few important rules are now presented below. Effective use of these rules can reduce the time taken to solve a problem.

(1) **Basic Proportionality Theorem:** A line parallel to one side of a triangle divides the other two sides proportionally i.e. it creates two similar triangles, one inside the other, sharing a common vertex.

In triangle ABC, DE is drawn parallel to BC.

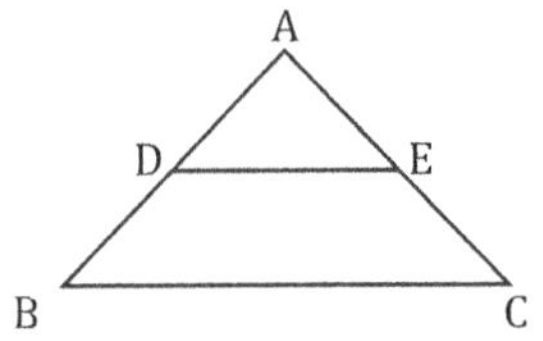

Thus, we have: $\triangle ABC \sim \triangle ADE$.

$$=> \frac{AD}{AB} = \frac{AE}{AC} => \frac{AD}{DB} = \frac{AE}{EC}$$

(2) **Mid-point Theorem:** The straight line joining the mid-points of two sides of a triangle is parallel to the third side and equal to half the third side.

(3) **Angle Bisector Theorem:** A bisector of any angle of a triangle will divide the opposite side in the same proportion as the two sides adjacent to the angle.

In $\triangle$ABC, AP is the angle bisector of $\angle$A. Thus, we have:

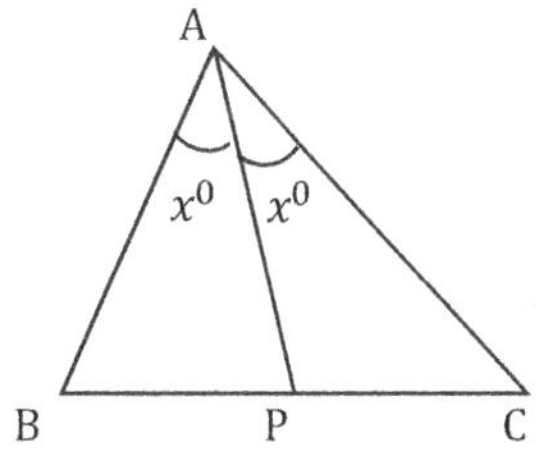

$$\frac{AB}{AC} = \frac{BP}{PC}$$

(4) **Right Triangles:**

- $30° - 60° - 90°$ Triangle: Basically, a $30° - 60° - 90°$ triangle is half of an equilateral triangle.

  It can be seen that half of triangle XBC is the $30° - 60° - 90°$ triangle ABC.

  The ratio of the sides of a $30° - 60° - 90°$ triangle is shown:

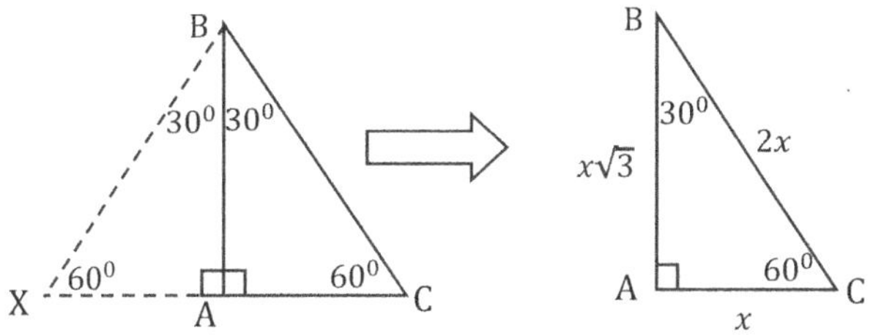

- $45° - 45° - 90°$ Triangle: Basically, a $45° - 45° - 90°$ triangle is half of a square.
  It can be seen that half of square ABXC is the $45° - 45° - 90°$ triangle ABC.

  The ratio of the sides of a $45° - 45° - 90°$ triangle is shown:

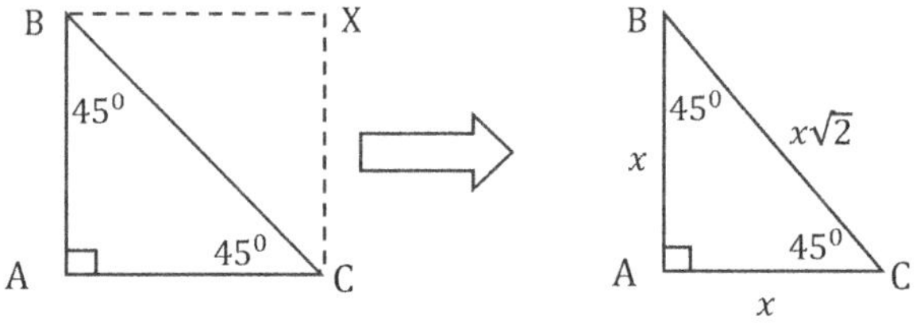

15.1.4 Polygons

A polygon is a closed figure bounded by line segments. Some common names of polygons:

A polygon with:

3 sides = **triangle**	6 sides = **hexagon**	9 sides = **nonagon**
4 sides = **quadrilateral**	7 sides = **heptagon**	10 sides = **decagon**
5 sides = **pentagon**	8 sides = **octagon**	Infinite sides = **circle**

Nomenclature:

- A polygon with all sides equal AND all angles equal is called a **regular polygon.**
- A polygon with no interior angle greater than 180° is called a **convex polygon.**
- A polygon with at least one interior angle greater than 180° is called a **concave polygon.**

Let us look at the important points associated with polygons: If the number of sides of a polygon be n, we have:

<table>
<tr><td>1. Sum of all interior angles
$= (n - 2) \times 180^\circ$</td><td>Any polygon of n sides can be broken down to $(n - 2)$ non-overlapping triangles as shown:
<table>
<tr><td>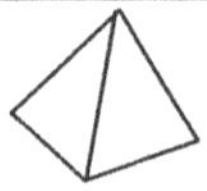</td><td>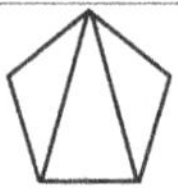</td><td>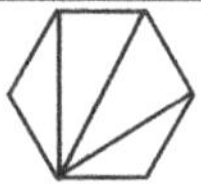</td></tr>
<tr><td>$n = 4$
of △s
$= 4 - 2 = 2$</td><td>$n = 5$
of △s
$= 5 - 2 = 3$</td><td>$n = 6$
of △s
$= 6 - 2 = 4$</td></tr>
</table>
Thus, sum of interior angles is simply the sum of the angles of all $(n - 2)$ triangles

$= (n - 2) \times 180^\circ$</td></tr>
<tr><td>2. Sum of all exterior angles
$= 360^\circ$</td><td>Sum of all exterior angles

= (Sum of all interior & exterior angles) - (Sum of interior angles)

$= n \times 180^\circ - (n - 2) \times 180^\circ$

$= 360^\circ$</td></tr>
<tr><td>3. Number of diagonals
$= \frac{n(n-3)}{2}$</td><td>Number of diagonals

= (# of ways in which any two vertices of the polygon can be joined) - (# of sides)

$= C_2^n - n = \frac{n(n-1)}{2} - n$

$= \frac{n(n-3)}{2}$</td></tr>
</table>

4. Each interior angle of a regular polygon $= \frac{(n-2) \times 180^\circ}{n}$	Each interior angle of a regular polygon $= \frac{\text{Sum of all interior angles}}{\text{\# of angles}}$ $= \frac{(n-2) \times 180^\circ}{n}$
5. Each exterior angle of a regular polygon $= \frac{360^\circ}{n}$	Each exterior angle of a regular polygon $= \frac{\text{Sum of all exterior angles}}{\text{\# of angles}}$ $= \frac{360^\circ}{n}$
6. Area of a regular polygon $= \frac{1}{2} \times (\text{perimeter}) \times (\perp \text{ from centre})$	Area of a regular polygon is simply the sum of the areas of n triangles formed by joining the vertices to the centre of the polygon as shown: 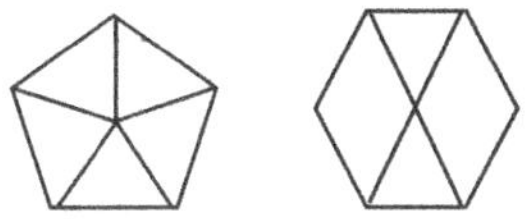 Thus, area of the polygon $= n \times \frac{1}{2} \times (\text{Side}) \times (\perp \text{ from centre})$ $= \frac{1}{2} \times (n \times \text{ Side}) \times (\perp \text{ from centre})$ $= \frac{1}{2} \times (\text{perimeter}) \times (\perp \text{ from centre})$

Two special polygons:

(1) **Regular Hexagon**: A regular hexagon is a polygon of six sides.

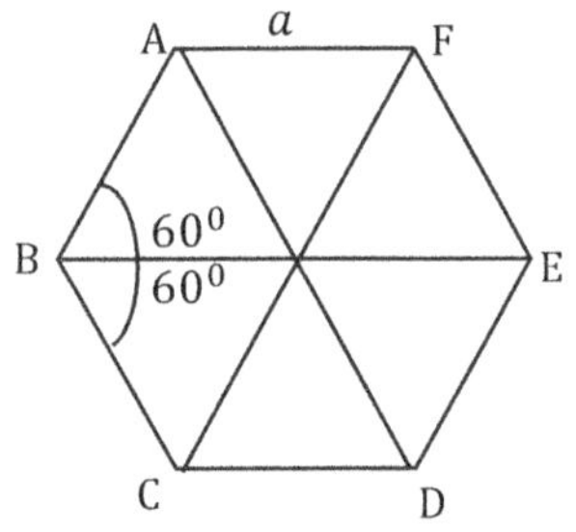

- All the six sides are of equal length
- Each interior angle is 120°
- Sum of Interior angles = (6 − 2) × 180° = 4 × 180° = 720°
- It can be broken into six equilateral triangles

Thus, if each side of the regular hexagon is a, we have:

- **Perimeter = 6*a***
- **Area** = $\frac{3\sqrt{3}}{2}a^2$

(2) **Regular Octagon**: A regular octagon is a polygon of eight sides.

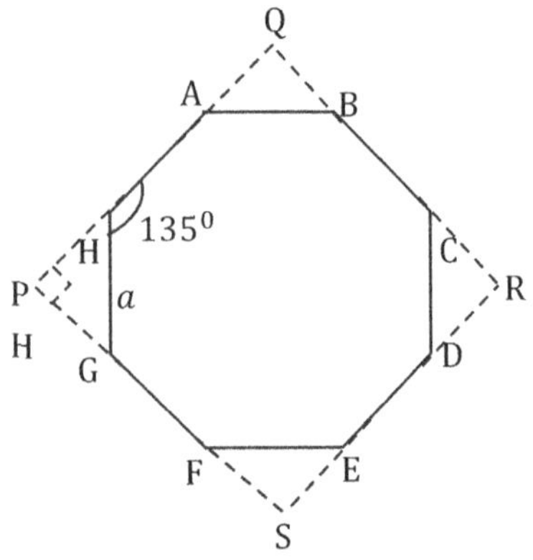

- All the eight sides are of equal length.
- Each interior angle is 135°
- Sum of Interior angles = (8 − 2) × 180° = 6 × 180° = 1080°

To find the area, the dotted lines are constructed.

△PHG is right-angled isosceles (45° − 45° − 90°).

Since GH = a, PH = PG = $\frac{a}{\sqrt{2}}$.

Thus, QA = PH = $\frac{a}{\sqrt{2}}$. Hence, QP = $a + \frac{2a}{\sqrt{2}} = a\left(1 + \sqrt{2}\right)$.

Area of octagon = Area of square PQRS - Area of 4 corner $\triangle$s

$$= PQ^2 - 4 \times \frac{1}{2} \times PH \times PG = \left(a\left(1 + \sqrt{2}\right)\right)^2 - 4 \times \frac{1}{2} \times \left(\frac{a}{\sqrt{2}}\right)^2 = 2\left(1 + \sqrt{2}\right)a^2$$

Thus, if each side of the regular hexagon is a, we have:

- **Perimeter = 8a**
- **Area = $2\left(1 + \sqrt{2}\right)a^2$**

15.1.5 Quadrilateral

A quadrilateral has four sides and sum of all four angles is 360°.

Let us look at the properties of the main quadrilaterals:

(1) **Parallelogram:** The properties are given below:

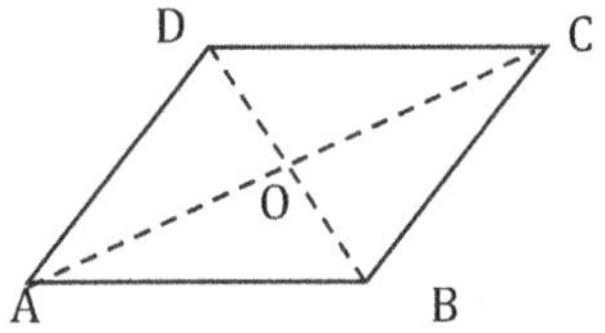

- Opposite sides are parallel and equal: AB || CD, AB = CD, AD || BC, AD = BC
- Opposite angles are equal: $\angle A = \angle C$, $\angle B = \angle D$
- Diagonals bisect each other: AO = OC, BO = DO
- Sum of any two adjacent angles = 180°: $\angle A + \angle D = 180°$, $\angle C + \angle B = 180°$
- Straight lines joining the midpoints of adjacent sides of any quadrilateral form a parallelogram
- Area = Base × Height

(2) **Rectangle:** All the above properties of a parallelogram hold true for a rectangle. However, there are few additional properties:

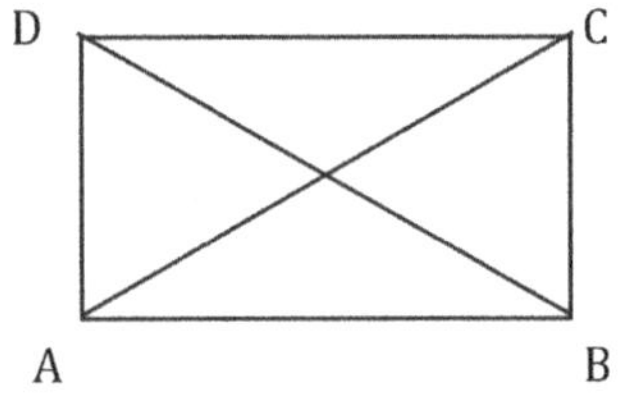

- All angles are equal and equal to a right angle: $\angle A = \angle B = \angle C = \angle D = 90°$
- Diagonals are equal: AC = BD
- Perimeter = 2 × (Length + Breadth)
- Area = Length × Width
- Length of Diagonal = $\sqrt{\text{Length}^2 + \text{Breadth}^2}$

(3) **Rhombus:** All the properties of a parallelogram hold true for a rhombus. However, there are some additional properties:

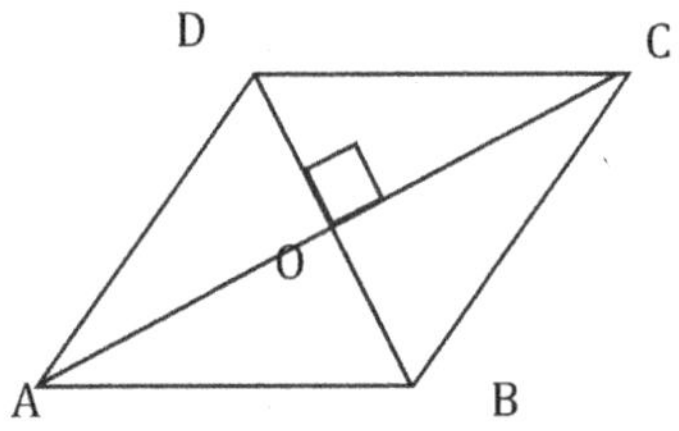

- All sides are equal: AB = BC = CD = DA
- The diagonals bisect each other at right angles: $\angle DOC = 90^\circ$
- The diagonals bisect the angles at the vertex: $\angle BAC = \angle DAC$, etc.
- Area = $\frac{1}{2} \times$ (Product of the diagonals) $= \frac{1}{2} \times AC \times BD$

Note: Area of any quadrilateral having diagonals perpendicular to one another can be calculated as $\frac{1}{2} \times$ (Product of the diagonals)

(4) **Square:** All the properties mentioned above for all the figures hold true for a square.

Thus, we have:

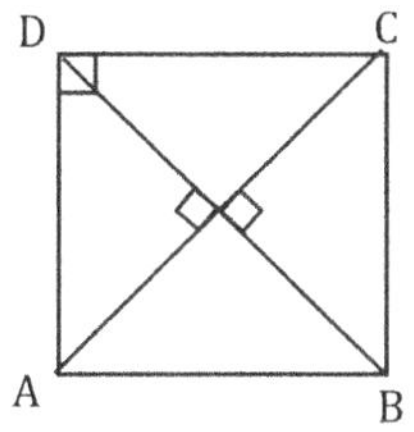

- All sides are equal: AB = BC = CD = DA
- All angles are right angles.
- The diagonals are equal and bisect each other at right angles.
- Perimeter of a square = $4a$, where a is the length of a side.
- Length of a diagonal = $a\sqrt{2}$
- Area = $(\text{Side})^2 = \frac{1}{2} \times (\text{Diagonal})^2 = a^2$

(5) **Trapezium:** A trapezium has only one pair of opposite sides parallel.

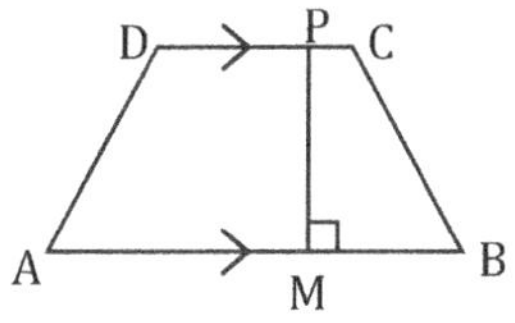

- Area = $\frac{1}{2} \times$ (Sum of || sides) $\times$ (Distance between || sides) $= \frac{1}{2} \times (AB + CD) \times MP$

15.1.6 Circles

Let us look at the important points associated with circles:

1. Tangent is perpendicular to the radius i.e. $AB \perp OA$ 2. Perpendicular from the centre bisects the chord i.e. OZ bisects XY 3. Similarly, if Z is the mid-point of XY, then OZ is perpendicular to XY	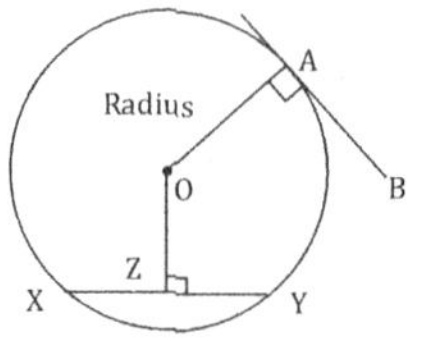 O: Centre of the circle OA: Radius AB: Tangent XY: Chord
4. Tangent segments drawn to a circle from an external point are equal i.e. PR = PQ	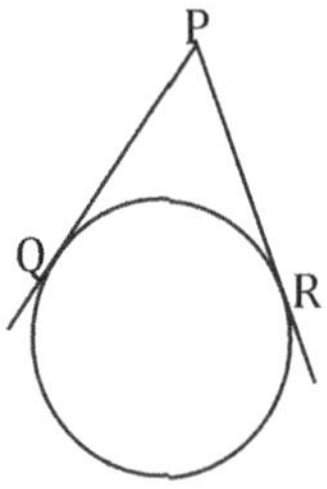
5. Angle subtended at the centre by an arc (QR) is equal to twice the angle made by the same arc at any other point on the remaining circumference $=> \angle QOR = 2 \times \angle QPR$	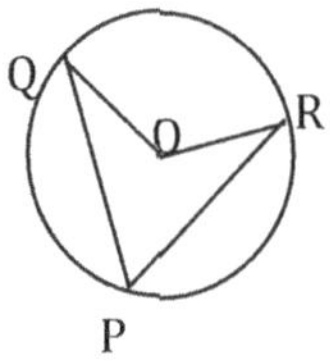
6. If two chords intersect: $AE \times BE = CE \times DE$	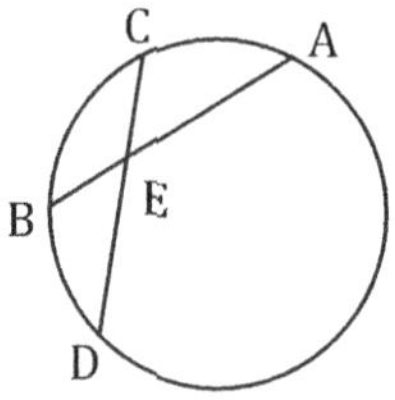

7. Area of a circle of radius $r = \pi r^2$ 8. Circumference of the above circle $= 2\pi r$ 9. Length of the arc subtending angle θ at the centre of a circle of radius r $= \left(\frac{\theta}{360}\right) \times 2\pi r$ 10. Area of the sector formed by the above arc $= \left(\frac{\theta}{360}\right) \times \pi r^2$ $= \frac{1}{2} \times r \times \left\{\left(\frac{\theta}{360}\right) \times 2\pi r\right\}$ $= \frac{1}{2} \times r \times$ (Length of arc) 11. Perimeter of the sector above = (Length of the arc) + (2 × Radius) $= \left(\frac{\theta}{360}\right) \times 2\pi r + 2r$	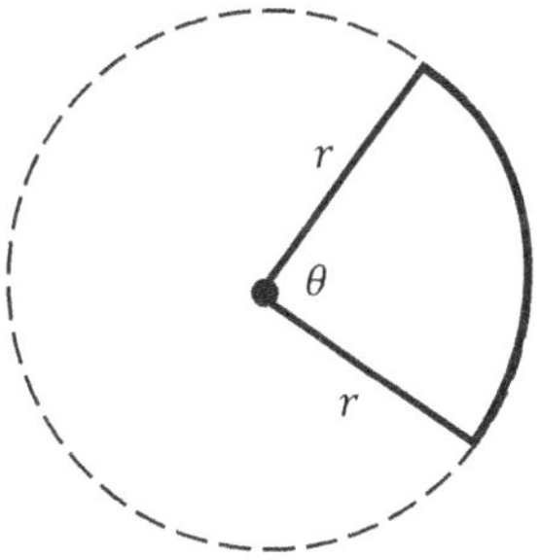

15.1.7 3-D Geometry

We will consider different solid shapes in this subsection.

Few important 3-D figures:

(1) **Cuboid:** A cuboid is a prism having six rectangular faces.

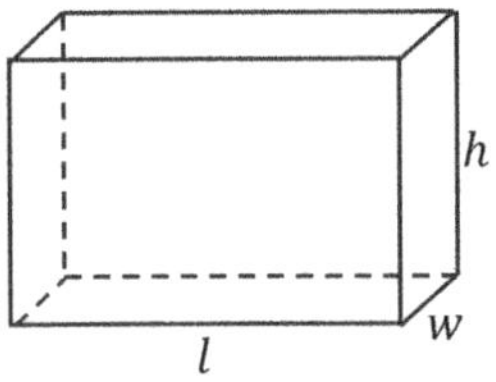

- **Volume** $= l \times w \times h$, where l = length, w = width, h = height

  $= \sqrt{\text{Product of areas of the mutually perpendicular faces}}$

  $= \sqrt{(lw)\,(wh)\,(lh)} = lwh$

- **Area of four walls** $= 2\,(l + w) \times h$
- **Total surface area** $= 2\,(lw + wh + lh)$
- **Body Diagonal** $= \sqrt{(l^2 + w^2 + h^2)}$

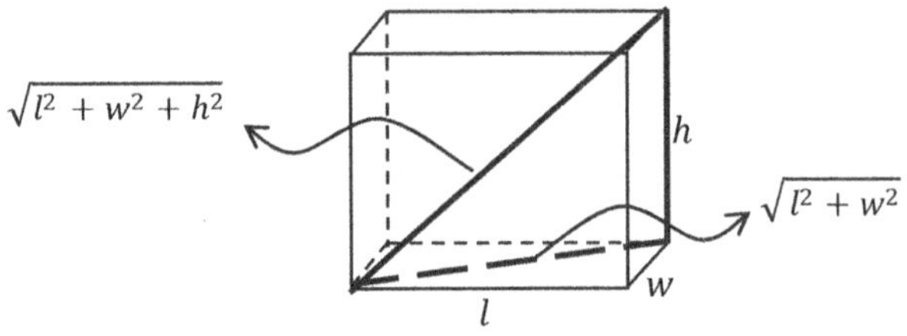

The 'Body Diagonal' is the longest line joining any two vertices of the cuboid.

(2) **Cube:** A prism whose every face is a square is called a cube.

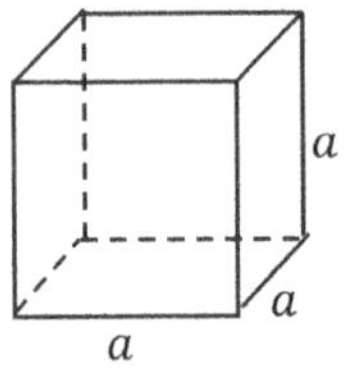

- **Volume** $= a^3$,where a = edge of the cube
- **Total surface area of the cube** = Area of 6 squares of area $a^2 = 6a^2$
- **Body Diagonal of a cube** $= a\sqrt{3}$

(3) **Cylinder:** If a rectangle is revolved about one of its sides as the axis, the solid thus formed is a right circular cylinder.

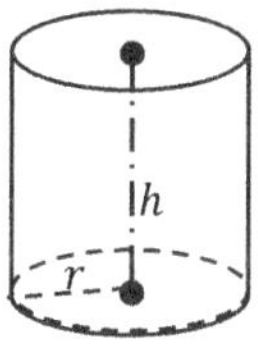

- **Volume** = $\pi r^2 h$, where r = radius of base, h = height
- **Curved surface area** = $2\pi r h$
- **Total surface area** = $2\pi r\ (r + h)$
- **Longest distance between two points** = $\sqrt{(2r)^2 + h^2}$

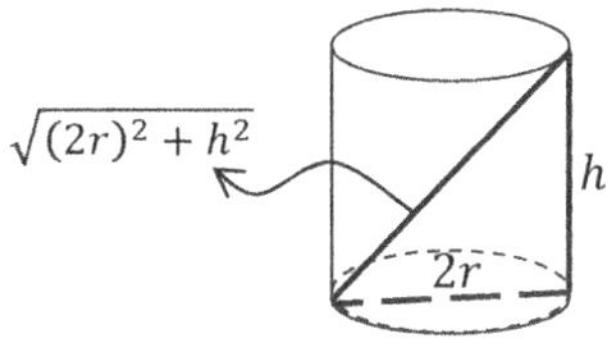

(4) **Cone:** If a right triangle is revolved about one of its sides containing the right angle as the axis the solid formed is called a cone.

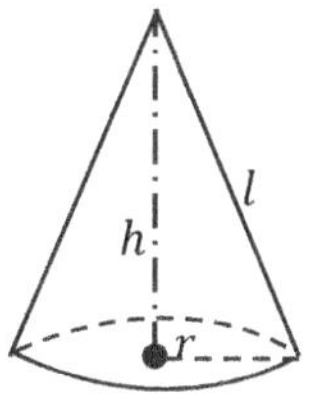

- **Volume** = $\frac{1}{3}\pi r^2 h$, where r = radius of base, h = height
- **Curved surface area** = $\pi r l$ where l = slant height = $\sqrt{(r^2 + h^2)}$
- **Total surface area** = $\pi r(r + l)$

(5) **Sphere:** When a circle is revolved about its diameter, the solid thus formed is called a sphere.

- **Volume** = $\frac{4}{3}\pi r^3$, where r = radius of the sphere
- **Surface area** = $4\pi r^2$

Note that the volume of a cone is proportional to $r^2 h$ and that for the sphere is proportional to r^3.

15.2 Trigonometry

The word "Trigonometry" comes from the Greek "trigonon" (meaning triangle) and "metron" (meaning measure). So, simply put, Trigonometry is the study of the measures of triangles. This includes the lengths of the sides, the measures of the angles and the relationships between the sides and angles.

Radians and Degrees

Angles in Trigonometry can be measured in either radians or degrees. There are 360 degrees (360°) in one rotation around a circle. Although there are various accounts of how a circle came to have 360 degrees, most of these are based on the fact that earlier, a complete year was considered to have 360 days.

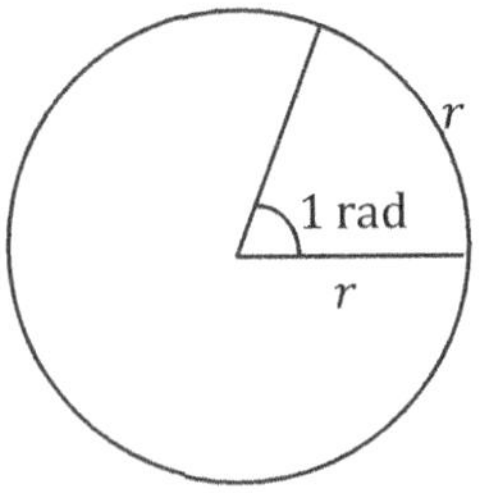

There are 2π (= 6.28) radians in one rotation around a circle. One radian is the measure of the angle made from wrapping the radius of a circle along the circle's circumference.

Measure of an Arc

One of the simplest and most basic formulas in Trigonometry provides the measure of an arc in terms of the radius of the circle, r, and the arc's central angle θ, expressed in radians.

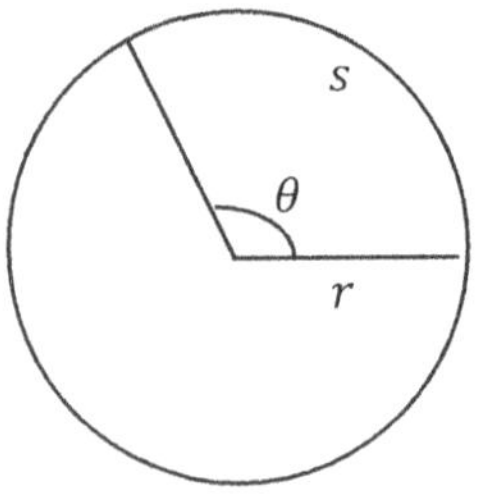

Since there are 2π radians in one full rotation around the circle, the measure of an arc, s, with central angle θ, expressed in radians, is:

$$s = 2\pi r \times \left(\frac{\theta}{2\pi}\right) = r\theta$$

If the angle θ were measured in degree, we would have:

$$s = 2\pi r \times \left(\frac{\theta^\circ}{360^\circ}\right)$$

Triangle ratios:

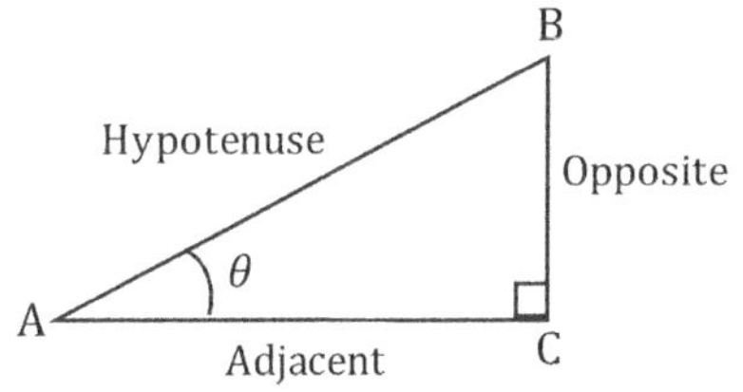

$\sin\theta = \dfrac{\text{Opposite}}{\text{Hypotenuse}} = \dfrac{BC}{AB}$; $\text{cosec}\,\theta = \dfrac{1}{\sin\theta}$

$\cos\theta = \dfrac{\text{Adjacent}}{\text{Hypotenuse}} = \dfrac{AC}{AB}$; $\sec\theta = \dfrac{1}{\cos\theta}$

$\tan\theta = \dfrac{\text{Opposite}}{\text{Adjacent}} = \dfrac{BC}{AC} = \dfrac{\sin\theta}{\cos\theta}$; $\tan\theta = \dfrac{1}{\cot\theta}$

Key Angles conversion:

- $0° = 0$ radians
- $30° = \dfrac{\pi}{6}$ radians
- $45° = \dfrac{\pi}{4}$ radians
- $60° = \dfrac{\pi}{3}$ radians
- $90° = \dfrac{\pi}{2}$ radians
- $180° = \pi$ radians

Height and Distance

Angle of Elevation: If point X is observed from point A, and X is at a higher level than A, then $\angle\theta$ is the angle of elevation.

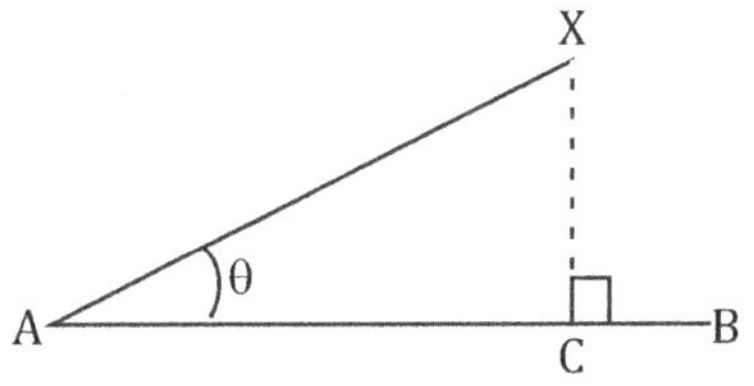

Angle of Depression: If point X be observed from point A, and X is at a lower level than A, then $\angle\theta$ is the angle of depression.

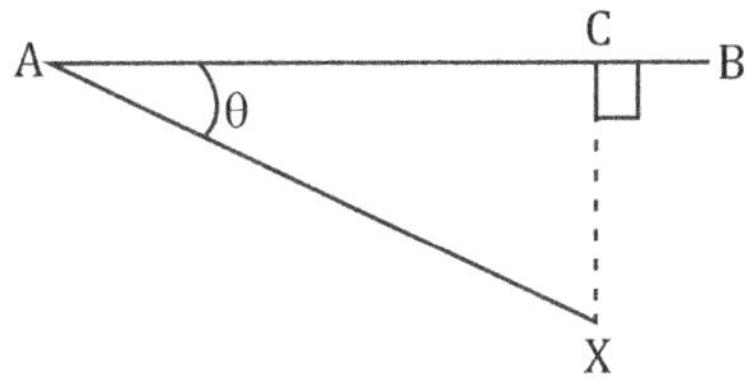

Let us look at a couple of examples:

(1) The length of the shadow of a pole increases by 1 foot when the altitude of the Sun changes from 45° to 30°. What is the height, in feet, of the pole?

Explanation:

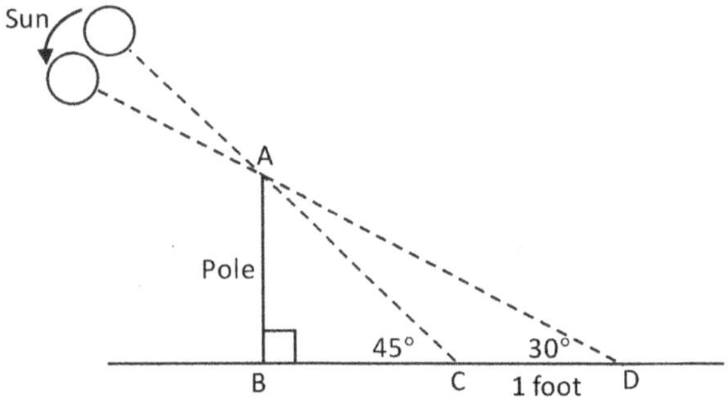

In the diagram, two positions of the Sun are shown - at 30° and at 45° altitudes. The lengths of the shadows of the pole AB at the two positions are BD and BC, respectively.

Let the length of the pole AB be l feet.

We have:

$$\tan \angle ACB = \tan 45^\circ = 1 = \frac{l}{BC}$$

$$=> BC = l \ldots \text{ (i)}$$

$$\tan \angle ADB = \tan 30^\circ = \frac{1}{\sqrt{3}} = \frac{l}{BD}$$

$$=> BD = l\sqrt{3} \ldots \text{ (ii)}$$

We know that: BD − BC = 1 foot

$$=> l\sqrt{3} - l = 1$$

$$=> l = \frac{1}{\sqrt{3}-1} \text{ feet} = 1.37 \text{ feet}$$

The correct answer is 1.37.

(2) A man approaching a building sees the angle of elevation of the top of the building change from 30° to 45° in 10 minutes. If the height of the building is 20 feet, what is the speed, in feet per minute, of the man?

Explanation:

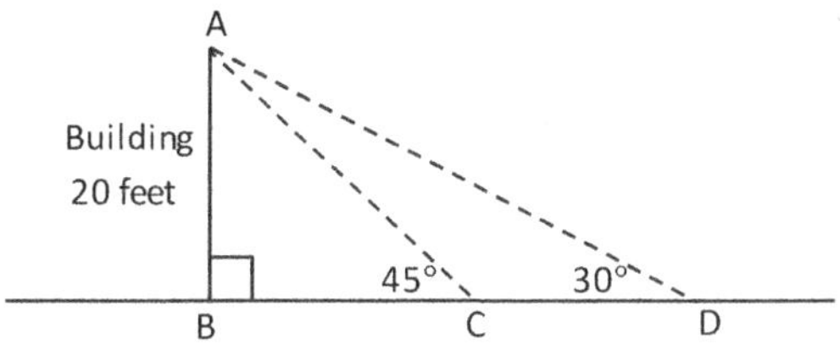

Let the distance the man moved CD be d feet.

We have:

$$\tan \angle ADB = \tan 30^\circ = \frac{1}{\sqrt{3}} = \frac{20}{BD}$$

$$=> BD = 20\sqrt{3} \ldots \text{ (i)}$$

$$\tan \angle ACB = \tan 45^\circ = 1 = \frac{20}{BC}$$

$$=> BC = 20 \ldots \text{ (ii)}$$

Thus, required distance

$$CD = BD - BC$$

$$=> 20\sqrt{3} - 20$$

$$= 20 \times 1.73 - 20$$

$$=> 34.6 - 20 = 14.6 \text{ feet}$$

The correct answer is 14.6.

Chapter 16

Questions on concepts

Many following questions may be out of scope from the test point of view, but you must understand them to develop your fundamentals.

16.1 Lines and Angles

(1) In the figure given below, line AB is parallel to line CG and $\angle FEC = 123°$. If BE = CE and $\angle GCD = 75°$, find $\angle ABF$.

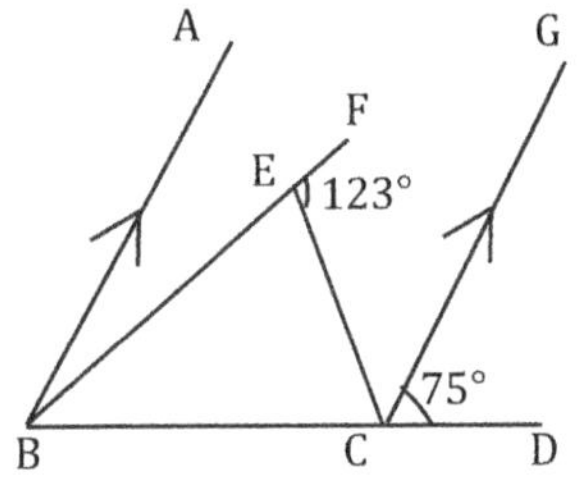

In the problem, we see a pair of parallel lines AB and CG.

Also, we see an external angle of triangle BEC, which is isosceles
(given that BE = CE)

Thus, the main three things we should focus on are:

- Are there any corresponding or alternate angles?
- Which angles of the isosceles triangle are equal?
- External angle of a triangle equals the sum of opposite interior angles

Let us now solve the problem.

We know that:

AB || CG
=> $\angle GCD = \angle ABC = 75°$ (Corresponding angles)

Now, BE = CE
=> $\angle EBC = \angle ECB$ (Angles opposites to equal sides would be equal)

Also, $\angle FEC = \angle CBE + \angle ECB$
(External angle of a triangle equals the sum of opposite interior angles)

=> $123° = 2\angle EBC$
=> $\angle EBC = 61.5°$
=> $\angle ABF = 75° - 61.5° = 13.5°$

(2) In the figure, if PQ = PR = RS, ST = RT and ∠STU = 120°, what is the value of x?

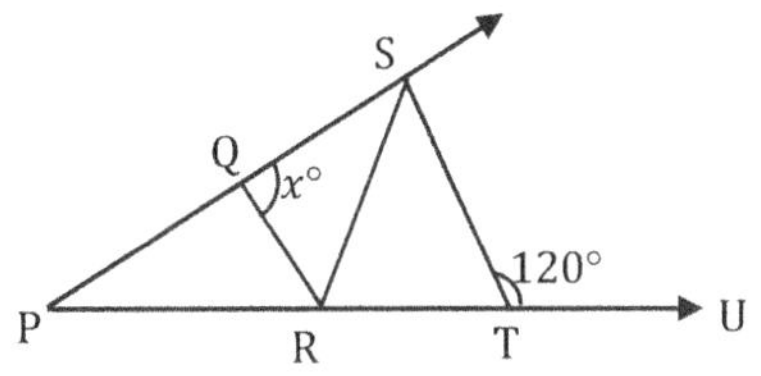

In the problem, we see a number of lines equal.

Thus, the main things we should focus on are:

- Are there any isosceles triangles?
- Are there any equilateral triangles?

Let us now solve the problem.

$\angle STR = 180^\circ - 120^\circ = 60^\circ$

Since ST = RT, triangle TSR is isosceles

$=> \angle TSR = \angle TRS = \frac{(180^\circ - 60^\circ)}{2} = 60^\circ$

Note: Triangle STR is thus equilateral

$\angle SRP = 180^\circ - 60^\circ = 120^\circ$

Since RS = PR, triangle SRP is isosceles

$=> \angle RPS = \angle RSP = \frac{(180^\circ - 120^\circ)}{2} = 30^\circ$

Again, since PQ = PR, triangle QRP is isosceles

$=> \angle RQP = \angle QRP = \frac{(180^\circ - 30^\circ)}{2} = 75^\circ$

Since x is the external angle of triangle PQR:

$=> x = \angle RPQ + \angle QRP = 30^\circ + 75^\circ = 105^\circ$

(3) In the given figure, AB is parallel to CD and BC is parallel to DF, if ∠ABC = 45° and EDF = 40°, then find the measure of CDE.

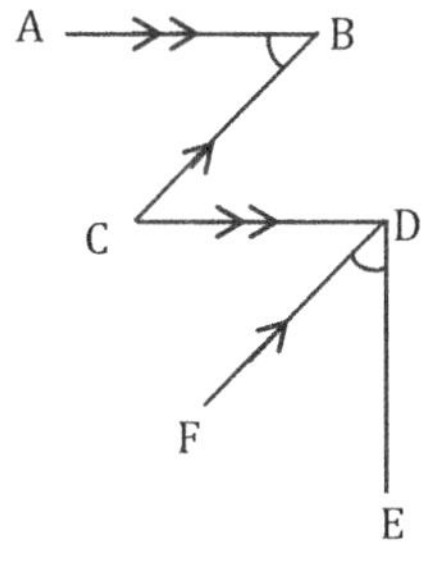

In the problem, we see two pairs of parallel lines.

Thus, the main thing we should focus on is:

- Are there any corresponding or alternate angles?

Let us now solve the problem.

It is given that: $\angle ABC = 45^\circ$

=> $\angle BCD = 45^\circ$ (Alternate angles, AB parallel to CD)

=> $\angle CDF = 45^\circ$ (Alternate angles, BC parallel to FD)

It is given that: $\angle EDF = 40^\circ$

=> $\angle CDE = 45^\circ + 40^\circ = 85^\circ$

16.2 Triangles

(4) If ray CX bisects ∠ACB, which of the following is true?

A) BC = BX B) BC < BX
C) BC > BX D) AX = BX

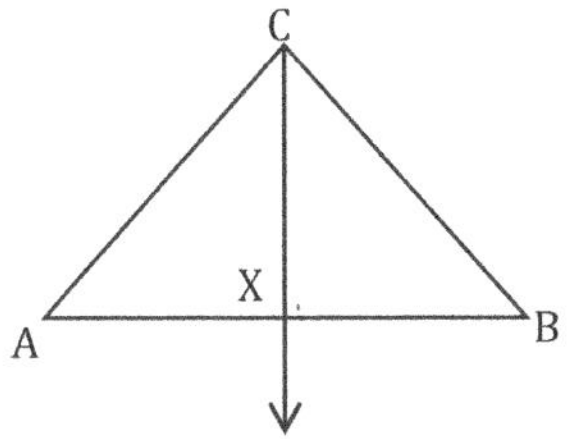

In the problem, we see that there is an angle bisector.

Also, we need to choose an option which states an inequality or equality.

Thus, the main three things we should focus on are:

- Is there any isosceles triangle which we can use to prove the equality?
- Since there is an inequality in the options, we should focus on the inequality regarding the sides of a triangle
- Can we compare the angles of a triangle and from there derive the inequality of the sides?

Let us now solve the problem.

∠CXB is the exterior angle of ΔACX

=> ∠ACX = ∠BCX

Hence, ∠CXB = ∠ACX + ∠CAX

=> ∠CXB > ∠ACX

=> ∠CXB > ∠BCX

=> BC > BX
(Since the largest side of a triangle is opposite to the largest angle of the triangle)

Hence, option C is the correct answer.

(5) If $AX_1 = X_1Y_1 = Y_1Z_1 = Z_1W_1 = W_1B$, $AB = 9$ and area of ΔABC is 150, what is the area of $Z_1W_1W_2Z_2$?

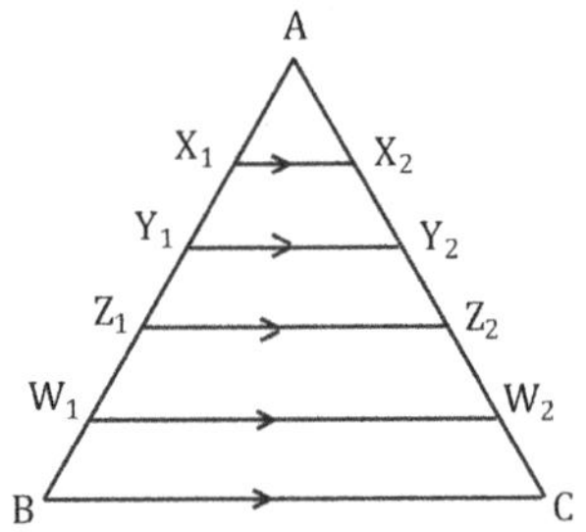

In the problem, we see that there are parallel lines drawn in a triangle.

Also, we need to find the area of a region in the triangle.

Thus, the main things we should focus on are:

- Which are the similar triangles?
- What is the ratio of the sides of the similar triangles?
- Area ratio equals the square of side ratio for similar triangles.

Let us now solve the problem.

$AX_1 = X_1Y_1 = Y_1Z_1 = Z_1W_1 = W_1B = \frac{1}{5}(9) = 1.8$

Area of $Z_1W_1W_2Z_2$ = (Area of AW_1W_2) − (Area of AZ_1Z_2)

Since X_1X_2 parallel to Y_1Y_2 parallel to Z_1Z_2 parallel to W_1W_2 parallel to BC, we have:

Triangles AZ_1Z_2 and ABC are similar

$=> \frac{AZ_1}{AB} = \frac{AZ_2}{AC} = \frac{3}{5}$

$=> \frac{(\text{Area of triangle } AZ_1Z_2)}{(\text{Area of triangle ABC})} = \left(\frac{3}{5}\right)^2 = \frac{9}{25}$

$=>$ Area of $AZ_1Z_2 = \frac{9}{25} \times 150 = 54$

Similarly, triangles AW_1W_2 and ABC are similar

$=> \frac{AW_1}{AB} = \frac{AW_2}{AC} = \frac{4}{5}$

$=> \frac{(\text{Area of triangle } AW_1W_2)}{(\text{Area of triangle ABC})} = \left(\frac{4}{5}\right)^2 = \frac{16}{25}$

$=>$ Area of $AW_1W_2 = \frac{16}{25} \times 150 = 96$

$=>$ Area of $Z_1W_1W_2Z_2 = 96 - 54 = 42$

(6) In the figure, BD =12. Find the length of the angle bisector of ∠A.

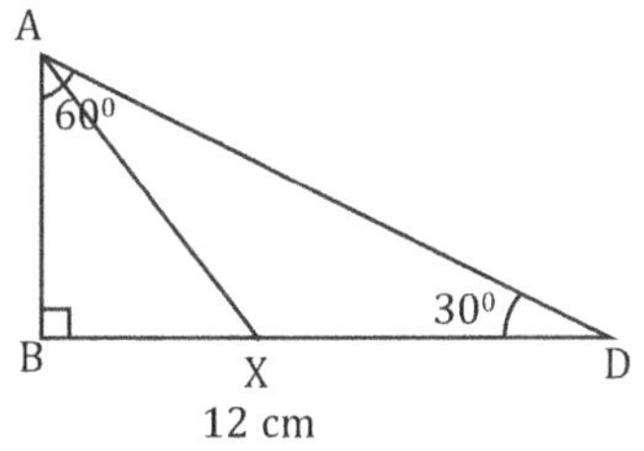

In the figure, we see that there is a 30-60-90 triangle.

Also, we need to find the length of a side in the triangle.

Thus, the main thing we should focus on is:

- Relation between the sides of a 30-60-90 triangle.

Let us now solve the problem.

In the 30-90-60 triangle ABD:

$AB = \frac{BD}{\sqrt{3}} = \frac{12}{\sqrt{3}} = 4\sqrt{3}$

Since AX is an angle bisector, $\angle BAX = \frac{60^\circ}{2} = 30^\circ$

$=> \angle AXB = 60^\circ$

Thus, ABX is a 30-60-90 triangle

$=> BX = \frac{4\sqrt{3}}{\sqrt{3}} = 4$

$=> AX = 2 \times BX = 8$

Note: If you observed carefully, there is a 30-60-90 triangle inside another 30-60-90 triangle!

16.3 Quadrilaterals

(7) ABCD is a the square of side 1 cm. Equilateral triangles AQD and BPC are drawn inside the square. Find the length of PQ.

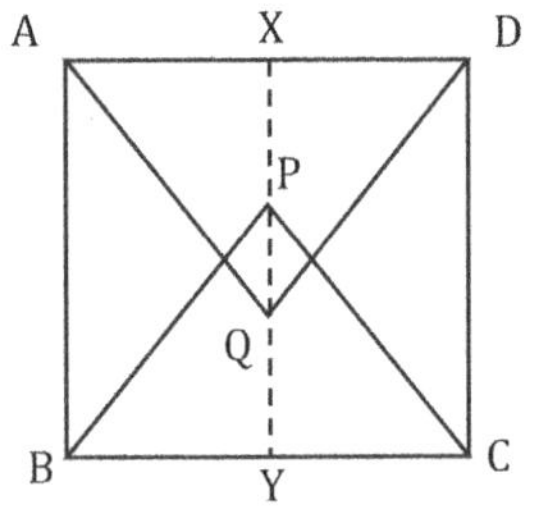

In the problem, we see information regarding a the square and two equilateral triangles.

Also, we need to find the length of a line.

Thus, the main things we should focus on are:

- Relation for the height of an equilateral triangle and its sides.
- How to relate the equilateral triangle to the square?

Let us now solve the problem.

We can see from the diagram that: QX + PY − PQ = XY = AB

=> Height of $\triangle$AQD + Height of $\triangle$BPC − PQ = Side of the square ABCD

=> PQ = 2 × (Height of one equilateral triangle) − (Side of square)

$= 2\left(\frac{\sqrt{3}}{2}\right)(1) - 1 = \sqrt{3} - 1$

Length of PQ = $\sqrt{3} - 1$

(8) In a quadrilateral ABDC as shown in the figure, diagonal AD is the perpendicular bisector of diagonal BC. If $\angle ABD = 105^\circ$, $\angle BDC = 60^\circ$ and BD = DC = 12, find the area of the quadrilateral ABDC.

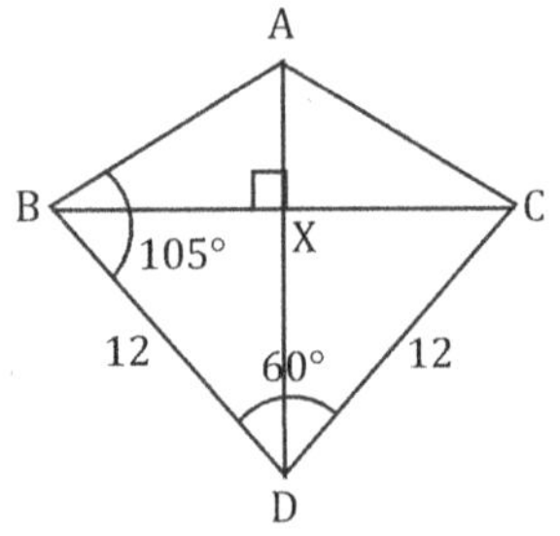

In the problem, we see that there is a 60° angle.

Also, we need to find the area of the quadrilateral.

Thus, the main things we should focus on are:

- Is there an equilateral triangle or a 30-60-90 triangle?
- Since the area of a quadrilateral cannot be directly determined, how do we break it into triangles whose areas can be easily determined?
- Since there is a 105° angle as well, and $105^\circ = 60^\circ + 45^\circ$, is there a 45-45-90 triangle too?

Let us now solve the problem.

Since BD = DC, and $\angle$BDC = 60°, ΔBDC is equilateral.

=> $\angle$ABC = $105^\circ - 60^\circ = 45^\circ$

Since AD is the perpendicular bisector of diagonal BC:

Triangles ABX and ACX are congruent
(AX is common, BX = CX, $\angle$AXB = $\angle$AXC = 90° => "SAS")

=> $\angle$ACB = $\angle$ABC = 45°

Hence, $\angle$BAC = $180^\circ - 45^\circ - 45^\circ = 90^\circ$

Thus, ABC is 90-45-45 triangle => AB = AC = $\frac{BC}{\sqrt{2}} = \frac{12}{\sqrt{2}} = 6\sqrt{2}$

Area ABDC = Area BDC + Area ABC

$= \frac{\sqrt{3}}{4}(12)^2 + \frac{1}{2}\left(6\sqrt{2}\right)\left(6\sqrt{2}\right)$

$= 36\sqrt{3} + 36$

Area of the quadrilateral ABDC = $36(\sqrt{3} + 1)$

(9) Two squares, with side lengths A and B ($B > A$), are placed together such that the right side of the square with side A touches the left side of the square with side B, and their bases are collinear. A diagonal is drawn from the bottom left corner of the square with side A to the top right corner of the square with side B. What is the area below the diagonal in the square with side A in terms of A and B?

A) $\frac{A^2B}{2(A+B)}$

B) $\frac{B}{2}(A+B)$

C) $\frac{A}{2}(A+B)$

D) $\frac{AB^2}{2(A+B)}$

In the problem, we see information regarding squares.
Also, we need to find the area of a triangle.

Thus, the main thing we should focus on is:

- How to determine the base and height of the triangle?

Let us now solve the problem.

Referring to the diagram given below,

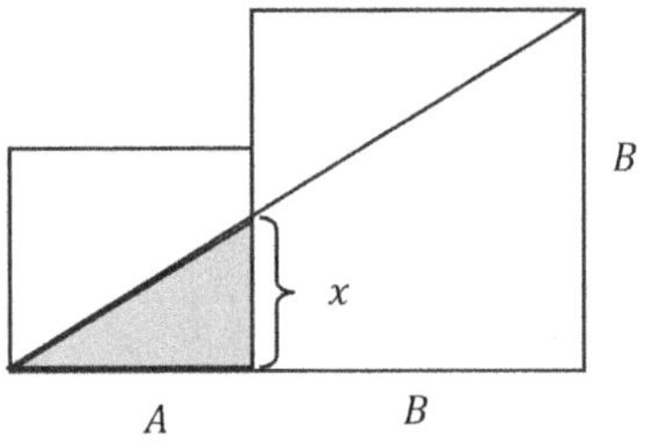

From similar triangles, we have:

$\frac{x}{B} = \frac{A}{A+B} => x = \frac{AB}{A+B}$

Required area $= \frac{1}{2}A \times x = \frac{1}{2}A\left(\frac{AB}{A+B}\right) = \frac{A^2B}{2\,(A+B)}$

Hence, option A is the correct answer.

(10) In the figure given below, if the perimeter of ΔABC is x, then what is the perimeter of the regular hexagon in terms of x?

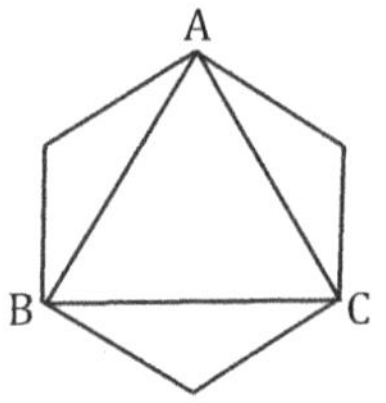

In the problem, we see information regarding a regular hexagon and an equilateral triangle.

Thus, the main things we should focus on are:

- Relate the equilateral triangle and the hexagon
- Find means of calculating the side of the hexagon, noting that a regular hexagon consists of six smaller equilateral triangles.

Let us now solve the problem.

Perimeter of equilateral triangle ABC = x

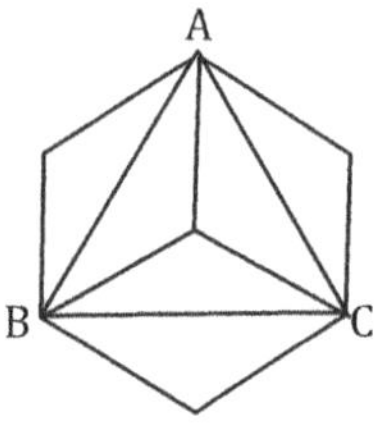

Each side of the triangle ABC = $\frac{x}{3}$

Thus, area of the equilateral triangle ABC = $\frac{\sqrt{3}}{4}\left(\frac{x}{3}\right)^2$

Area of each of six small isosceles triangles = $\frac{1}{3}\left[\frac{\sqrt{3}}{4}\left(\frac{x}{3}\right)^2\right]$

Area of the hexagon = 6 × Area of small triangle = $6 \times \frac{\sqrt{3}}{3 \times 4}\left(\frac{x}{3}\right)^2$

Let us calculate the perimeter.

From the following diagram:

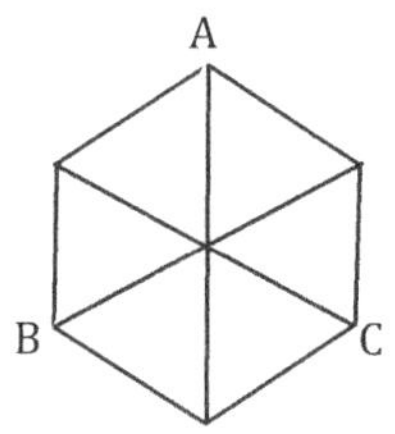

Let the side of the hexagon be y

Area of the hexagon

Area of six small equilateral triangles with side y = Area of six small isosceles triangles

$$=> 6 \times \frac{\sqrt{3}}{4}y^2 = 6 \times \frac{\sqrt{3}}{3 \times 4}\left(\frac{x}{3}\right)^2 => y = \frac{x}{3\sqrt{3}}$$

$$=> \text{Perimeter of the hexagon} = 6y = 6 \times \frac{x}{3\sqrt{3}} = \frac{2x}{\sqrt{3}}$$

(11) Six horses are tethered at six corners of a regular hexagonal plot of side 14 meter using equal length of ropes so that the adjacent horses can just reach one another. Find the area of the plot that is not grazed.

The situation of the horses and their reaches is depicted in the diagram given below. We see information regarding a regular hexagon and six sectors of a circle.

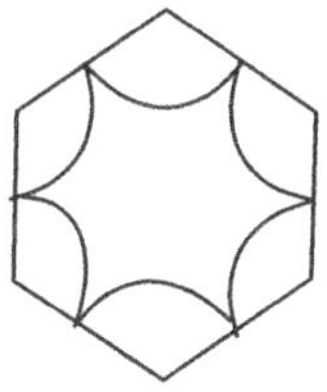

Thus, the main things we should focus on are:

- Find means of calculating the area of the hexagon, noting that a regular hexagon consists of six smaller equilateral triangles.

- What is the interior angle of a regular hexagon?
- Relation for the area of a sector of a circle.

Let us now solve the problem.

In a regular polygon of n sides, measure of each interior angle = $\frac{(n-2)\times 180^\circ}{n}$

Thus, in a regular hexagon, the measure of each interior angle

$$= \frac{(6-2)\times 180^\circ}{6} = 120^\circ$$

Thus, the angle of each of the six sectors = 120°

Since the horses are tethered using equal lengths of rope and the adjacent horses just reach each other, the length of rope = radius of the sector

$$= \frac{(\text{Side of hexagon})}{2} = \frac{14}{2} = 7 \text{ m.}$$

Area of the sector of a circle having angle θ and radius r is given by $\pi r^2\left(\frac{\theta}{360}\right)$.

Thus, area not grazed = Area of hexagon − Area of six sectors

$$= \left[6\times\frac{\sqrt{3}}{4}\times 14^2\right] - \left[6\times\frac{120}{360}\times\pi\times 7^2\right] = 499.80 - 307.7 = 192.1$$

16.4 Circles

(12) Three equal circles having radius of 1 cm each, touch each other as shown in the figure. Find the area enclosed by the circles (approx. in sq. cm).

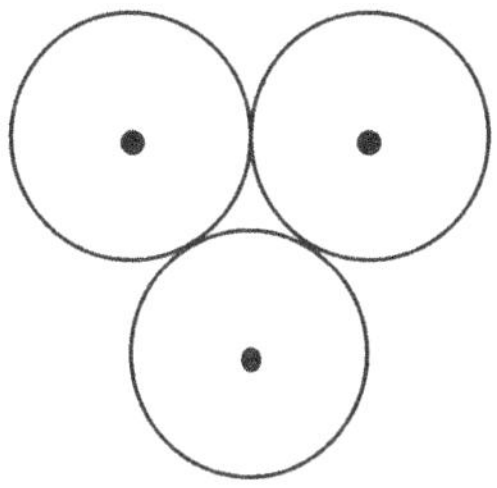

In the problem, we see information regarding three equal circles.

We need to determine the area enclosed by the circles.

Thus, the main things we should focus on are:

- Since the circles are equal, their centers form an equilateral triangle.
- Relation for the area of an equilateral triangle
- Relation for the area of a sector of a circle

Let us now solve the problem.

We join the centers X, Y, Z of the three circles to form an equilateral triangle of side 2cm.

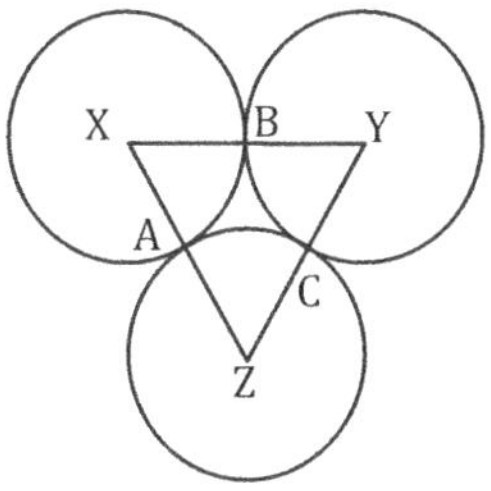

Let the points where the circles touch each other be A, B, C.

Area of each of the sectors XAB, YBC and ZAC $= \pi\,(1^2)\left(\frac{60°}{360°}\right) = \frac{\pi}{6}$

Thus, area of enclosed region

= Area of equilateral triangle XYZ − Area of the three sectors

$$= \frac{\sqrt{3}}{4}\,(2^2) - 3\left(\frac{\pi}{6}\right) = \sqrt{3} - \frac{\pi}{2} = 1.73 - \frac{3.14}{2} = 1.73 - 1.57 = 0.16$$

(13) The outline of a tear-drop sign is made by placing $\frac{3}{4}$ of the circumference of a circle of radius 2 feet on top of an isosceles triangle of height 5 feet. What is the perimeter, in feet, of the sign?

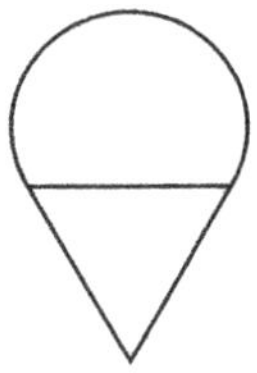

In the problem, we see information regarding a sector of a circle and an isosceles triangle.

Thus, the main things we should focus on are:

- Relation of the length of arc of a sector of a circle

Let us now solve the problem.

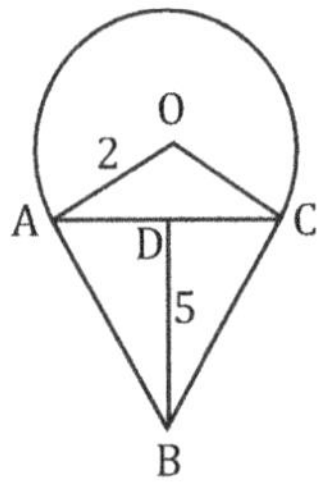

Reflex angle AOC $= \frac{3}{4} \times 360^\circ = 270^\circ$

Thus, $\angle AOC = 90^\circ \Rightarrow AC = \sqrt{AO^2 + CO^2} = 2\sqrt{2} \Rightarrow AD = \sqrt{2}$

$\Rightarrow AB = BC = \sqrt{5^2 + \left(\sqrt{2}\right)^2} = \sqrt{27} = 3\sqrt{3}$

Perimeter of the sign $= \frac{3}{4} \times (2\pi \times 2) + 2 \times 3\sqrt{3} = 3\pi + 6\sqrt{3}$

(14) The triangle PAB is formed by three tangents to a circle with center at O. What is the measure of $\angle AOB$?

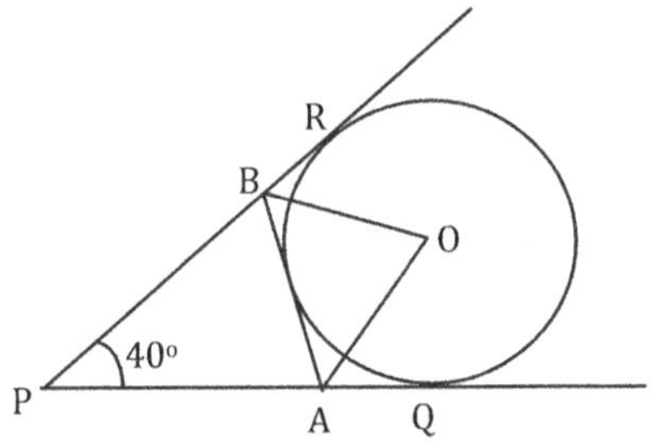

In the problem, we see information regarding a circle and tangents.

We need to determine an angle.

Thus, the main things we should focus on are:

- Radius $\perp$ Tangent
- How to relate $\angle AOB$ to $\angle BPA$ using suitable constructions?

Let us now solve the problem.

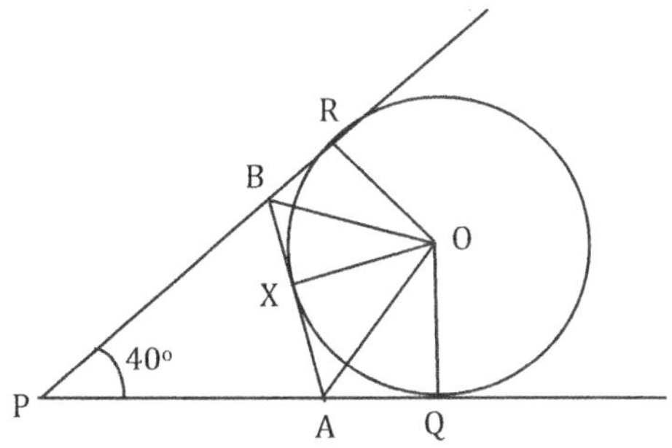

We drop perpendicular OX and join OR and OQ as shown.

$\angle ORB = \angle OQA = \angle OXA = 90°$

Also: BOX =ROB
(Triangles ROB and BOX are congruent: BR = BX, $\angle ORB = \angle OXA$, OB is common)

Similarly, AOX =QOA

In PROQ, $\angle ROQ = 360° - 40° - 90 - 90° = 140°$
=> 2 (BOX + AOX) = 140° => BOX + AOX = AOB = 70°

16.5 3-D geometry

(15) A cone is cut parallel to its base from a height equal to $\frac{1}{3}$ of the total height, $AS' = \frac{1}{3}AS$. Find the ratio of the volume of cone ADE to the frustum DECB.

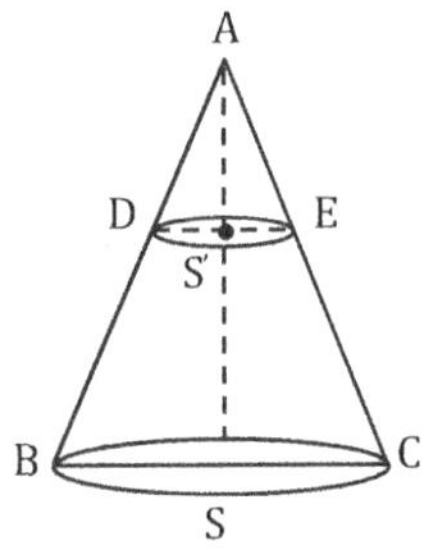

In the problem, we see information regarding a cone.

We need to determine the ratio of the volumes of two parts of the cone.

Thus, the main things we should focus on are:

- There are no dimensions given, so we do NOT need to calculate the exact volume.
- The smaller and larger cones seem similar; so do we use the property that the volume ratio of two similar solids equals the cube of the side ratio?

Let us now solve the problem.

The upper small cone and the entire cone (uncut) are similar solids.

Thus, ratio of their volume is cube of the ratio of their dimensions

$$=> \frac{\text{(Volume of cone ADE)}}{\text{(Volume of cone ABC)}} = \left(\frac{\text{(Height of come ADE)}}{\text{(Height of come ABC)}}\right)^3 = \left(\frac{1}{3}\right)^3 = \frac{1}{27}$$

$$=> \frac{\text{(Volume of cone ADE)}}{\text{(Volume of DECB)}} = \frac{1}{27-1} = \frac{1}{26}$$

(16) From a solid cone of height 6 cm. and radius $6\sqrt{2}$ cm., a cube with the maximum possible volume is cut out. What is the volume of the remaining portion of the cone? Use the relation: Volume of a cone of radius r and height $h = \frac{1}{3}\pi r^2 h$.

In the problem, we see information regarding a cube and a cone.

We need to determine the volume of the cube.

Thus, the main things we should focus on are:

- Since the volume of a cube depends only on its side (volume is the cube of its side), we need to determine the length of the side of the cube.

- Since the cone, represented in 2-D form, is basically a triangle, do we need to use similarity to determine the side of the cube?

Let us now solve the problem.

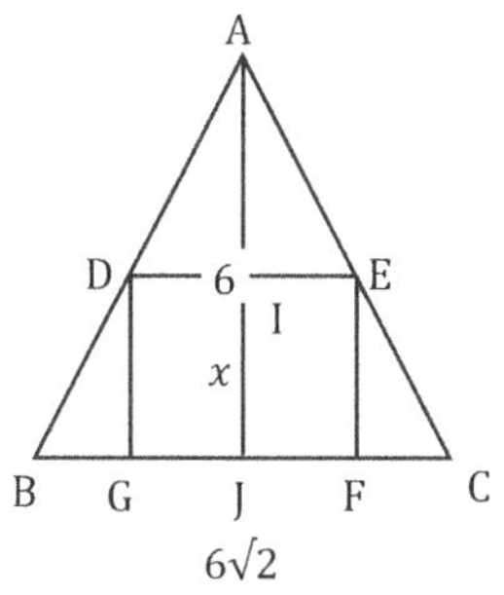

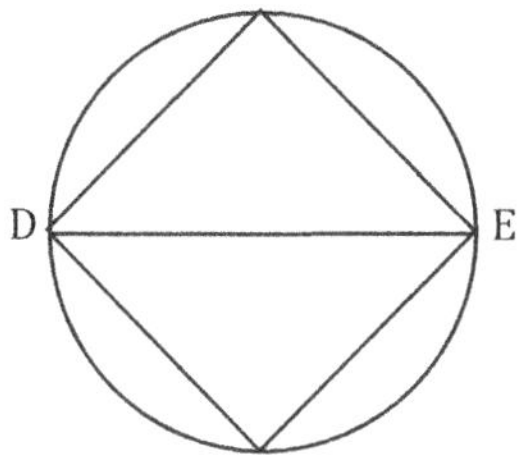

Let each side of the cube be x.

$=> \text{AI} = 6 - x$

DE is the diagonal of the top face of the cube $= x\sqrt{2}$

$=> \text{IE} = \frac{\text{DE}}{2} = \frac{x}{\sqrt{2}}$

Note: Do NOT, by mistake, assume that DE is the side of the cube.

Since the corners of the cube touch the surface of the cone, DE must be the diagonal of the square face on top of the cube, as shown beside:

ΔAIE is similar to ΔAJC.

$$=> \frac{6 - x}{6} = \frac{\left(\frac{x}{\sqrt{2}}\right)}{6\sqrt{2}} => 12 - 2x = x => x = 4$$

Volume of remaining portion of cone

$$= \frac{\pi}{3}\left(6\sqrt{2}\right)^2 \times 6 - 4^3 = 144\pi - 64$$

Chapter 17

Geometry & Trigonometry: Practice Questions

17.1 Geometry

1. A chord is drawn in a circle at a distance of 4 units from the center of the circle, as shown in the figure below. If the radius of the circle is 5 units, what is the length of the chord?

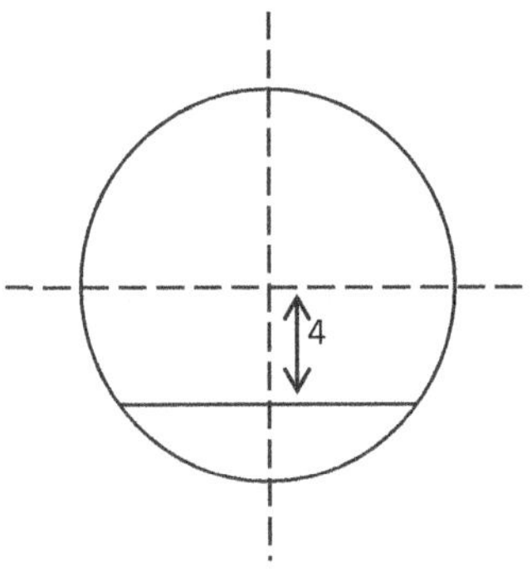

(A) 3
(B) 4
(C) 6
(D) 8

2. **SPR:**

A rectangular lawn of dimensions 10 meters by 20 meters has a 2-meter wide pathway along three of its sides. What is the largest possible straight-line distance, in meters, between any two points of the pathway?

3. **SPR:**

A rectangular lawn of dimensions 10 meters by 20 meters has a 2-meter wide pathway along three of its sides. What is the largest possible area, in square meters, of the pathway surrounding the lawn?

4. In the diagram shown below, what is the value of x?

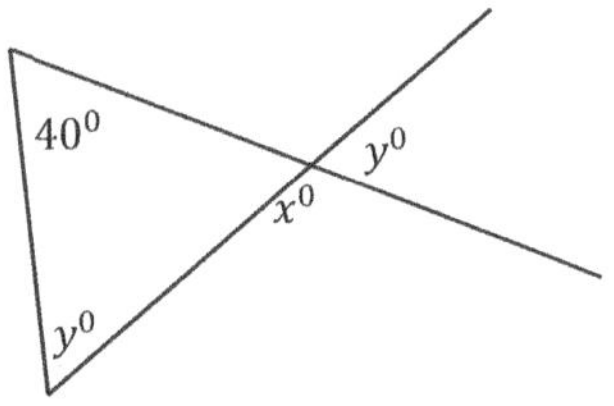

(A) 50°
(B) 70°
(C) 90°
(D) 110°

5. **SPR:**

ABFE is a rectangle with AB = 20, and AE = 10, as shown in the figure below. C is any point on AB. DE = DF and EG = GC. What is the ratio of the shaded area to the area of the rectangle ABFE?

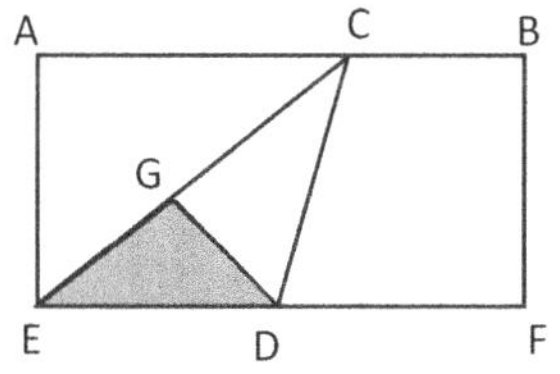

6. If ABC is a straight line as shown in the figure below, and x, y are integer multiples of 10 with $x > y$, what is the value of $(x + y)$?

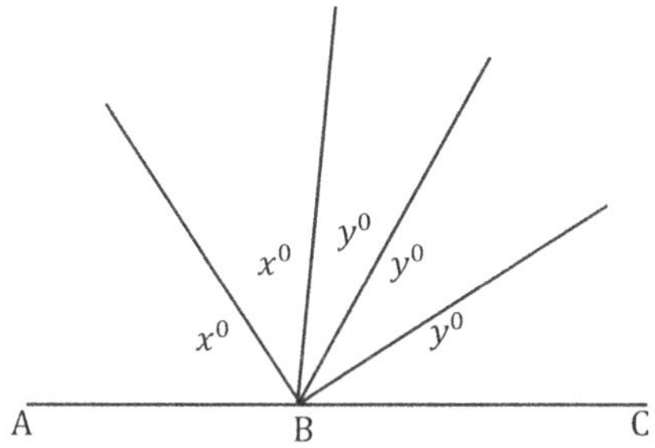

(A) 50°
(B) 65°
(C) 80°
(D) 90°

7. In the figure below, each small square measures 1 unit by 1 unit. What is the area of the shaded region?

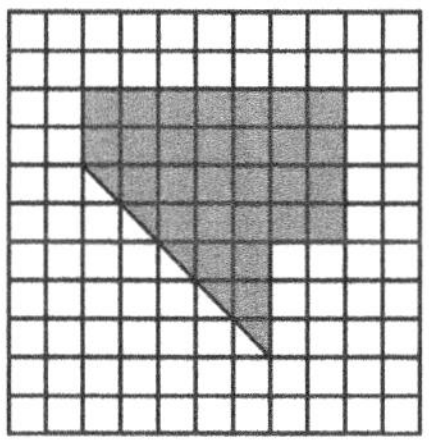

(A) 28.5 unit^2
(B) 30 unit^2
(C) 30.5 unit^2
(D) 32 unit^2

8. In the figure shown below, what is the value of x?

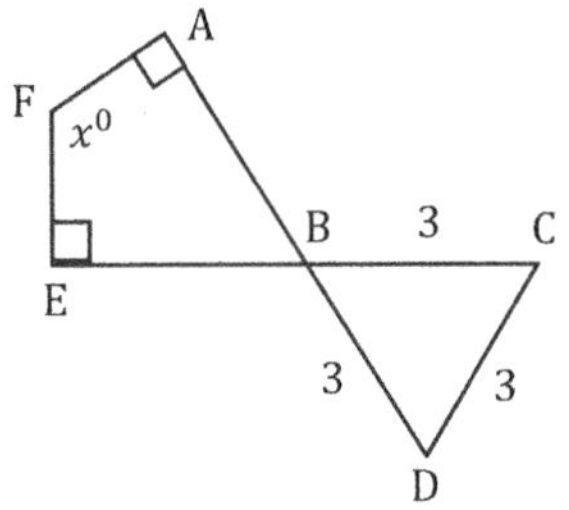

(A) 60°
(B) 90°
(C) 120°
(D) 150°

9. In the figure below, $\angle A = 80°$, AC = BC, FG = FE, and $\angle D = 90°$. What is the measure of $\angle G$?

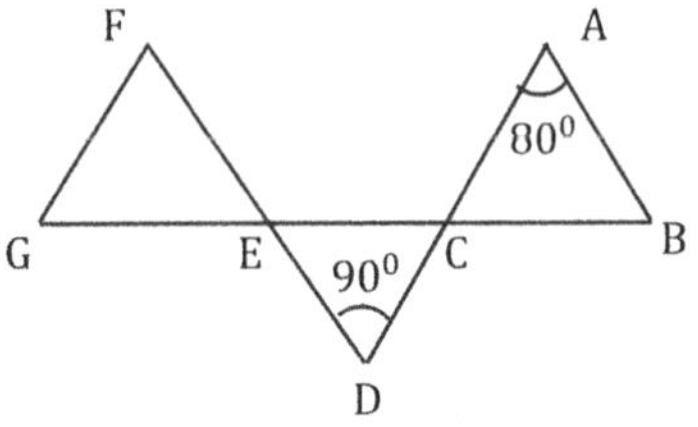

(A) 40°
(B) 45°
(C) 70°
(D) 75°

10. **SPR:**

In the figure shown below, ABCD and DBEF are squares. What is the ratio, to the nearest tenth, of the perimeters of the squares ABCD and DBEF?

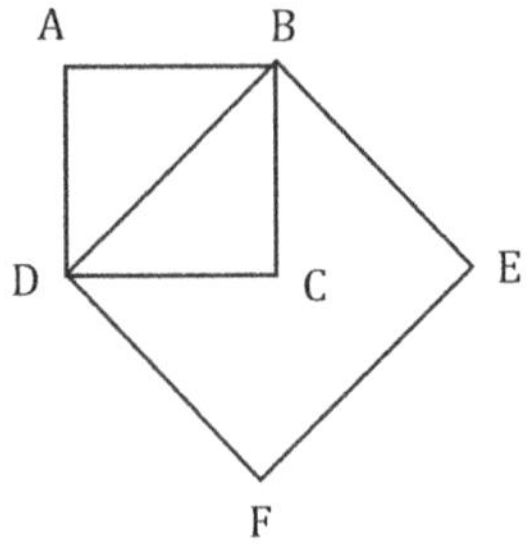

11. In the figure below, AD is the median to side BC. If AD = x, AC = y, AB = z, and CD = w, which of the following relations is NOT true?

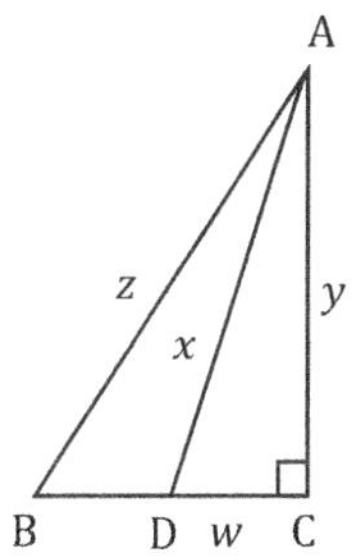

(A) $z^2 = x^2 + 3w^2$

(B) $z^2 = 4x^2 - 3y^2$

(C) $z^2 = 4w^2 + y^2$

(D) $z^2 = 4x^2 + y^2$

12. The perimeter of an isosceles triangle is x units. Each of the two equal sides is y units shorter than the third unequal side. What is the length of one of the equal sides, when expressed in terms of x and y?

(A) $\frac{x + 2y}{3}$

(B) $\frac{x - y}{3}$

(C) $\frac{x + y}{3}$

(D) $\frac{x - 2y}{3}$

13. In the figure shown below, ABCD is a square of side 2. If O is the center of the circle, what is the radius of the circle?

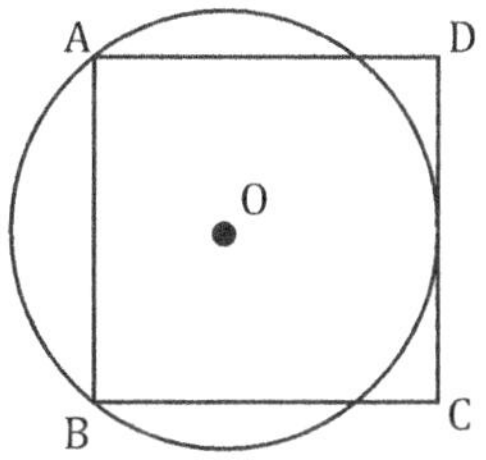

14. A triangle has a base equal to half the circumference of a circle with a radius of r. If the area of the triangle is twice that of the circle, what is the height of the triangle in terms of r?

(A) $\frac{r}{2}$

(B) $2r$

(C) $4r$

(D) $8r$

15. SPR:

The perimeters of a square and an equilateral triangle are the same. The area of the triangle forms what percent, to the nearest integer, of the area of the square? Ignore the 'percent' symbol while posting your answer in the grid.

16. In the figure given below, $w + z = 170°$. Which of the following is true?

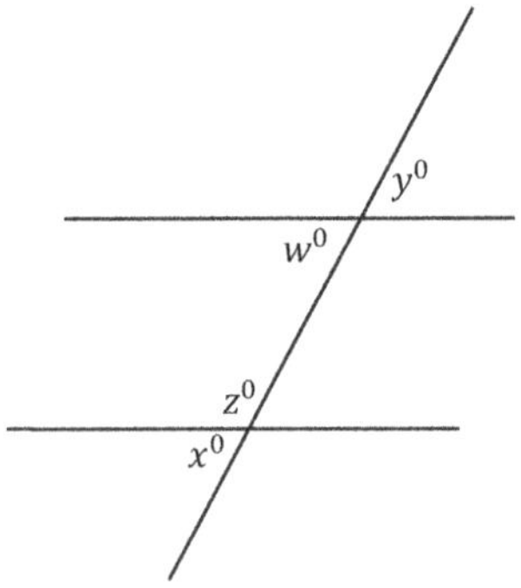

I. $x > y$
II. $x + w = 180°$
III. $x = w$

(A) Only I
(B) Only II
(C) I and II
(D) All three

17. In a right triangle ABC with $\angle B = 90°$, BC = 6 and the perimeter of the triangle is 24. What is the area of triangle?

(A) 6
(B) 12
(C) 24
(D) Insufficient data

18. The sides of a triangle are x, $(2x - 1)$, and $(5 - x)$, where x is an integer. What is the perimeter of the triangle?

(A) 5
(B) 6
(C) 8
(D) 9

19. SPR:

In the figure shown below, ABCD is a rectangle. E is a point on CD such that BE = 18. F is a point on BE such that AF is perpendicular to EB and AF = 15. What is the area of the rectangle ABCD?

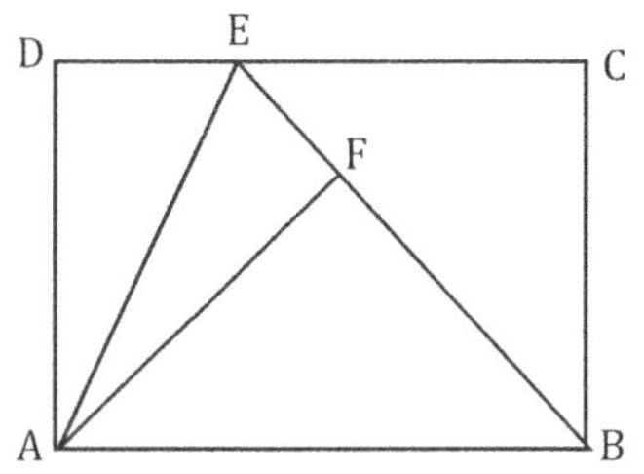

20. If the areas of triangles AMC and HMF are $36\sqrt{3}$ and $9\sqrt{3}$, respectively, what is the area of rectangle BDGK?

(A) $36\sqrt{3}$

(B) 81

(C) $81\sqrt{3}$

(D) $162\sqrt{3}$

21. What is the difference between the areas KLMH and CDEM?

(A) 0

(B) $6\sqrt{3}$

(C) $12\sqrt{3}$

(D) 24

22. In the figure shown below, ACEG, BCDM and GFMH are squares. If the area of BCDM is 16 times that of GFMH, and perimeter of ACEG is 40, what is the difference between the perimeters of BCDM and GFMH?

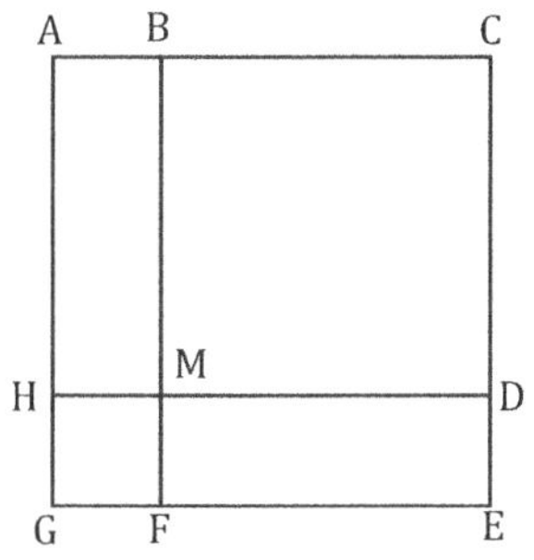

(A) 6

(B) 18

(C) 24

(D) 60

23. **SPR:**

The side of Square A is k times the diagonal of Square B. If the area of Square A is k times the area of Square B, what is the value of k?

24. In the given figure, AB = 6, BC = 7, and CA = 8. AD is the perpendicular from vertex A to the side BC.

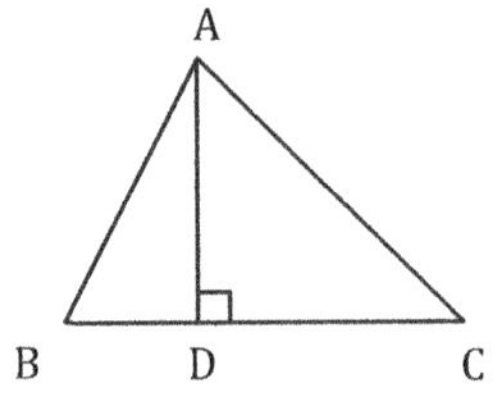

What is the length of BD?

(A) $\frac{1}{2}$

(B) 1

(C) $\frac{3}{2}$

(D) 2

25. In the given figure, AB = 6, BC = 7 and CA = 8. AD is the perpendicular from vertex A to the side BC.

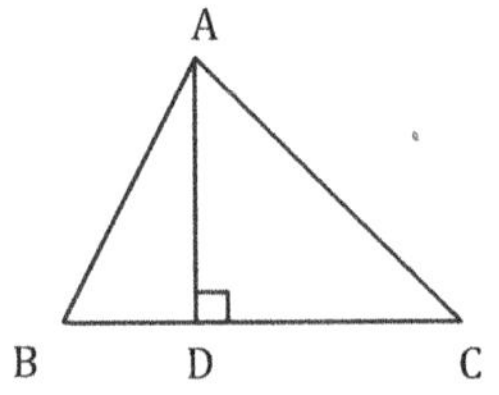

What is the ratio of the lengths of BD and CD?

(A) 3 : 11

(B) 3 : 7

(C) 4 : 7

(D) 3 : 4

26. In the given figure, AB is parallel to CG and $\angle FEC = 92^\circ$. If BE = CE and $\angle GCD = 60^\circ$, what is the measure of $\angle ABF$?

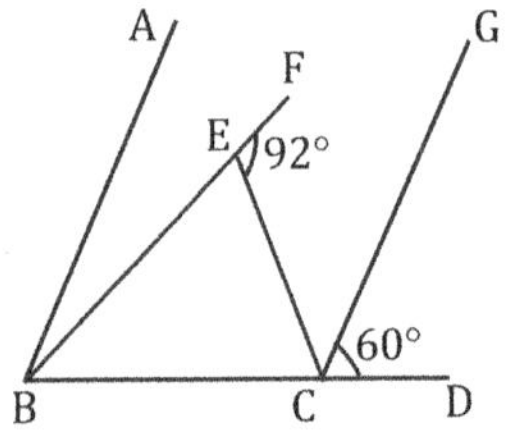

(A) 14°

(B) 16°

(C) 24°

(D) 46°

27. Three circles, each having radius 1 cm, touch each other as shown in the given figure.

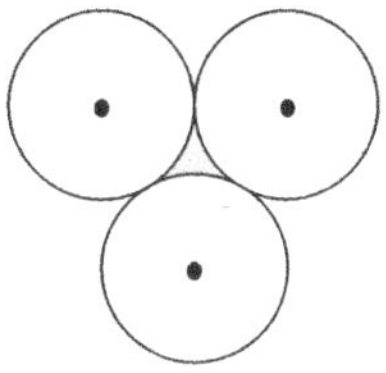

What is the perimeter of the shaded region?

(A) $\frac{\pi}{3}$

(B) $\frac{2\pi}{3}$

(C) π

(D) 2π

28. Three circles, each having radius 1 cm, touch each other as shown in the given figure.

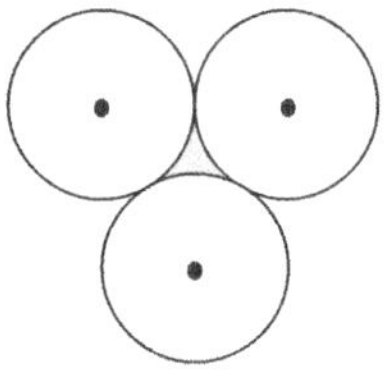

What is the OUTER perimeter of the region formed by the three circles?

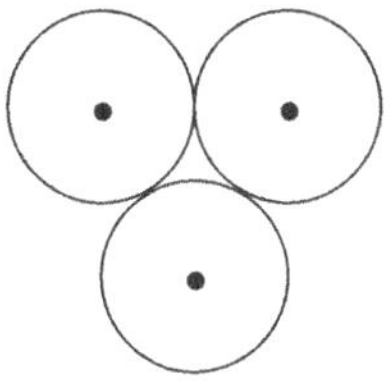

(A) 4π

(B) $\frac{9}{2}\pi$

(C) $\frac{13}{3}\pi$

(D) 5π

29. Three circles, each having radius 1 cm, touch each other as shown in the given figure.

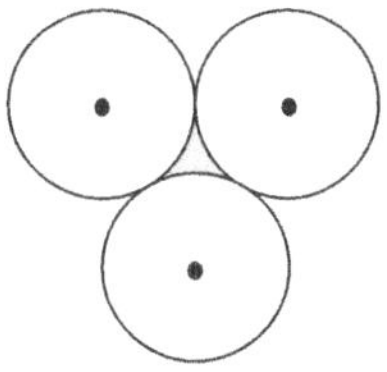

If the area of the shaded region is given by $(a\sqrt{3} + b\pi)\ cm^2$, what is the value of $(a + b)$?

(A) $\frac{1}{2}$

(B) $\frac{2}{3}$

(C) $\frac{3}{4}$

(D) 1

30. SPR

The areas of the three distinct faces of a rectangular wooden beam are 15 m^2, 3 m^2, and 20 m^2.

What is the volume, in cubic meters, of the rectangular beam?

31. SPR

The areas of the three distinct faces of a rectangular wooden beam are 15 m^2, 3 m^2, and 20 m^2.

What is the sum of the lengths of all the edges of the rectangular wooden beam?

32. SPR

In the given figure, AP = PR = RB, and the lines PQ, RS and BC are parallel to one another. Also, BC = 9 and the area of triangle ABC is 90 square units.

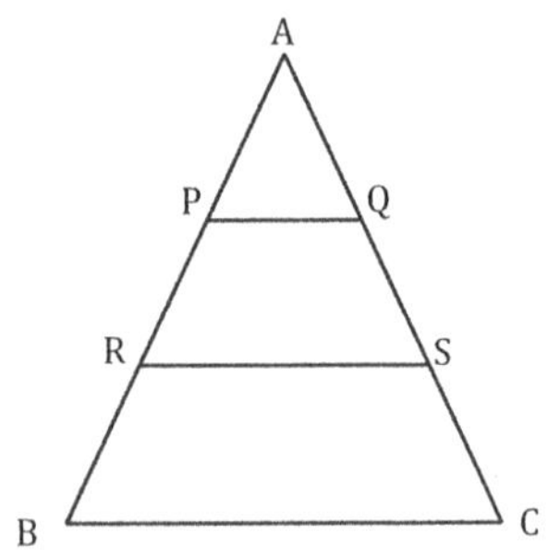

What is the length of PQ?

33. SPR

In the given figure, AP = PR = RB, and the lines PQ, RS and BC are parallel to one another. Also, BC = 9 and the area of triangle ABC is 90 square units.

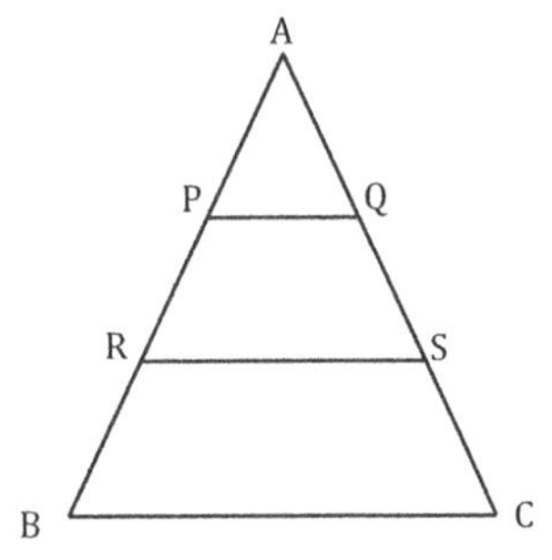

What is the area of triangle APQ?

34. **SPR**

In the given figure, AP = PR = RB, and the lines PQ, RS and BC are parallel to one another. Also, BC = 9 and the area of triangle ABC is 90 square units.

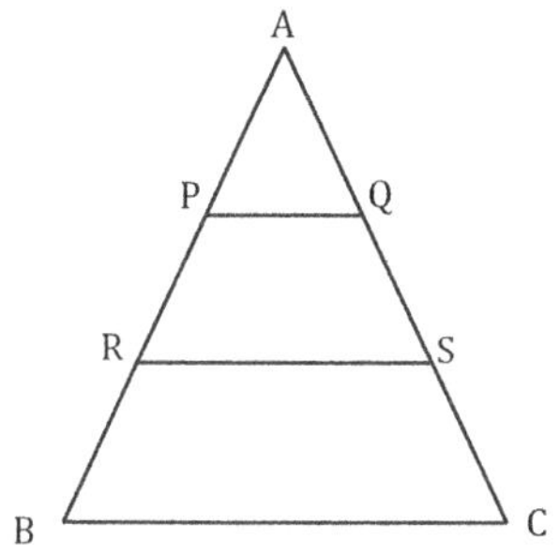

What is the area of PQSR?

35. **SPR**

ABCD is a trapezium whose parallel sides AB and CD are 8 and 14, respectively and the oblique sides AD and BC are 5 and 7, respectively.

What is the height of the trapezium, to the nearest tenths?

36. **SPR**

ABCD is a trapezium whose parallel sides AB and CD are 8 and 14, respectively and the oblique sides AD and BC are 5 and 7, respectively.

What is the longest distance, to the nearest integer, between any two vertices of the trapezium?

37. In the given figure, AB and CD represent two parallel chords of a circle. The shortest distance between the two chords is 14. If the radius of the circle is 10 and the length of the chord AB is 12, what is the length of the chord CD?

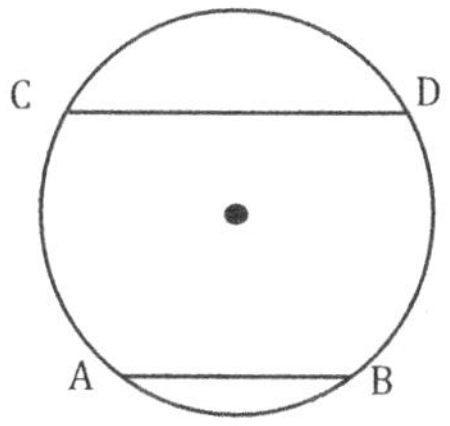

(A) 8
(B) 12
(C) 16
(D) 24

38. A cylinder of length 18 cm is melted down and re-cast into a solid sphere of radius 6 cm. What is the radius of the original cylinder?

(A) 2 cm

(B) 4 cm

(C) 5 cm

(D) 6 cm

39. In the given figure, what is the measure of ∠ACD?

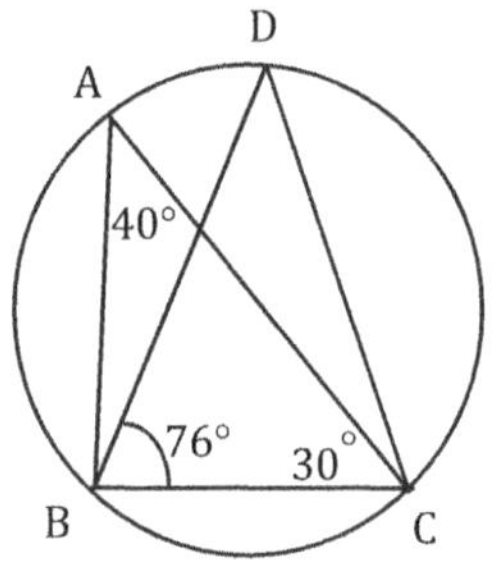

(A) 14°

(B) 34°

(C) 36°

(D) 60°

40. In the given figure, P is the center of the circle. What is the area, to the nearest tenth, of the shaded region?

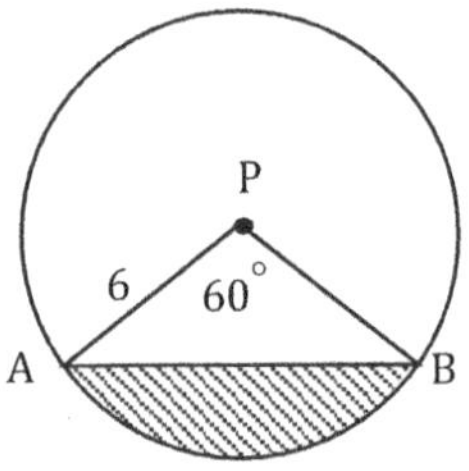

(A) 3.3

(B) 7.8

(C) 12.7

(D) 15.6

41. In the given figure, a ladder, 25 feet long, represented by the line AC, is kept leaning on a wall, AO, with its foot 7 feet away from the foot of the wall. The ladder slips to a new position, denoted by the line BD. If AB = 19 feet, how much farther away, in feet, did the foot of the ladder slip from the foot of the wall?

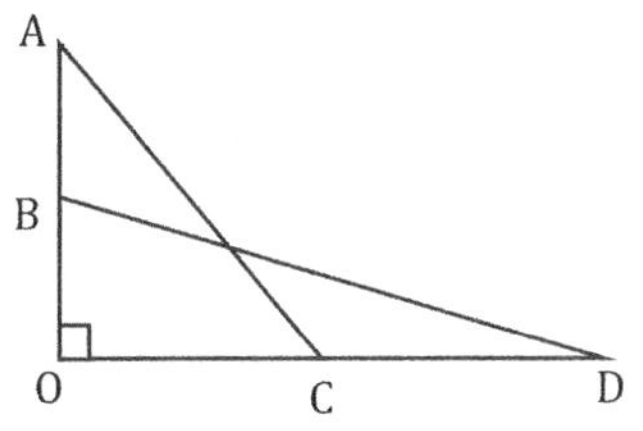

(A) 9.8

(B) 11.3

(C) 17.5

(D) 24.5

42. SPR

In the given figure, PQ = PR = RS, QS = QR, and ∠QPR = 20°. What is the measure, in degrees, of ∠SRU?

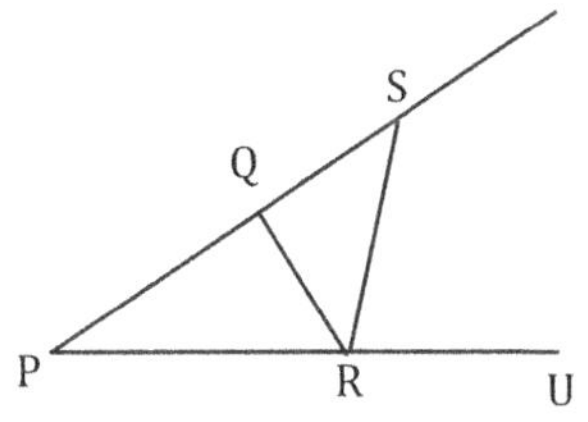

43. In the given figure, AB = ED = 3, AE = 10, and CD = 4. If the measure of ∠AEC is 90, what is the length, to the nearest tenth, of BC?

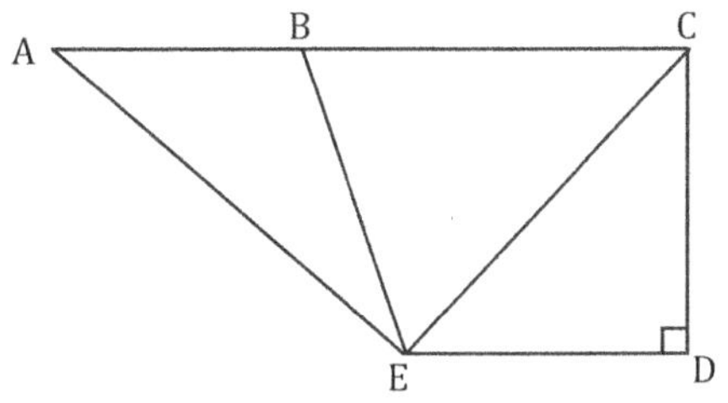

(A) 6.4

(B) 8.2

(C) 10.6

(D) 11.2

44. In the given figure, AB = BC and CE is parallel to AB. If CE bisects ∠ACD, what is the measure of ∠ABC?

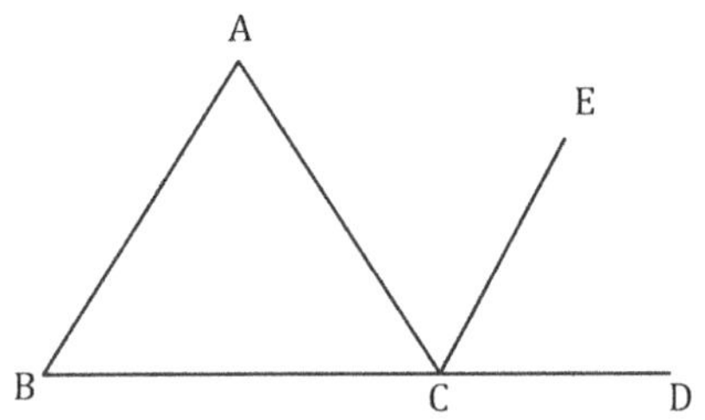

(A) 30°
(B) 60°
(C) 75°
(D) 90°

45. In the given figure, AB = AF and BC = CD. If the measure of ∠DBF is 60°, what is the measure of ∠AEC?

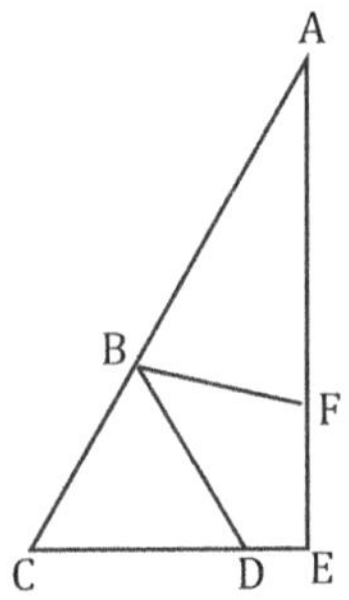

(A) 30°
(B) 45°
(C) 60°
(D) 90°

46. In the given figure, PQRS is a rectangle which has been divided into six congruent rectangles. What is the ratio of the sides PQ and QR?

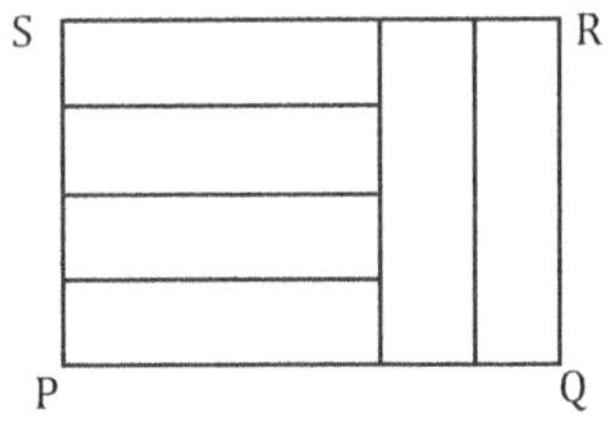

(A) 3 : 1
(B) 3 : 2
(C) 4 : 3
(D) 1 : 1

47. SPR

In the given figure, O is the center of a semicircle with a radius of 3 units. Points D and E lie on the arc AC such that the arcs CD and AE are $\frac{1}{10}$ of the arc AC. What is the area, to the nearest integer, in square units, of the sector DOE?

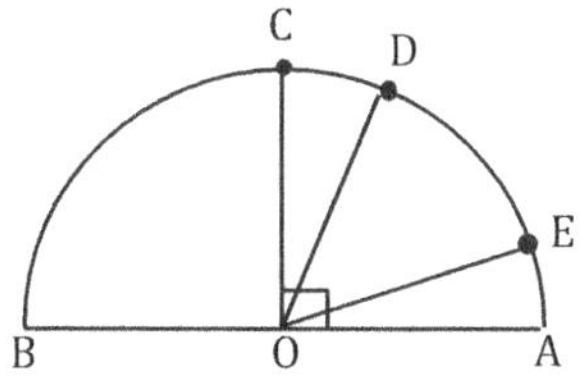

48. In the given figure, DF is parallel to AB and BC is perpendicular to AB. If BC = 4, BG = 1, AB = 3 and AD = 2, what is the length of CE?

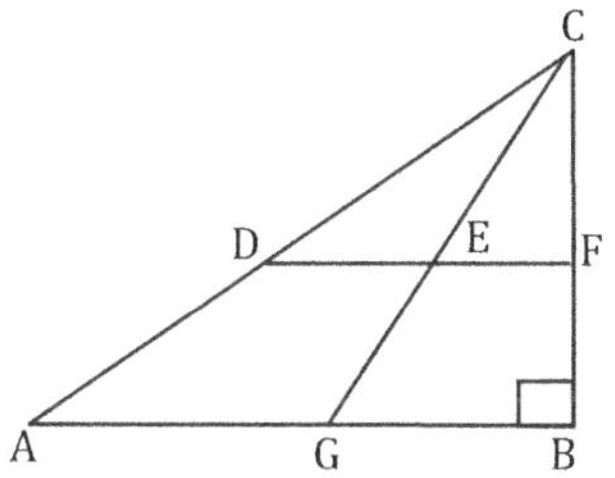

(A) $\frac{\sqrt{17}}{5}$

(B) $\frac{2\sqrt{17}}{5}$

(C) $\frac{3\sqrt{17}}{5}$

(D) $\frac{2\sqrt{17}}{3}$

49. In the right-triangle PQR, the smaller sides are 10 and 48. The triangle is divided into two triangles by the side PS, such that QS = SR = 24. By what percent is the sum of the perimeters of the two triangles thus generated greater than the perimeter of the triangle PQR?

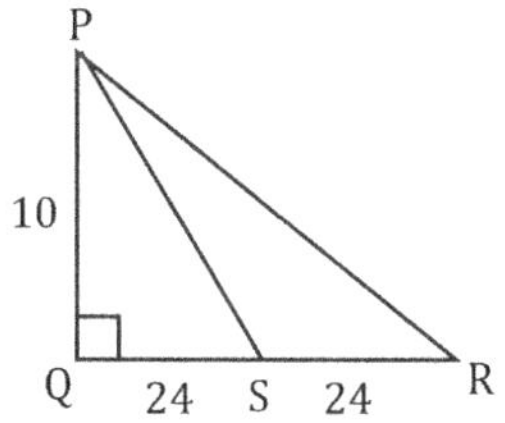

(A) 23.5%

(B) 34.6%

(C) 41.1%

(D) 48.6%

50. PQRS is a quadrilateral in which PQ and SR are parallel. Further, PQ = 10, QR = PS = 5, and RS = 4. What is the area of the quadrilateral PQRS?

(A) 20

(B) 28

(C) 32

(D) 36

51. **SPR**

A 21 × 14 × 11 cubic centimeters rectangular water container holds water up to $\frac{1}{3}$ its height. How many spherical marbles of diameter 1.4 centimeters can be placed in the container so that water just reaches the brink?

52. In the given figure, XY is parallel to BC, XY = 3 and BC = 5. If the area of BCYX is 32 square units, what is the area, in square units, of triangle ABC?

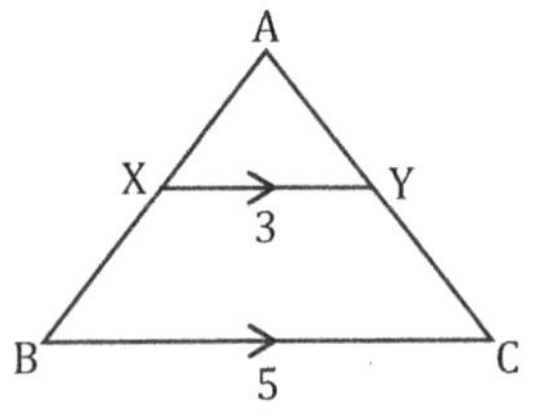

(A) 32

(B) 40

(C) 50

(D) 64

53. A, B, and C are three cities, as shown in the diagram below. A man, driving a car, leaves city A for city C at 8 a.m., while a train leaves city B for city C at a speed of 60 miles per hour at 12:00 p.m. The distance between the cities A and B is 273 miles.

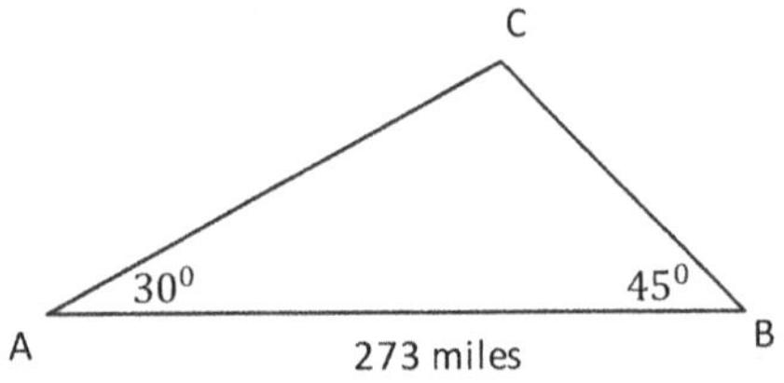

At what speed, in miles per hour, should the man drive his car so that he reaches city C at the same time as when the train from city B arrives?

(A) 18.8

(B) 26.7

(C) 31.5

(D) 37.5

54. A circular table is placed in the corner of a room, touching two adjacent walls perpendicular to each other, as shown in the top view below.

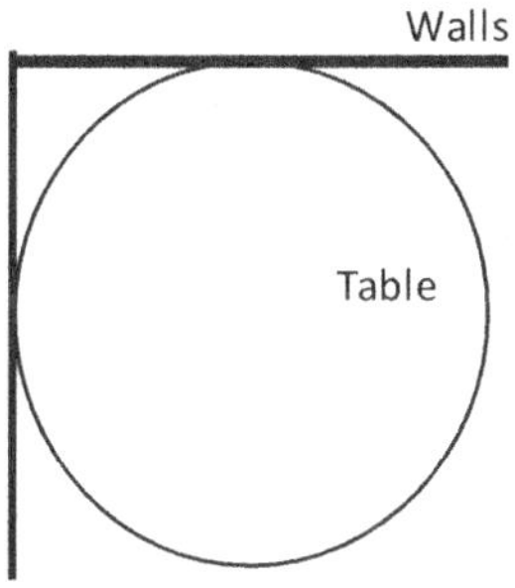

The table touches the walls at the points A and B. A random point P is chosen on the major arc AB. What is the measure of $\angle APB$?

(A) 30°

(B) 45°

(C) 60°

(D) There is not enough information to determine the answer

55. A circular table is placed in the corner of a room, touching two adjacent walls perpendicular to each other, as shown in the top view below.

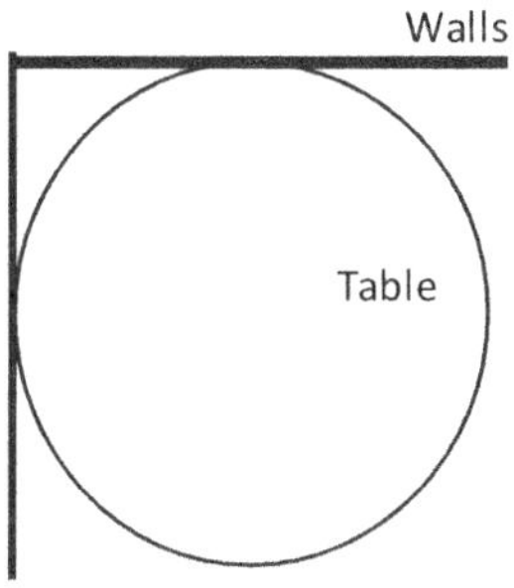

In the above question, another point Q, on the minor arc AB is chosen. What is the measure of $\angle AQB$?

(A) 120°

(B) 135°

(C) 150°

(D) There is not enough information to determine the answer

56. A circular table is placed in the corner of a room, touching two adjacent walls perpendicular to each other, as shown in the top view below.

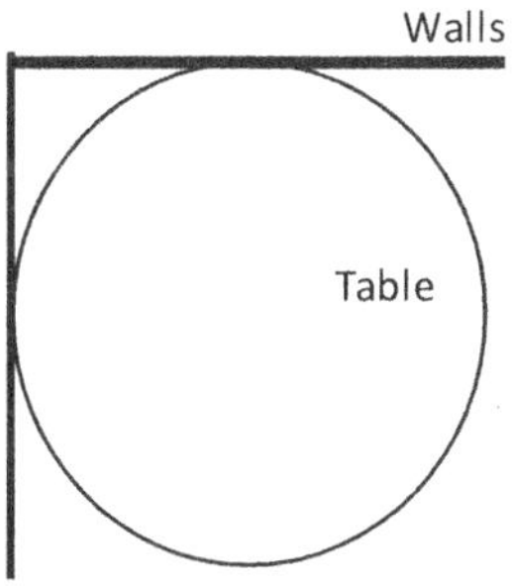

What is the ratio of the length of the line connecting the two points of the table in contact with the walls to the radius of the table?

(A) $2 : \sqrt{3}$
(B) $\sqrt{3} : \sqrt{2}$
(C) $\sqrt{2} : 1$
(D) $\sqrt{3} : 1$

57. From the top view of the scenario shown, what is the volume, to the nearest integer cubic meter, of the space enclosed between the table and the two walls? Assume that the circular table is a cylinder of radius 2.5 m and radius 3 m.

(A) 4
(B) 5
(C) 6
(D) 8

17.2 Trigonometry

58. **SPR:**

What is the perimeter, to the nearest integer, of the quadrilateral ABCD as shown below?

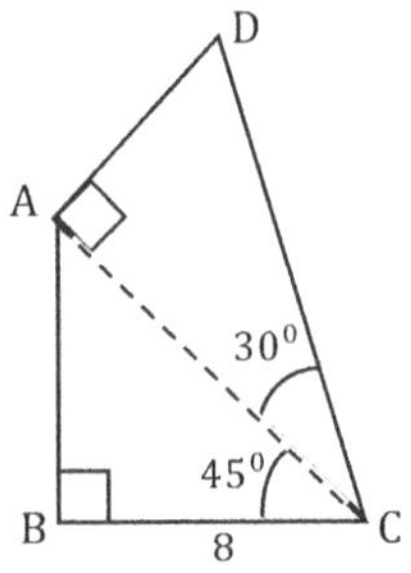

59. **SPR:**

In the figure below, what is the sum of the lengths of AD and CD to the nearest tenth?

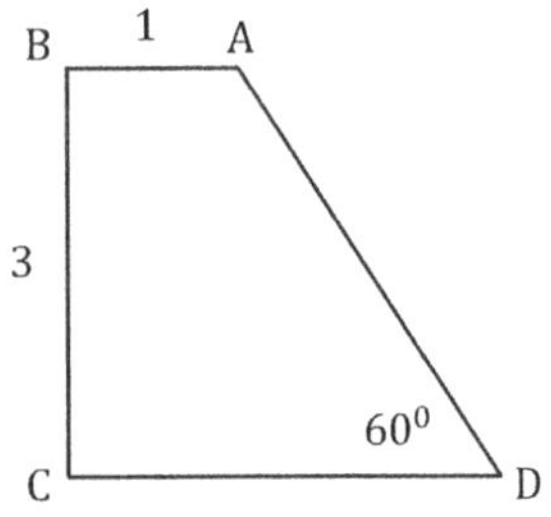

60. SPR

In triangle ABC, AB = 6, ∠BAC = 45° and ∠ABC = 75°. What is the length, to the nearest integer, of BC?

61. SPR

In the given figure (not to scale), the length of each side of triangle ACD is 20 units. If E is the midpoint of AD and angle B is 90 degrees, what is the area in square units, to the nearest integer, of the quadrilateral BCDE?

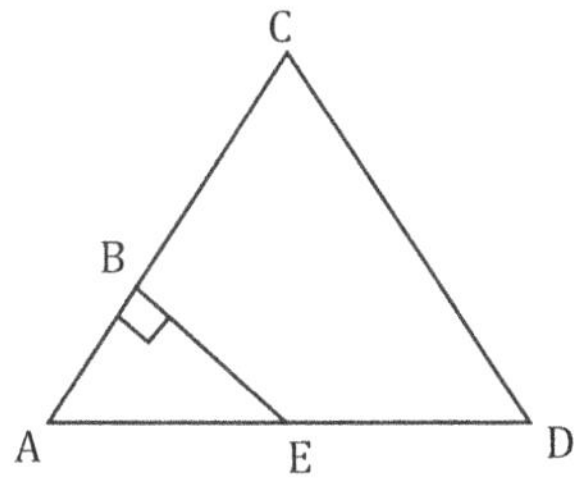

62. Which of the following is equivalent to the value of $\left(\frac{4\pi^c}{5} + \frac{11\pi^c}{6} - \frac{13\pi^c}{3}\right)$, when converted to degrees?

(A) 54°

(B) 124°

(C) 214°

(D) 306°

63. In the given figure, AB is perpendicular to BC and BD is perpendicular to AC. The angle at vertex A is 30° and the length of AB = 5.

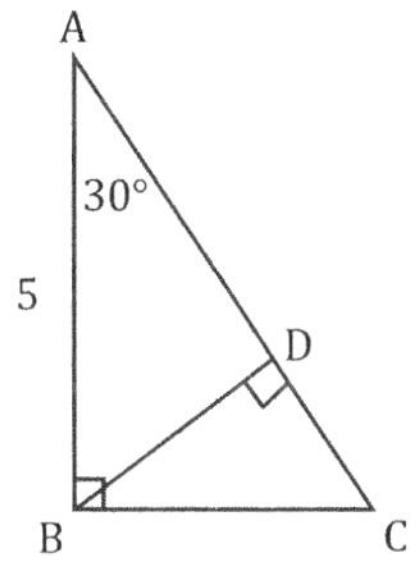

What is the length of BC?

(A) $\frac{5}{3}$

(B) $\sqrt{5}$

(C) 2.5

(D) $\frac{5}{\sqrt{3}}$

64. In the given figure, AB is perpendicular to BC and BD is perpendicular to AC. The angle at vertex A is 30° and the length of AB = 5.

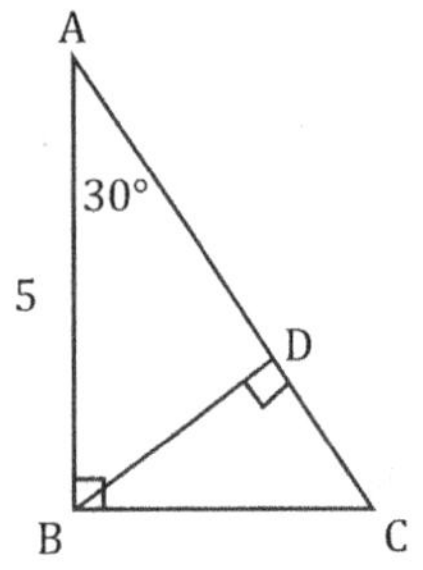

What is the length of CD?

(A) $\frac{5}{2\sqrt{5}}$

(B) $\frac{5}{2\sqrt{3}}$

(C) 2.5

(D) $\frac{5}{\sqrt{3}}$

65. In the given figure, AB is perpendicular to BC and BD is perpendicular to AC. The angle at vertex A is 30° and the length of AB = 5.

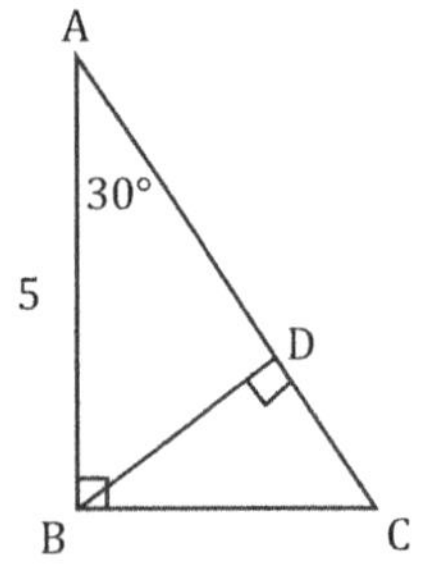

What is the perimeter of triangle ABC?

(A) 8.65

(B) 10.65

(C) 11.65

(D) 13.65

66. **SPR**

In the given figure, if AB = 5, what is the length, to the nearest integer, of CD?

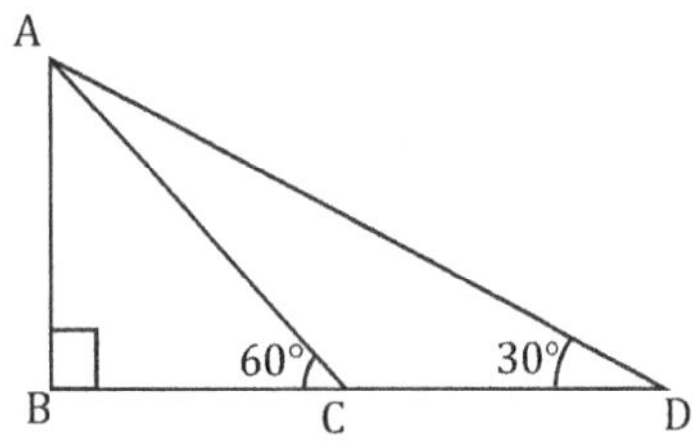

67. A, B, and C are three cities, as shown in the diagram below. A man, driving a car, leaves city A for city C at 8 a.m., while a train leaves city B for city C at a speed of 60 miles per hour at 12:00 p.m. The distance between the cities A and B is 273 miles.

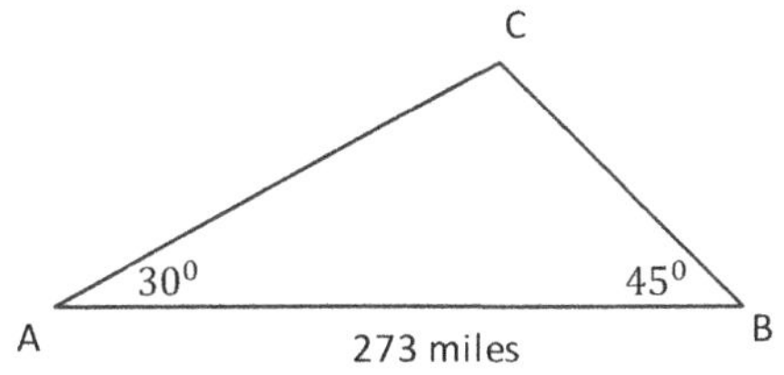

What is the distance between the cities A and C?

(A) 75 miles

(B) 100 miles

(C) 200 miles

(D) 225 miles

68. A, B, and C are three cities, as shown in the diagram below. A man, driving a car, leaves city A for city C at 8 a.m., while a train leaves city B for city C at a speed of 60 miles per hour at 12:00 p.m. The distance between the cities A and B is 273 miles.

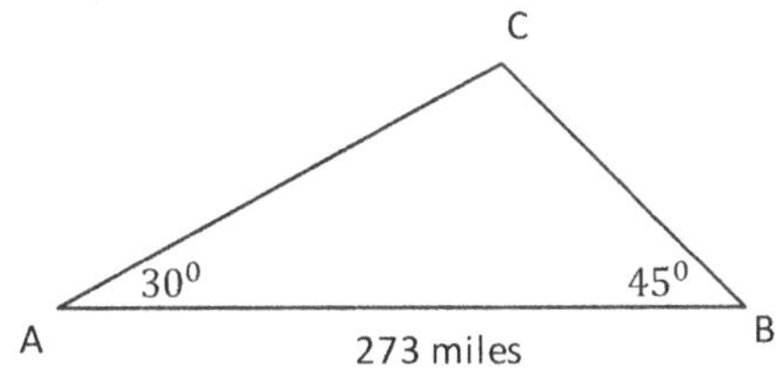

What is the distance between the cities B and C?

(A) 50 miles

(B) 87 miles

(C) 100 miles

(D) 141 miles

Chapter 18

Geometry & Trigonometry Answer-key

(1) C

(2) 26.83

(3) 108

(4) D

(5) 1/8 or .125

(6) C

(7) C

(8) C

(9) C

(10) 0.71

(11) D

(12) B

(13) 5/4 or 1.25

(14) C

(15) 77

(16) A

(17) C

(18) C

(19) 270

(20) D

(21) A

(22) C

(23) 1/2 or 0.5

(24) C

(25) A

(26) A

(27) C

(28) D

(29) A

(30) 30

(31) 54

(32) 3

(33) 10

(34) 30

(35) 4.9

(36) 14

(37) C

(38) B

(39) B

(40) A

(41) C

(42) 60

(43) B

(44) B

(45) C

(46) B

(47) 6

(48) C

(49) D

(50) B

(51) 1500

(52) C

(53) C

(54) B

(55) B

(56) C

(57) A

(58) 36

(59) 6.2

(60) 5

(61) 152

(62) A

(63) D

(64) B

(65) D

(66) 6

(67) C

(68) D

Chapter 19

Geometry & Trigonometry: Solutions to Practice Questions

19.1 Geometry

1. Let us bring out the image.

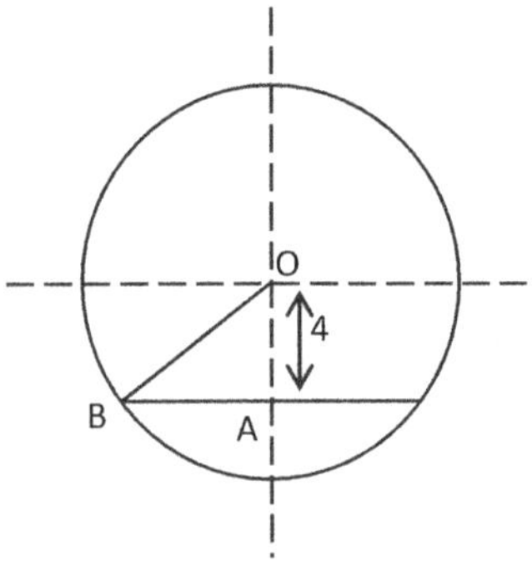

Let us join OA.

We have OB = radius = 5

Thus, we have a right-angled triangle OAB

=> $OB^2 = OA^2 + AB^2$

=> $AB^2 = 5^2 - 4^2 = 9$

=> AB = 3

Thus, the length of the chord = 2AB = 6

The correct answer is choice C.

2. The possible scenarios are (the longest distance is shown by a dotted line):

Case I represents a scenario in which the two parallel sides of 10-meter length have the pathways, while only one side of length 20 meter has pathways.

Case II represents a scenario in which the two parallel sides of 20-meter length have the pathways, while only one side of length 10 meter has pathways.

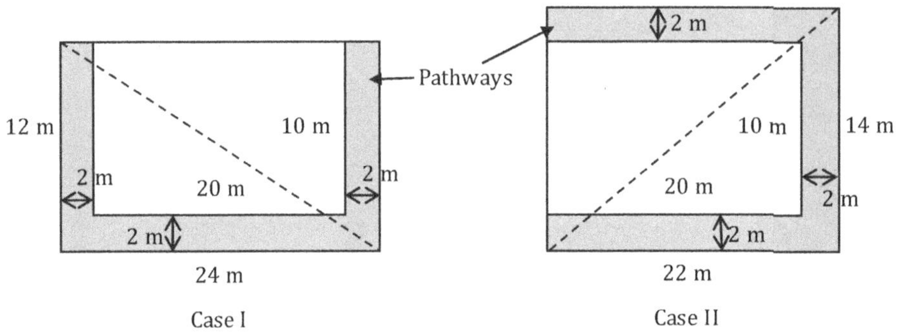

The longest distance:

- Case I: $\sqrt{12^2 + 24^2}$ = **26.83 m**

- Case II: $\sqrt{14^2 + 22^2}$ = 26.07 m

The correct answer is 26.83.

3. Referring to the two cases above, we have, area of the path:

- Case I: $24 \times 12 - 20 \times 10 = 88$ square meters

- Case II: $22 \times 14 - 20 \times 10 =$ **108 square meters**

The correct answer is 108.

Note: The longest distance between any two points occurs in Case I, while the largest area occurs in Case II.

4. Let us bring out the image.

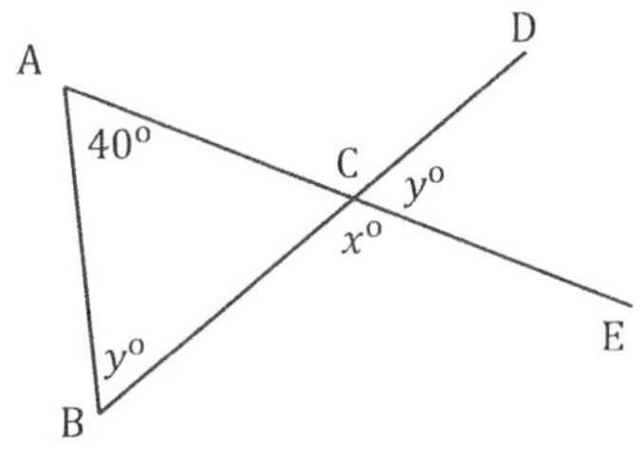

$\angle ACB = \angle DCE = y$

Thus, in triangle ABC:

$40^\circ + 2y = 180^\circ$

$=> y = 70^\circ$

Since BCD is a straight line:

$\angle BCE + \angle DCE = 180^\circ$

$=> x = 180^\circ - 70^\circ = 110^\circ$

The correct answer is choice D.

5. Let us bring out the image.

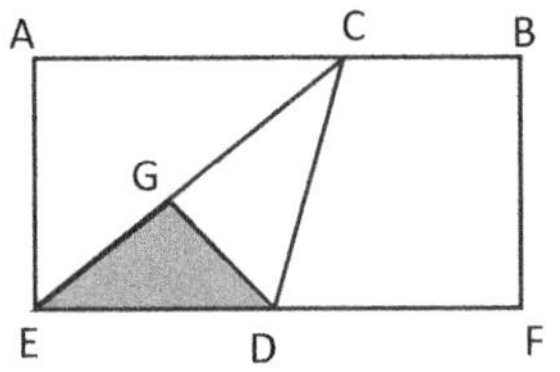

Area of triangle ECD = $\frac{1}{2} \times ED \times Height = \frac{1}{2} \times \frac{EF}{2} \times BF$

$= \frac{1}{4} \times$ Area of ABFE

Area of triangle EGD = $\frac{1}{2} \times GE \times Height = \frac{1}{2} \times \frac{CE}{2} \times Height$

$= \frac{1}{2} \times$ Area of triangle CED

(Since height of triangles DCE and GED are the same, taking their bases as CE and GE, respectively)

Thus, we have:

Area of triangle GED = $\frac{1}{2} \times \frac{1}{4} = \frac{1}{8}$ of Area of ABFE

The ratio of the shaded area to the area of the rectangle ABFE

$= \frac{\text{Area of triangle GED}}{\text{Area of ABFE}} = \frac{1}{8} = 0.125.$

The correct answer is $\frac{1}{8}$ or .125.

Note: The information about the measures of length AB = 20 and AE = 10 is redundant.

6. Let us bring out the image.

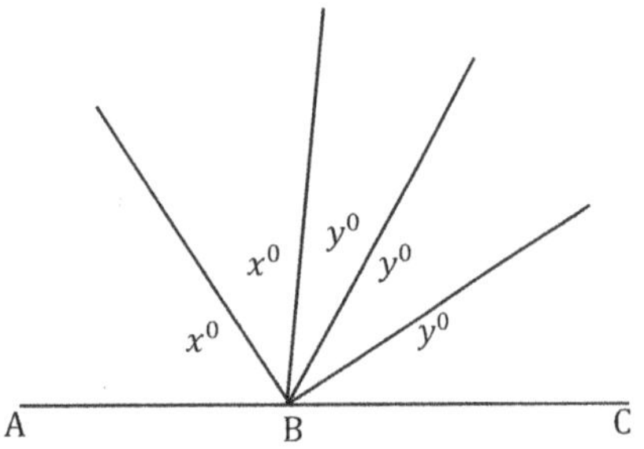

We have:

$2x + 3y = 180^\circ$

Let $x = 10p$ and $y = 10q$, where p and q are positive integers

$\Rightarrow 20p + 30q = 180^\circ$

$\Rightarrow 2p + 3q = 18$

Since $2p$ and 18 are even numbers, the value of q must also be even.

Thus, the possible solutions of p and q are:

$q = 2 \Rightarrow p = 6$ OR $q = 4 \Rightarrow p = 3$

Since $x > y \Rightarrow p > q$

$\Rightarrow p = 6$ and $q = 2$

$\Rightarrow x + y = 10p + 10q = 80$

The correct answer is choice C.

7. The above region can be broken down as shown below:

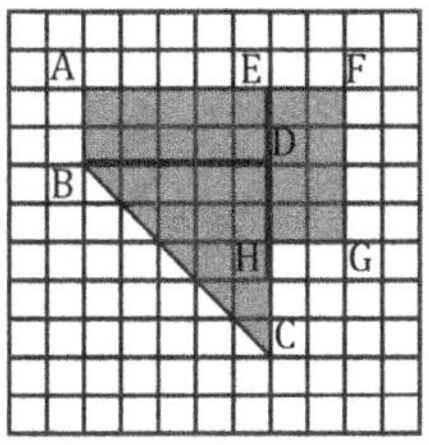

Area of the triangle BDC $= \frac{1}{2} \times \text{Base} \times \text{Height} = \frac{1}{2} \times 5 \times 5 = 12.5$

Area of the bigger rectangle AEDB $= 5 \times 2 = 10$

Area of the smaller rectangle EFGH $= 4 \times 2 = 8$

Thus, total area $= 12.5 + 10 + 8 = 30.5$

The correct answer is choice C.

8. Let us bring out the image.

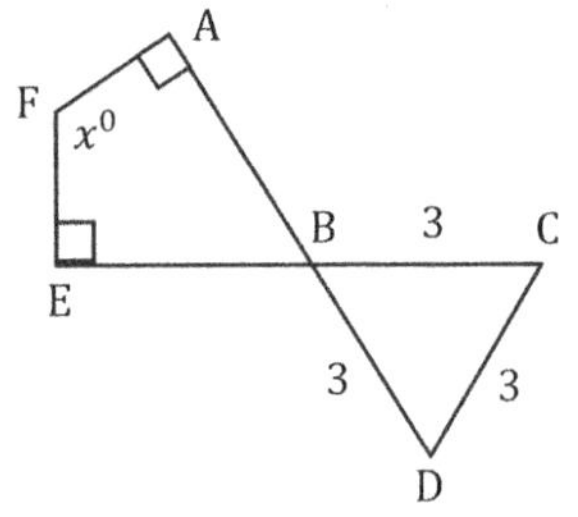

Since triangle BCD is equilateral, $\angle CBD = 60^\circ = \angle ABE$

In quadrilateral ABEF, we have:

$x = 360^\circ - (90^\circ + 90^\circ + 60^\circ) = 120^\circ$

The correct answer is choice C.

9. Let us bring out the image.

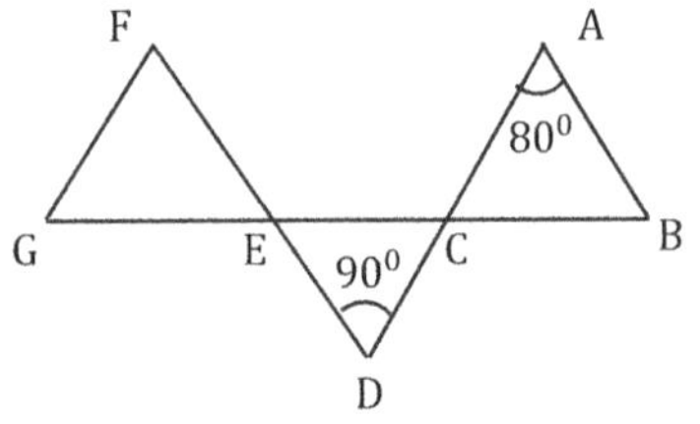

Since AC = BC, we have: $\angle A = \angle B = 80^\circ$

=> $\angle ACB = 180^\circ - (80^\circ + 80^\circ) = 20^\circ$

=> $\angle DCE = 20^\circ$

=> $\angle CED = \angle FEG = 180^\circ - (20^\circ + 90^\circ) = 70^\circ$

Since FG = FE, we have: $\angle G = \angle FEG = 70^\circ$

The correct answer is choice C.

10. Let us bring out the image.

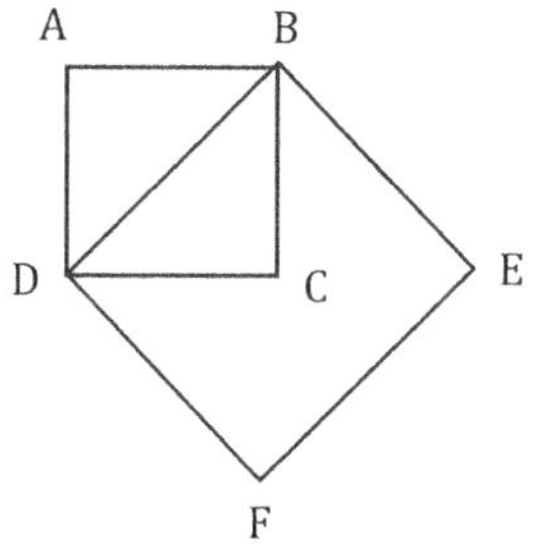

Let the side of square ABCD be x

Thus, in the right angled isosceles triangle ABD, from Pythagoras' theorem, we have:

BD = $x\sqrt{2}$

Thus, perimeter ratio of the two squares

$$= \frac{4x}{4x\sqrt{2}} = \frac{1}{\sqrt{2}}$$

$= 0.707 \approx= 0.71$

The correct answer is 0.71.

11. Let us bring out the image.

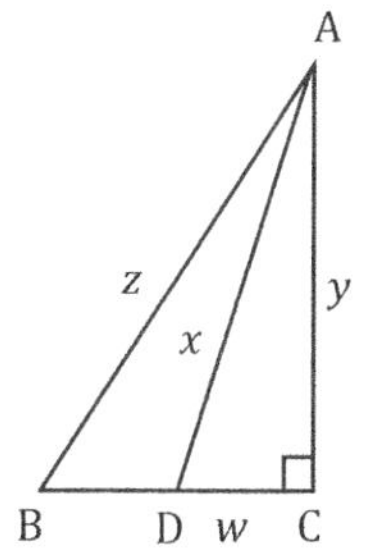

From right angled triangle ACD:

$x^2 = y^2 + w^2 \ldots$ (i)

From right angled triangle ABC:

$z^2 = (2w)^2 + y^2$

$=> z^2 = 4w^2 + y^2$ - Choice C is correct

Using (i), we have:

$\Rightarrow z^2 = x^2 + 3w^2$ - Choice A is correct

$\Rightarrow z^2 = 4x^2 - 3y^2$ - Choice B is correct

The correct answer is choice D.

12. Let the unequal side be a.

Thus, each of the equal sides $= a - y$

Thus, perimeter of the triangle

$= a + 2(a - y) = 3a - 2y$

$\Rightarrow 3a - 2y = x$

$\Rightarrow a = \dfrac{x + 2y}{3}$

Thus, the length of the equal side $= a - y$

$= \dfrac{x + 2y}{3} - y = \dfrac{x - y}{3}$

The correct answer is choice B.

13. Let us bring out the image.

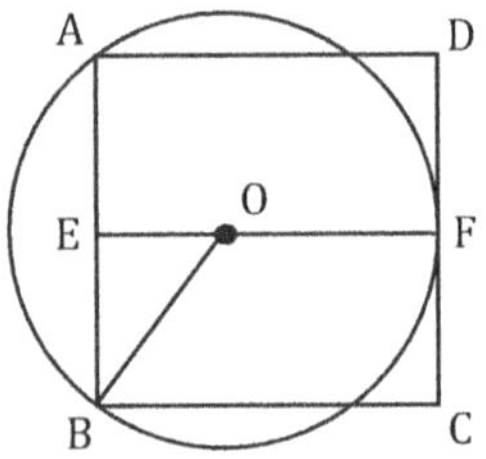

Assuming the radius of the circle to be r, we have:

$OE = EF - OF = 2 - r$

$OB = r$

$EB = \dfrac{1}{2} \times 2 = 1$

Thus, from the right triangle OEB:

$(2 - r)^2 + 1^2 = r^2$

$=> 4 - 4r + r^2 + 1 = r^2$

$=> r = \frac{5}{4} = 1.25$

The correct answer is $\frac{5}{4}$ or 1.25.

14. Circumference of the circle = $2\pi r$

Thus, base of the triangle = $\frac{1}{2} \times 2\pi r = \pi r$

Area of the circle = πr^2

Thus, area of the triangle = $2\pi r^2$

Thus, we have:

$2\pi r^2 = \frac{1}{2} \times \pi r \times \text{Height}$

=> Height = $4r$

The correct answer is choice C.

15. Let the perimeter of the square and the triangle be $12x$ each (we take 12 since it is divisible by 3 and 4).

Thus, each side of the square = $\frac{12x}{4} = 3x$

=> Area of the square = $(3x)^2 = 9x^2$

Again, each side of the equilateral triangle = $\frac{12x}{3} = 4x$

=> Area of the equilateral triangle = $\frac{\sqrt{3}}{4} \times (4x)^2 = 4\sqrt{3}x^2 = 6.93x^2$

Thus, required percent = $\frac{6.93x^2}{9x^2} \times 100 = 76.98 \approx= 77\%$

The correct answer is 77.

16. Let us bring out the image.

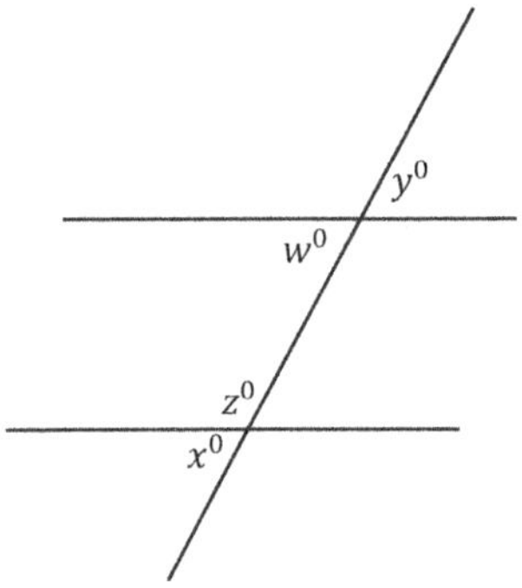

We have:

$w = y \ldots$ (i)

$x + z = 180^\circ \ldots$ (ii)

$w + z = 170^\circ \ldots$ (iii)

From (ii) – (iii): $x - w = 10^\circ \ldots$ ($=> x \neq w$) - Statement III is not true

From (i): $x - y = 10^\circ => x > y$ - Statement I is true

There is no information about the value of $(x + w)$ - Statement II is not true

The correct answer is choice A.

17. We have AB + BC + CA = 24 and BC = 6

=> AB + CA = 24 – 6 = 18

Let AB = x => CA = 18 – x

From Pythagoras' theorem:

$AC^2 = AB^2 + BC^2$

$=> (18 - x)^2 = x^2 + 6^2$

$=> 324 - 36x + x^2 = x^2 + 36$

$=> x = 8 = AB$

Thus, area of triangle ABC $= \frac{1}{2} \times AB \times BC = \frac{1}{2} \times 8 \times 6 = 24$

The correct answer is choice C.

18. The perimeter of the triangle = $x + (2x - 1) + (5 - x) = 2x + 4$

However, the value of x cannot be determined. Let us apply another approach to answer the question.

Since sum of any two sides is greater than the third side, we have:

- $x + (2x - 1) > 5 - x \Rightarrow x > 1.5$
- $x + (5 - x) > 2x - 1 \Rightarrow x < 3$
- $(5 - x) + (2x - 1) > x \Rightarrow 4 > 0$ - Always true (for all values of x)

Thus, we have:

$1.5 < x < 3$

Since x is an integer, the only possible value of $x = 2$

Thus, the sides are

$x = 2$

$2x - 1 = 3$

$5 - x = 3$

Thus, the perimeter of the triangle = 2 + 3 + 3 = 8

The correct answer is choice C.

19. Let us bring out the image.

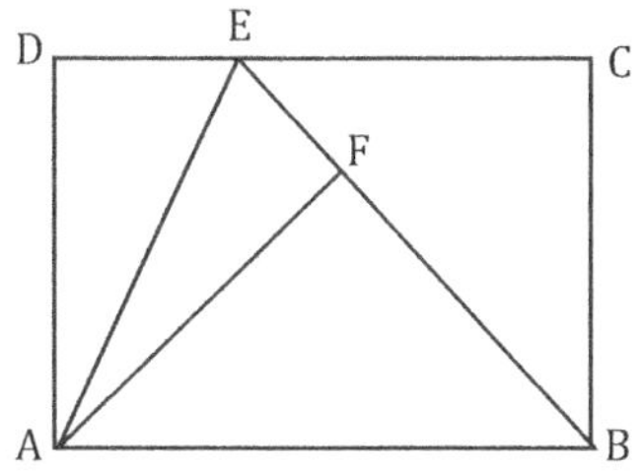

Area of triangle ABE = $\frac{1}{2} \times BE \times AF = \frac{1}{2} \times 18 \times 15 = 135$

Again: Area of triangle ABE = $\frac{1}{2} \times AB \times AD$ (since the height of the triangle, taking base as AB, is AD)

=> Area of triangle ABE = $\frac{1}{2} \times$ Area of ABCD

=> Area of ABCD = $2 \times 135 = 270$

The correct answer is 270.

20. Continuing from the previous solutions, we have:

Area of triangle AMC = $\frac{\sqrt{3}}{4}x^2 = 36\sqrt{3}$

=> $x = 12$

Area of triangle HMF = $\frac{\sqrt{3}}{4}y^2 = 9\sqrt{3}$

=> $y = 6$

Thus, we have:

GD = $y + x = 18$

GK = $\frac{\sqrt{3}}{2}(y + x) = 9\sqrt{3}$

Thus, area of rectangle BDGK = GD × GK = $18 \times 9\sqrt{3} = 162\sqrt{3}$

The correct answer is choice D.

21. Continuing with the solutions of the previous two questions, we have:

Area of rectangle KLMH = LM × MH = $\left(\frac{\sqrt{3}}{2}x\right) \times y$

Area of rectangle CDEM = ME × CM = $\left(\frac{\sqrt{3}}{2}y\right) \times x$

Thus, the two areas are equal.

The correct answer is choice A.

22. Let us bring out the figure.

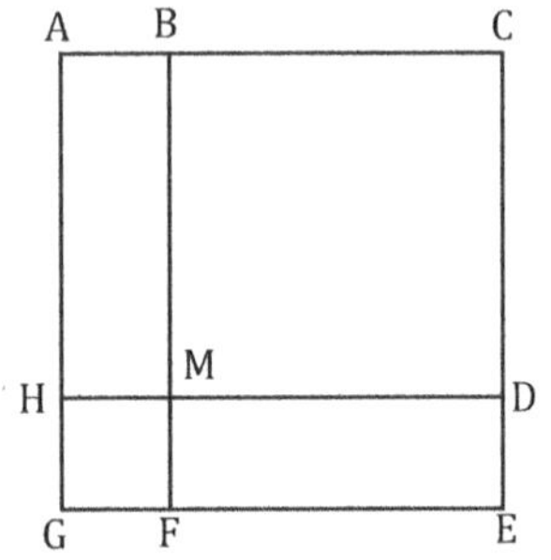

Let HM = MF = x

=> Area of GFMH = x^2

=> Area of BCDM = $16x^2$

=> BC = CD = $4x$

=> GE = MH + MD = $x + 4x = 5x$

Thus, perimeter of ACEG = $4 \times 5x = 20x$

=> $20x = 40$ => $x = 2$

Thus, required difference = $4 \times 4x - 4 \times x = 12x = 24$

The correct answer is choice C.

23. Let the side of Square B be x

=> Length of the diagonal of Square B = $x\sqrt{2}$

=> Side of Square A = $x\sqrt{2} \times k$

=> Area of Square A = $\left(x\sqrt{2} \times k\right)^2 = 2x^2k^2$

Thus, we have:

$2x^2k^2 = k \times x^2$

=> $k = \frac{1}{2} = 0.5$

The correct answer is $\frac{1}{2}$ or 0.5.

24. Let BD be x

=> CD = $(7 - x)$

Since AD is perpendicular to BC, the Pythagoras theorem can be applied in right triangles ADB and ADC.

Thus, in triangle ADB:

$AD^2 = AB^2 - BD^2 = 36 - x^2 \ldots$ (i)

Also, in triangle ADC:

$AD^2 = AC^2 - CD^2 = 64 - (7 - x)^2 \ldots$ (ii)

Thus, from (i) and (ii): $36 - x^2 = 64 - (7 - x)^2 \Rightarrow 36 - x^2 = 64 - 49 + 14x - x^2 \Rightarrow x = \frac{3}{2}$

The correct answer is choice C.

25. From the previous question, we have:

$BD = \frac{3}{2}$

$\Rightarrow CD = 7 - \frac{3}{2} = \frac{11}{2}$

Thus, the required ratio $= \frac{3}{2} : \frac{11}{2} = 3 : 11$

The correct answer is choice A.

26. Let's bring out the figure.

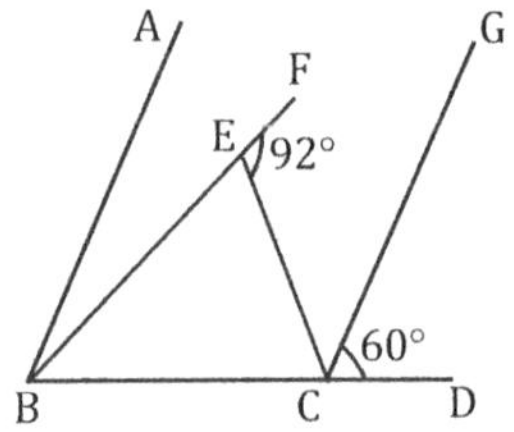

We have AB parallel to CG

$\Rightarrow \angle GCD = \angle ABC = 60°$

Now, in triangle BEC, it is given that BE = CE

$\Rightarrow \angle EBC = \angle ECB \ldots$ (i)

Again, in triangle BEC,

$\angle EBC + \angle ECB + \angle BEC = 180° \ldots$ (ii)

Combining (i) and (ii), we get:

$\angle EBC = \angle ECB = \frac{180 - BEC}{2} = \frac{CEF}{2} = 46$

$\Rightarrow \angle ABF = 60° - 46° = 14°$

The correct answer is choice A.

27. We join the centers X, Y, Z of the three circles.

Let the points where the circles touch each other be A, B and C.

Thus, the required perimeter is the sum of the lengths of the minor arcs AB, BC and CA.

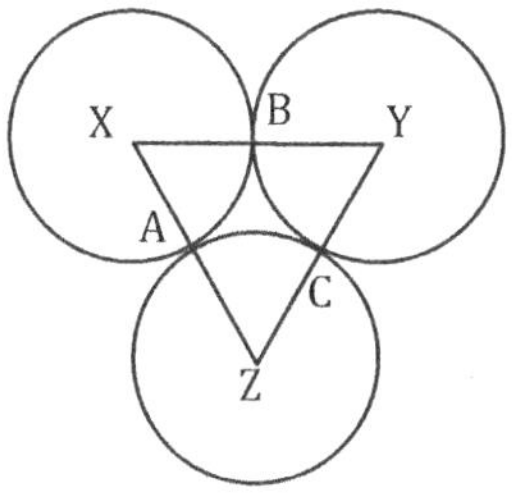

In triangle XYZ, side XY = XB + BY = 1 + 1 = 2 cm.

Similarly, sides YZ and ZX too measure 2 cm each.

Thus, triangle XYZ is equilateral.

Therefore, we have: $\angle XYZ = \angle YZX = \angle ZXY = 60$

So, (Length of minor arc AB) = (Length of minor arc BC) = (Length of minor arc CA) $= 2\pi r \times \frac{60}{360} = 2\pi\left(\frac{1}{6}\right) = \frac{\pi}{3}$

Thus, required perimeter $= 3 \times \frac{\pi}{3} = \pi$ cm.

The correct answer is choice C.

28. Let's bring out the figure.

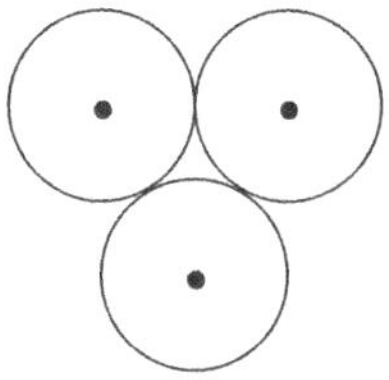

Total perimeter of the three equal circles

$= 3\,(2\pi r) = 6\pi$ cm (since radius r = 1 cm)

Thus, the required outer perimeter

= (Sum of perimeters of the three circles) – (Perimeter of the shaded region) $= 6\pi - \pi = 5\pi$ cm.

The correct answer is choice D.

29. Let's bring out the figure.

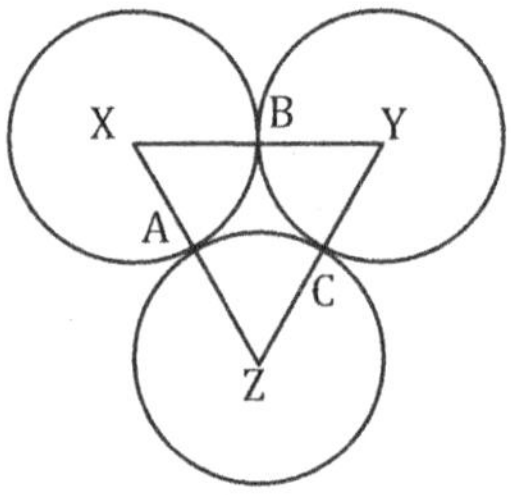

Area of each of the minor sectors XAB, YBC and ZAC = $\pi\ (1^2)\left(\frac{60}{360}\right) = \frac{\pi}{6}$ cm^2

Thus, area of enclosed region

= (Area of triangle XYZ) − (Sum of the areas of the above three sectors)

$= \frac{\sqrt{3}}{4}\ (2^2) - 3\left(\frac{\pi}{6}\right) = \sqrt{3} - \frac{\pi}{2}$ cm^2 ... (i)

Given: Area of this region = $(a\sqrt{3} + b\pi)\ cm^2$ ... (ii)

Comparing (i) and (ii): => $a = 1,\ b = -\frac{1}{2}$ => $a + b = 1 - \frac{1}{2} = \frac{1}{2}$

The correct answer is choice A.

30. Let the three edges (length, width and height) of the beam be l, w and h. So, the areas of the three distinct faces of the beam will be $lw,\ wh$ and hl square meters.

We are given the numerical values of these three areas.

Let us assume that,

$lw = 15$ m^2 and $wh = 3$ m^2

Then, $lh = 20$ m^2

Multiplying the above:
$(lw) \times (wh) \times (lh) = 15 \times 3 \times 20$

$l^2h^2w^2 = 900$

=> $lwh = 30$

Thus, the volume is 30 cubic meters.

The correct answer is 30.

31. We have:

$lw = 15\ m^2 \ldots$ (i)

$wh = 3\ m^2 \ldots$ (ii)

$lh = 20\ m^2 \ldots$ (iii)

$lwh = 30\ m^3 \ldots$ (iv)

Dividing (iv) by (i): $\frac{lwh}{lw} = \frac{30}{15} => h = 2\ m$

Similarly, Dividing (iv) by (ii): $l = 10\ m$

Dividing (iv) by (iii): $w = 1.5\ m$

In the rectangular beam, each edge is present four times. For example, there are four 'lengths' - two at the bottom face and two at the top face.

Thus, the required sum of edges = 4 (2 + 10 + 1.5) = 54 m

The correct answer is 54.

32. In triangles APQ, ARS and ABC, ∠A is common.

Since the lines PQ, RS and BC are parallel to one another,
∠APQ = ∠ARS = ∠ABC

And, ∠AQP = ∠ASR = ∠ACB

Thus, triangles APQ, ARS and ABC are similar to one another.

We know that: AP = PR = RB.

Therefore, AR = AP + PR = 2PR

And, AB = AP + PR + RB = 3PR

Thus, we see that the sides of similar triangles APQ, ARS and ABC are in the ratio

= AP : AR : AB = 1 : 2 : 3 = PQ : RS : BC

Since BC = 9, we have: PQ = $\frac{1}{3}$ BC = $\frac{9}{3}$ = 3 units.

The correct answer is 3.

33. We know that:

Triangles APQ, ARS and ABC are similar to one another with ratio of sides

= AP : AR : AB = 1 : 2 : 3

Thus, ratio of the areas of the triangles = Square of the ratio of sides

= 1 : 4 : 9

Thus, (Area of triangle APQ) = $\frac{1}{9}$ (Area of triangle ABC) = $\frac{90}{9}$ = 10 square units.

The correct answer is 10.

34. Area of PQSR = (Area of triangle ARS) – (Area of triangle APQ)

From the previous question, we know that the area of triangle APQ is 10 square units. So, we now need to find the area of triangle ARS.

We know that the ratio of the areas of the triangles APQ, ARS and ABC

= Square of the ratio of sides = 1 : 4 : 9

Thus, area of triangle ARS = $\frac{4}{9}$ (Area of triangle ABC) = $\frac{4\,(90)}{9}$ = 40 square units.

Thus, area of PQSR = 40 – 10 = 30 square units.

The correct answer is 30.

35. ABCD is the trapezium, where AB = 8, CD = 14, AD = 5, BC = 6.

Since the parallel sides are AB and CD, the height of the trapezium is equal to the length of the perpendiculars AE and BF, dropped from vertices A and B respectively to the base CD.

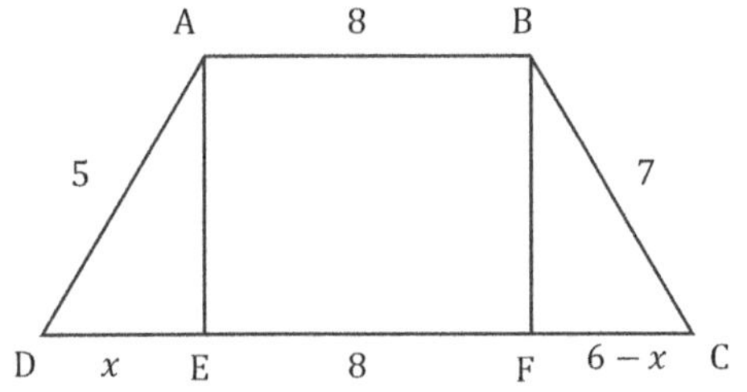

Let DE be x

=> CF = 14 – 8 – x = (6 – x)

In right triangle AED, $AE^2 = AD^2 - DE^2$

In right triangle BFC, $BF^2 = BC^2 - FC^2$

Equating AE and BF, we have:

$AD^2 - DE^2 = BC^2 - FC^2$

$=> 5^2 - x^2 = 7^2 - (6 - x)^2$

$=> 25 - x^2 = 49 - 36 - x^2 + 12x$

$=> x = 1$

$=> AE = \sqrt{5^2 - 1^2} = \sqrt{24} = 4.899 =\approx 4.9$

The correct answer is 4.9.

36. The longest distance between two vertices of the trapezium would be one of the diagonals of the trapezium; hence either AC or BD.

$AC = \sqrt{AE^2 + EC^2} = \sqrt{24 + (8 + 6 - 1)^2} = \sqrt{24 + 169} = 13.9 =\approx 14$

$BD = \sqrt{BF^2 + FD^2} = \sqrt{AE^2 + FD^2} = \sqrt{24 + (8 + 1)^2} = \sqrt{24 + 81} < AC$

The correct answer is 14.

37. Let's label the center of the circle as O.

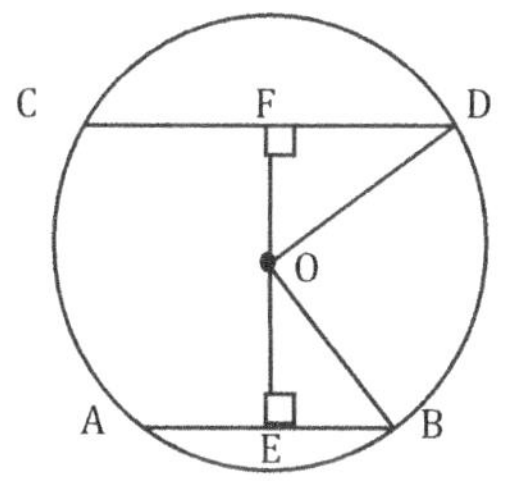

The shortest distance between two parallel lines is the perpendicular distance between them. So, let's drop a perpendicular FE from line CD to the line AB that also passes through the center of the circle, O.

We have:

EF = 14, AB = 12 and OD = OB = Radius = 10.

OE is the perpendicular from the center of the circle to the chord AB. So, it will bisect chord AB.

Therefore, AE = EB = $\frac{12}{2} = 6$

=> In right triangle OEB, OE = $\sqrt{10^2 - 6^2} = 8$

=> OF = EF − OE = 14 − 8 = 6

=> In right triangle OFD, FD = $\sqrt{10^2 - 6^2} = 8$

=> CD = 2 × FD = 16.

The correct answer is choice C.

38. Assuming the radius of the cylinder to be r cm, its volume

$= \pi r^2 \times 18$

Thus, equating the volume of the sphere and the cylinder, we have:

$\frac{4}{3}\pi \times 6^3 = \pi r^2 \times 18$

=> $r^2 = \frac{4}{3} \times \frac{6^3}{18} = 16$

=> $r = 4$ cm

(Taking the square root on both sides of the equation, the value $r = -4$ will be rejected because the radius of a physical object cannot be negative.)

The correct answer is choice B.

39. Let us bring out the figure.

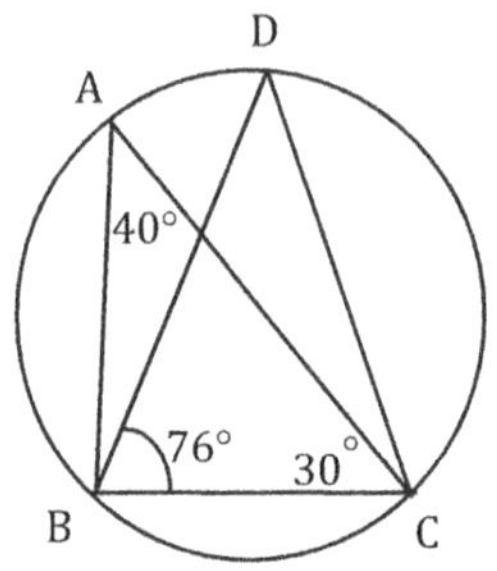

In triangle ABC:

$\angle ABC = 180° - (40° + 30°) = 110°$

=> $\angle ABD = 110° - 76° = 34°$

We know that angles in the same segment of a circle arc equal.

=> $\angle ACD = \angle ABD = 34^\circ$

The correct answer is choice B.

40. Let us bring out the figure.

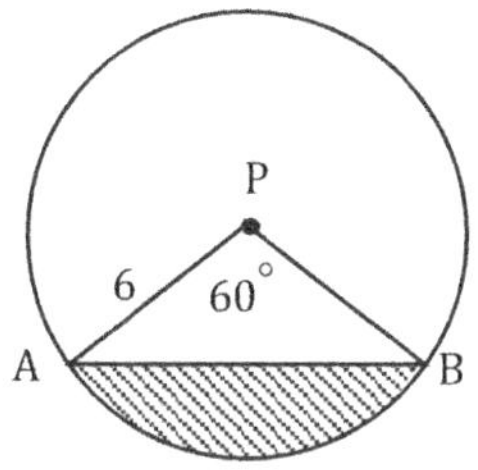

The area of the sector PAB = $\pi\ (6^2)\left(\frac{60^\circ}{360^\circ}\right) = 6\pi = 6 \times 3.14 = 18.84$ square units

In triangle APB:

$\angle APB = 60^\circ$

And, AP = PB = 6 (radius)

$=> \angle PAB = \angle PBA = \frac{180^\circ - 60^\circ}{2} = 60^\circ$

Thus, triangle ABP is equilateral with side 6.

=> Area of triangle PAB

$= \frac{\sqrt{3}}{4} \times 6^2 = 9\sqrt{3}$

$= 9 \times 1.73 = 15.57$ square units

Thus, the area of the shaded region

= (Area of sector PAB) – (Area of triangle PAB)

$= 18.84 - 15.57$

$= 3.27 \approx= 3.3$ square units

The correct answer is choice A.

41. Let us bring out the figure.

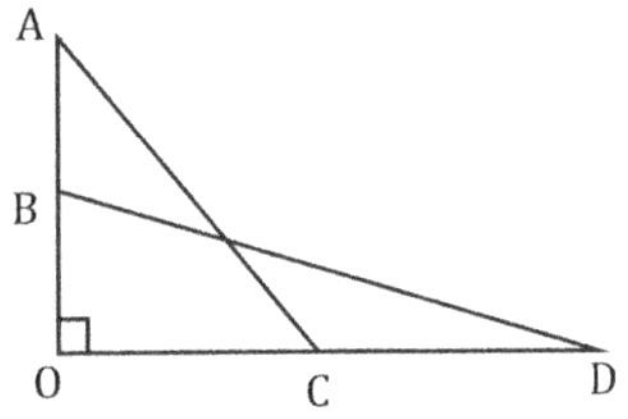

We have AC = BD = 25 feet (length of the ladder), OC = 7 feet and AB = 19 feet.

The distance by which the foot of the ladder slipped farther away from the wall = OD – OC = CD

We need to determine the length of CD.

From right-angled triangle AOC:

$AO = \sqrt{AC^2 - OC^2} = \sqrt{25^2 - 7^2} = 24$ feet

Since AB = 19 feet, we have:

OB = 24 – 19 = 5 feet

From right-angled triangle BOD:

$OD = \sqrt{BD^2 - OB^2} = \sqrt{25^2 - 5^2} = \sqrt{600} = 10\sqrt{6} = 24.5$ feet

Since OC = 7 feet, we have:

CD = 24.5 – 7 = 17.5 feet

The correct answer is choice C.

42. Let us bring out the figure.

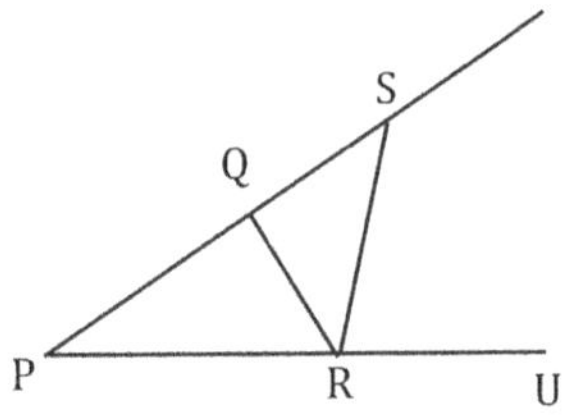

We have: In triangle PQR,

$\angle QPR = 20^\circ$ and PQ = PR

Thus: $\angle PQR = \angle PRQ = \dfrac{180^\circ - 20^\circ}{2} = 80^\circ$

=> $\angle SQR = 180^\circ - 80^\circ = 100^\circ$

=> In triangle QSR, $\angle QRS = \angle QSR = \dfrac{180^\circ - 100^\circ}{2} = 40^\circ$

=> $\angle SRU = 180^\circ - QRS - QRP = 180^\circ - 40^\circ - 80^\circ = 60^\circ$

The correct answer is 60.

43. Let us bring out the figure.

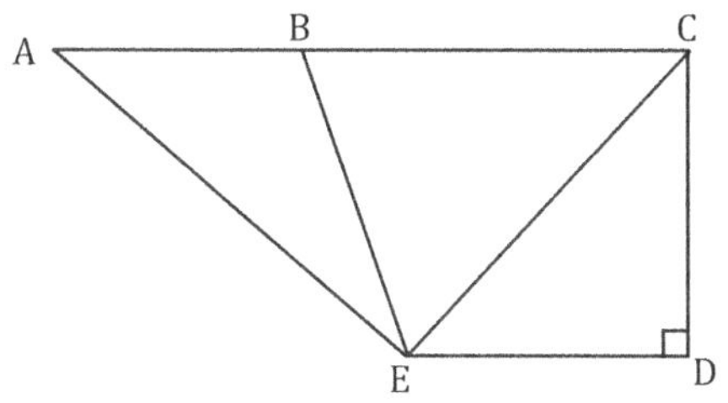

ED = 3, and CD = 4

Thus, in right-angled triangle CDE:

$CE = \sqrt{CD^2 + DE^2} = \sqrt{4^2 + 3^2} = 5$

Since AE = 10, in right-angled triangle AEC:

$AC = \sqrt{CE^2 + AE^2} = \sqrt{5^2 + 10^2} = \sqrt{125} = 11.2$

Since AB = 3, we have:

BC = 11.2 - 3 = 8.2

The correct answer is choice B.

44. Let us bring out the figure.

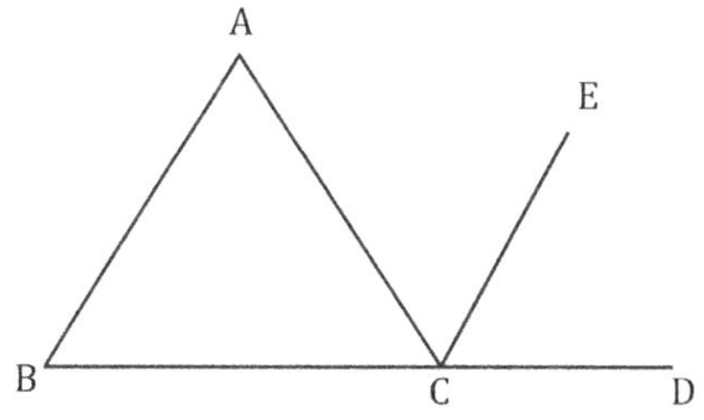

In triangle ABC, AB = BC

=> $\angle BAC = \angle BCA$

Let $\angle BAC = \angle BCA = x$

=> $\angle ACD = 180° - x$

Since CE bisects $\angle ACD$, we have:

$\angle ACE = \angle DCE = \dfrac{180° - x}{2}$

Since CE is parallel to AB, we have:

$\angle BAC = \angle ACE$ (alternate angles)

=> $x = \dfrac{180° - x}{2}$

=> $x = 60°$

Thus, in triangle ABC:

$\angle ABC = 180° - \angle BAC - \angle BCA = 180° - 2x$

$= 60°$

The correct answer is choice B.

45. Let us bring out the figure.

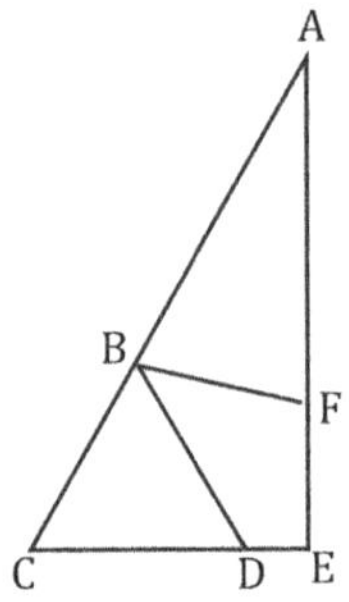

Since AB = AF in ΔABF, let $\angle ABF = \angle AFB = x$

=> $\angle BAF = 180° - 2x$

Since $\angle DBF = 60°$, we have:

$\angle CBD = 180° - (x + 60°) = 120° - x$

Since in ΔCBD, BC = CD, we have:

$\angle$BDC = $\angle$CBD = 120° − x

=> $\angle$BCD = 180° − 2 (120° − x) = 2x − 60°

Thus, in triangle AEC:

$\angle$AEC = 180° − (ACE + CAE) = 180° − (180° − 2x) − (2x − 60°) = 60°

The correct answer is choice C.

46. In the given figure, we have:

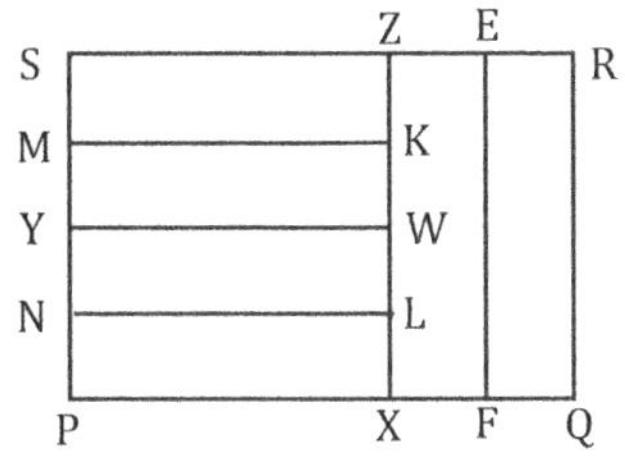

The six rectangles are congruent, that is, they have equal lengths and equal breadths.

=> PX = QR (lengths of the congruent rectangles)

And SM = MY = YN = NP = XF = FQ (breadths of the congruent rectangles)

Thus, we have:

QR = PX = 4SM

PQ = PX + QX = 4SM + 2SM = 6SM

Thus, we have:

$$\frac{PQ}{QR} = \frac{6SM}{4SM} = \frac{3}{2}$$

The correct answer is choice B.

47. Let us bring out the figure.

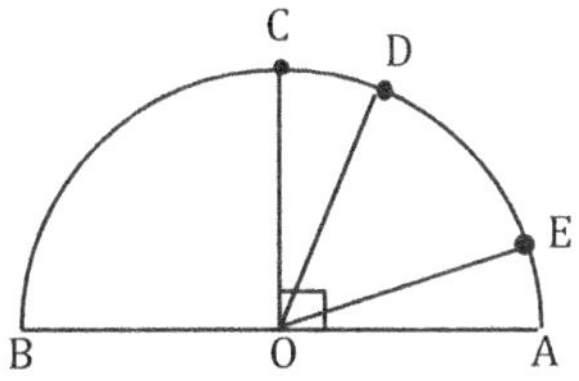

We have: (Length of Arc AE) = (Length of Arc CD) $= \frac{1}{10}$(Length of Arc AC)

Since the length of an arc depends on the angle subtended by that arc at the center of the circle, from the above equation, we can infer that:

$\angle AOE = \angle COD = \frac{1}{10}\angle AOC = \frac{1}{10} \times 90^\circ = 9^\circ$

$=> \angle DOE = 90^\circ - (9^\circ + 9^\circ) = 72^\circ$

$=>$ Area of the sector DOE $= \pi\ (3^2)\left(\frac{72^\circ}{360^\circ}\right) = 9\pi\left(\frac{1}{5}\right) = 1.8\pi$

$= 1.8 \times 3.14 = 5.65 \approx= 6$ square units.

The correct answer is 6.

48. Let us bring out the figure.

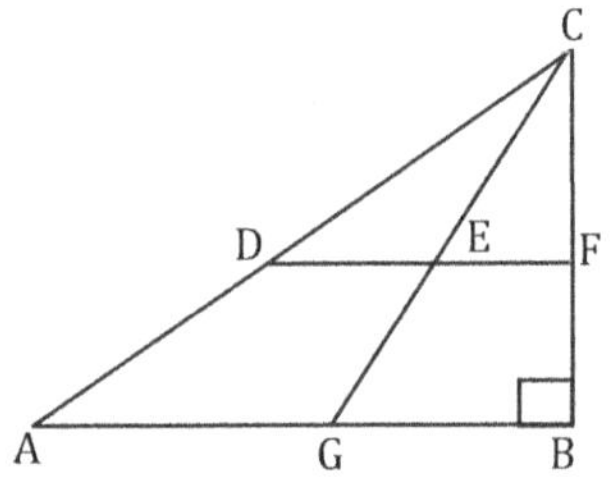

In right triangle ABC, we have:

$AC^2 = BC^2 + AB^2 = 4^2 + 3^2 => AC = 5$

Since AD = 2, we have: CD = 5 − 2 = 3

$=> \frac{CD}{AC} = \frac{3}{5}$

In right triangle CBG, we have:

$CG^2 = BC^2 + BG^2 = 4^2 + 1^2 => CG = \sqrt{17}$

For similar triangles CDE and CAG (The two triangles are similar because: ∠DCE is common to both triangles, and since DE is parallel to AG, ∠CDE = ∠CAG):

$\frac{CD}{AC} = \frac{CE}{CG} => \frac{3}{5} = \frac{CE}{\sqrt{17}}$

$=> CE = \frac{3\sqrt{17}}{5}$

The correct answer is choice C.

49. Let us bring out the figure.

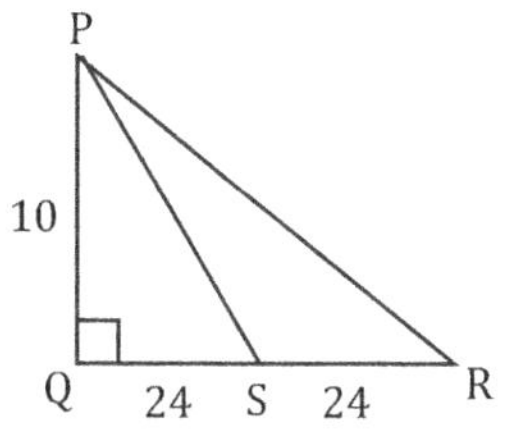

In the diagram shown, triangle PQR is right-angled at Q.

PS is the side that is drawn on the side QR of length 48 units.

PR = $\sqrt{48^2 + 10^2} = 49.03 \approx 49$

Also, we have PS = $\sqrt{24^2 + 10^2} = 26$

Perimeter of the triangle PQR = 10 + 48 + 49 = 107

Sum of perimeters of triangles PQS and PSR

= (24 + 10 + 26) + (26 + 24 + 49) = 159

Thus, required percent increase

$$= \frac{159 - 107}{107} \times 100 = 48.6\%$$

The correct answer is choice D.

50. In the diagram below of PQRS, SA and RB are drawn perpendicular to PQ. Since the lines SR and PQ are parallel, the perpendicular distance between them remains constant. So, SA = RB.

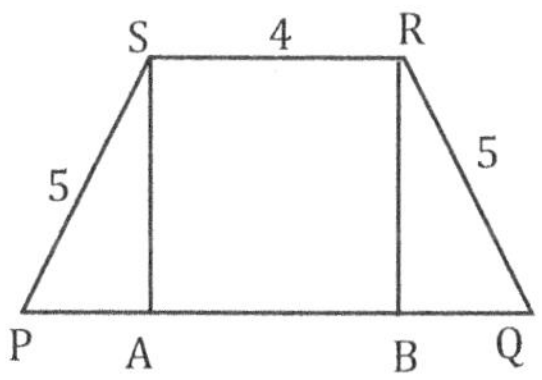

Thus, we have: AB = SR = 4 (since ABRS forms a rectangle)

Since PS = RQ, SA = RB and $\angle SAP = \angle RBQ = 90^\circ$, we have:

Triangle SAP is congruent to triangle RBQ

=> AP = BQ

$$=> AP = \frac{PQ - AB}{2} = \frac{10 - 4}{2} = 3$$

Since SAP is a right-angled triangle, we have:

$$AS = \sqrt{PS^2 - AP^2} = \sqrt{25 - 9} = 4$$

Area of PQRS

= (Area of triangle APS) + (Area of triangle BQR) + (Area of rectangle ABRS)

$$= \frac{1}{2}(3 \times 4) + \frac{1}{2}(3 \times 4) + (4 \times 4)$$

$$= 28$$

The correct answer is choice B.

51. Volume of the marbles to be added should equal the volume of additional water the container can hold.

Since the container holds water till $\frac{1}{3}$ of its height, it can hold another $\frac{2}{3}$ of its total volume of water. It does not matter which of the three sides, 21, 14, and 11 cm, is the height since we are concerned with volume.

Thus, assuming that the container can accommodate n spheres (spherical marbles), we have:

$$\frac{2}{3} \times 21 \times 14 \times 11 = n \times \frac{4}{3} \times \pi \times \left(\frac{1.4}{2}\right)^3$$

$$=> 14 \times 14 \times 11 = n \times \frac{4}{3} \times \frac{22}{7} \times (0.7)^3$$

$$=> n = \frac{6}{4 \times 0.001} = 1,500$$

The correct answer is 1500.

52. Let us bring out the figure.

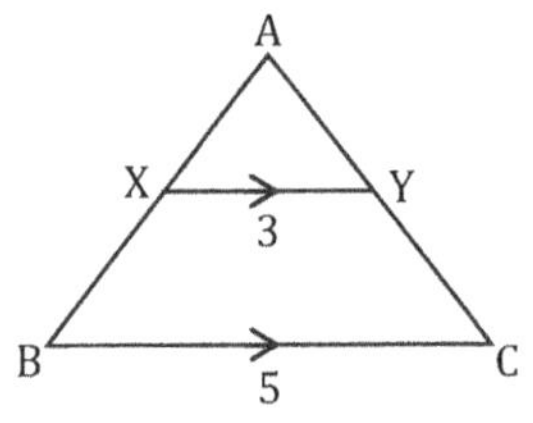

In Triangles AXY and ABC,

$\angle XAY = \angle BAC$ (same angle)

$\angle AXY = \angle ABC$ (corresponding angles of parallel lines XY and BC)

Thus, Triangles AXY and ABC are similar

Since the ratio of areas of similar triangles equals the square of ratio of sides, we have:

$$\frac{\text{Area of triangle AXY}}{\text{Area of triangle ABC}} = \left(\frac{3}{5}\right)^2 = \frac{9}{25}$$

$$=> \frac{(\text{Area of triangle ABC}) - (\text{Area of BCYX})}{\text{Area of triangle ABC}} = \frac{9}{25}$$

$$=> 1 - \frac{32}{\text{Area of triangle ABC}} = \frac{9}{25} => \frac{32}{\text{Area ABC}} = \frac{16}{25}$$

=> Area of triangle ABC = 50 square units.

The correct answer is choice C.

53. From the above solutions, we have:

BC = 141 miles and AC = 200 miles.

Time taken by the train to reach C = $\frac{141}{60}$ = 2 hours 21 minutes

Thus, the train, which leaves city B at 12 noon, reaches city C at 2:21 pm.

Thus, time the man has to reach city C, after starting from city A at 8 am

= 6 hours 21 minutes = $6\frac{21}{60}$ hours = 6.35 hours

Thus, his required speed (the distance being 200 miles)

$= \frac{200}{6.35}$ = 31.496 miles per hour

=≈ 31.5 miles per hour

The correct answer is choice C.

54. Let us bring out the image.

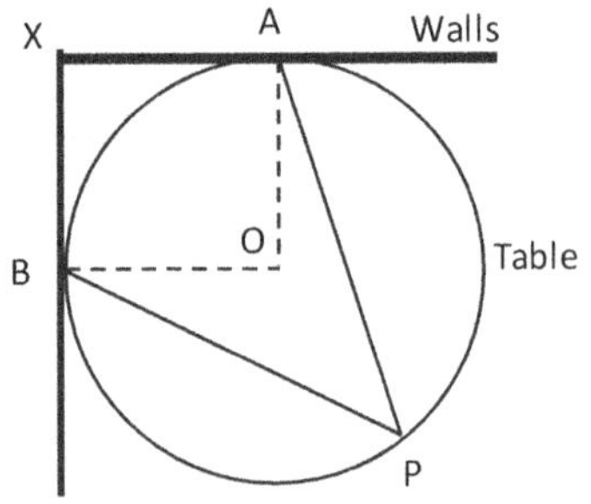

Let us join AP, BP, AO, and BO, where O is the center of the circle.

OA and OB are perpendicular to the walls since those are tangent to the circle (tangents are perpendicular to the radii).

Thus, $\angle AOB = 90^\circ$ (since in XAOB, all the other 3 angles are 90° each).

We know that the angle subtended by an arc at the center is twice the angle subtended by the same arc at the circumference.

Thus, the angle subtended by minor arc AB at the circumference, i.e. $\angle APB$

$$= \frac{1}{2} \times \angle AOB$$

$$= \frac{90^\circ}{2} = 45^\circ$$

The correct answer is choice B.

55. From the previous solution, we already know that $\angle APB = 45^\circ$

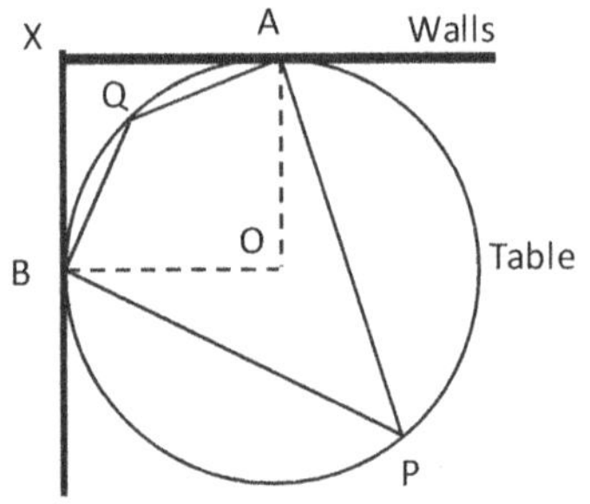

From the diagram, it is clear that quadrilateral APBQ is cyclic (since all the vertices lie on the circumference of the circle).

Thus, sum of opposite angles equals 180°

Thus, $\angle AQB = 180^\circ - \angle APB$

$= 180^\circ - 45^\circ$

$= 135^\circ$

The correct answer is choice B.

56. Let us bring out the image.

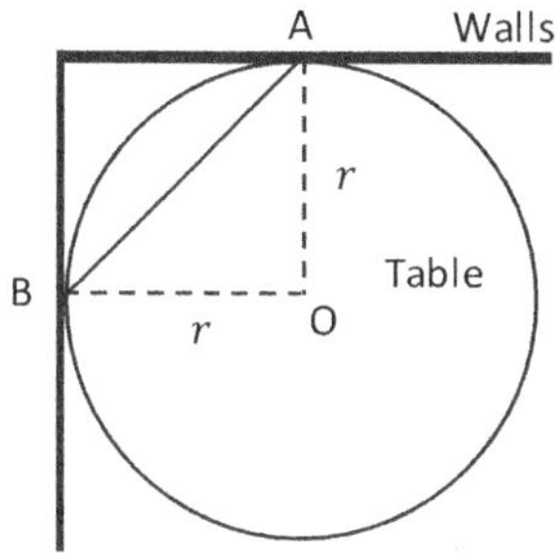

Let the table be in contact with the walls at points A and B. Also, let O be the center of the circular table.

From the previous solutions, we already know that $\angle AOB = 90^\circ$

Let the radius of the circle be r.

Thus, in right-angled triangle AOB, we have, from Pythagoras' theorem:

$AO^2 + BO^2 = AB^2$

$\Rightarrow AB^2 = 2r^2$

$\Rightarrow AB = r\sqrt{2}$

Here, AB represents the length of the shortest line connecting the two points of the table in contact with the walls.

Thus, the required ratio

$= \frac{AB}{r} = \frac{r\sqrt{2}}{r}$

$= \sqrt{2} : 1$

The correct answer is choice C.

57. Let us bring out the image.

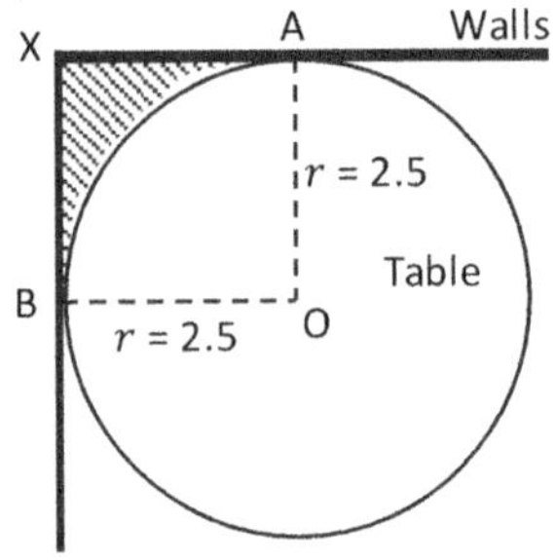

Let the table be in contact with the walls at points A and B. Also, let O be the center of the circular table.

From the previous solutions, we already know that $\angle AOB = 90^\circ$

The shaded area

= (Area of square AOBX) – (Area of quadrant AOB)

$= r^2 - \frac{\pi r^2}{4}$

$= 2.5^2 - 3.14 \times \frac{2.5^2}{4}$

$= 1.344 \text{ m}^2$

Thus, the required volume

= (Area of shaded region) × (Height)

$= 1.344 \times 3$

$= 4.032 \text{ m}^3 \approx = 4 \text{ m}^3$

The correct answer is choice A.

19.2 Trigonometry

58. Let us bring out the image.

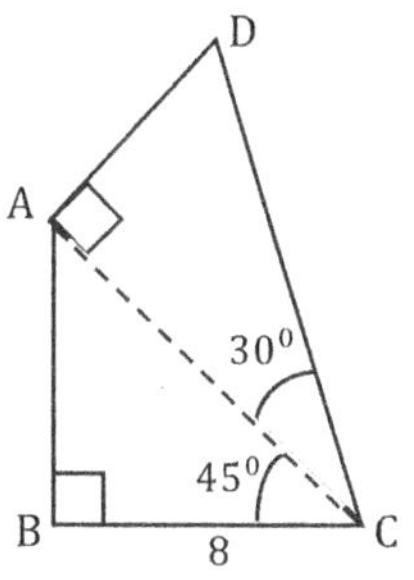

Since triangle ABC is right angled isosceles, we have:

AB = BC = 8

AC = $8\sqrt{2}$

In 30-60-90 triangle ACD:

$\tan 30^\circ = \frac{AD}{AC}$

$=> AD = \frac{1}{\sqrt{3}} \times 8\sqrt{2} = 6.53$

$\sin 30^\circ = \frac{AD}{DC}$

=> DC = 2AD = 13.06

Thus, the perimeter = AB + BC + CD + DA = 8 + 8 + 6.53 + 13.06 = 35.59 $\approx$= 36

The correct answer is 36.

59. We drop a line perpendicular to CD from A.

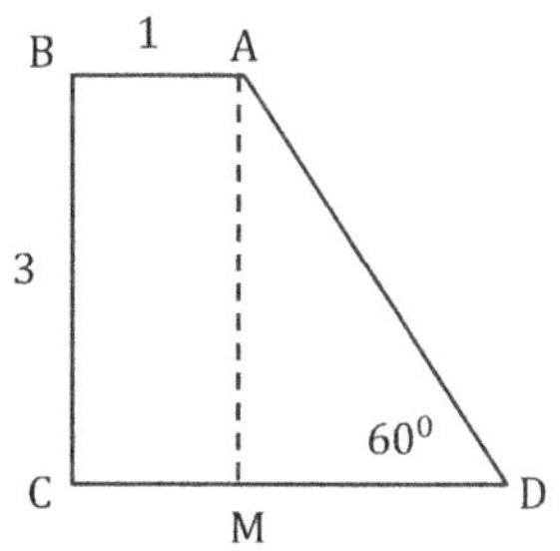

Thus, CM = AB = 1; also, AM = BC = 3

In right triangle AMD, with one angle 60°, we have:

$\tan 60^\circ = \frac{AM}{MD}$

$=> \sqrt{3} = \frac{3}{MD}$

$=> MD = \sqrt{3} \ldots$ (i)

Also, we have:

$\sin 60^\circ = \frac{AM}{AD}$

$=> \frac{\sqrt{3}}{2} = \frac{3}{AD}$

$=> AD = 2\sqrt{3}$

$=> AD + CD = AD + MD + CM$

$= 2\sqrt{3}+\sqrt{3} + 1 = 3\sqrt{3} + 1 = 6.196 \approx= 6.2$

The correct answer is 6.2.

60. The figure for triangle ABC is as shown.

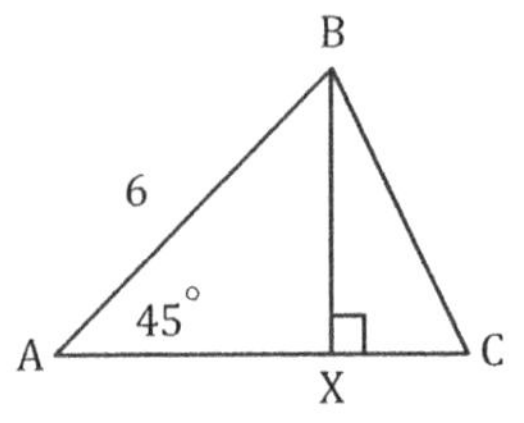

We draw a perpendicular BX on AC.

Thus, right-angled triangle AXB is a $45^\circ - 45^\circ - 90^\circ$ triangle

$=> AX = BX = \frac{AB}{\sqrt{2}} = \frac{6}{\sqrt{2}} = 3\sqrt{2}$

Also, since $\angle ABX = 45^\circ$, we have:

$\angle CBX = 75^\circ - 45^\circ = 30$

Thus, right-angled triangle BCX is a $30^\circ - 60^\circ - 90^\circ$ triangle

$=> \cos 30^\circ = \frac{BX}{BC}$

$=> BC = \frac{BX}{\left(\frac{\sqrt{3}}{2}\right)} = \frac{3\sqrt{2}\times 2}{\sqrt{3}} = 2\sqrt{6} = 4.89 \approx= 5$

The correct answer is 5.

61. Let us bring out the figure.

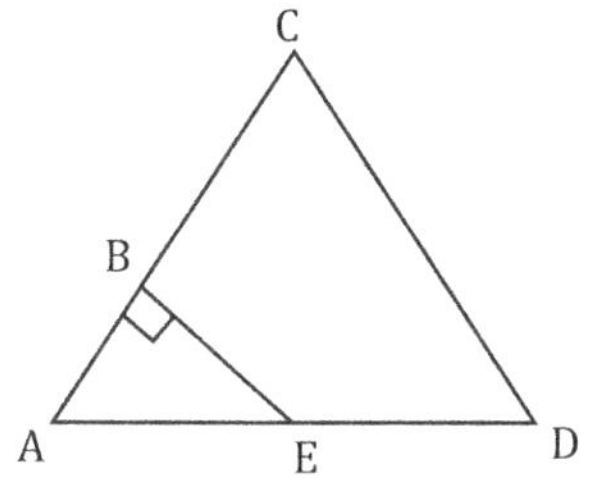

We have

AC = CD = DA = 20 units

Thus, area of equilateral triangle ACD

$= \frac{\sqrt{3}}{4} \times 20^2 = 100\sqrt{3} = 173.2$ square units

Now, in triangle ABE, we have:

$\angle ABE = 90°$; $\angle BAE = 60°$

and $\angle AEB = 180° - 90° - 60° = 30°$

Thus, triangle ABE is a 30-60-90 triangle with side $AE = \frac{AD}{2} = 10$ (Since AE = DE)

$\sin 30° = \frac{AB}{AE} \Rightarrow AB = AE \times \frac{1}{2} = 5$ units

$\cos 30° = \frac{BE}{AE} \Rightarrow BE = AE \times \frac{\sqrt{3}}{2} = 5\times\sqrt{3} = 8.66$ units

Thus, area of triangle ABE

$= \frac{1}{2}(AB)(BE) = \frac{1}{2}(5 \times 8.66) = 21.65$ square units

Thus, area of BCDE = (Area of ΔACD) - (Area of ΔABE)
$= 173.2 - 21.65 = 151.55 \approx= 152$ square units

The correct answer is 152.

62. We know that π radians are equivalent to 180°.

Thus, converting all radians to degrees, we have:

- $\frac{4\pi}{5}$ radians are equivalent to $180° \times \frac{4}{5} = 144°$.
- $\frac{11\pi}{6}$ radians are equivalent to $180° \times \frac{11}{6} = 330°$.

- $\frac{13\pi}{3}$ radians are equivalent to $180^\circ \times \frac{13}{3} = 780^\circ$.

Thus, the required value is $(144^\circ + 330^\circ - 780^\circ) = -306^\circ$.

A negative value of an angle θ implies that the angle has been measured in the counter clockwise direction. Thus, the equivalent positive value is $(\theta + 360)$. Thus, we have:

$-306^\circ = (360^\circ - 306^\circ) = 54^\circ$

The correct answer is choice A.

63. In the right triangle ABC, we know that $\angle BAC = 30^\circ$

$=> \angle ACB = 90^\circ - 30^\circ = 60^\circ$

We have: $\tan \angle ACB = \frac{AB}{BC}$

$=> \tan 60^\circ = \sqrt{3} = \frac{5}{BC}$

$=> BC = \frac{5}{\sqrt{3}}$

The correct answer is choice D.

64. From the previous solution, we have: $BC = \frac{5}{\sqrt{3}}$

In right triangle BDC, we know that $\cos \angle BCD = \frac{CD}{BC}$

$=> \cos 60^\circ = \frac{1}{2} = \frac{CD}{\left(\frac{5}{\sqrt{3}}\right)}$

$=> CD = \frac{5}{2\sqrt{3}}$

The correct answer is choice B.

65. We have: $AB = 5$

$\cos \angle BAC = \frac{AB}{AC}$

$=> \cos 30^\circ = \frac{\sqrt{3}}{2} = \frac{5}{AC}$

$=> AC = \frac{10}{\sqrt{3}}$

From the previous solutions, we have: $BC = \frac{5}{\sqrt{3}}$

Thus, perimeter of triangle ABC = AB + BC + AC

$= 5 + \frac{5}{\sqrt{3}} + \frac{10}{\sqrt{3}}$

$= 5 + \frac{15}{\sqrt{3}} = 5 + 5\sqrt{3}$

$= 5 + 5 \times 1.73 = 13.65$

The correct answer is choice D.

66. Let us bring out the figure.

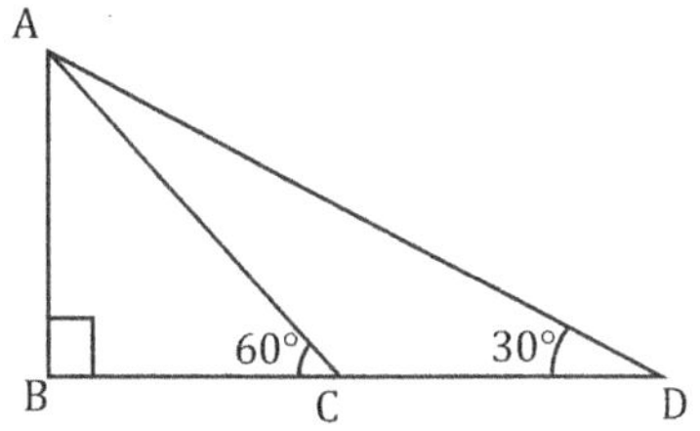

In the 30-60-90 triangle ABC, we have:

$\tan 60° = \sqrt{3} = \frac{AB}{BC} = \frac{5}{BC}$

$=> BC = \frac{5}{\sqrt{3}}$

In the 30-60-90 triangle ABD, we have:

$\tan 30° = \frac{1}{\sqrt{3}} = \frac{AB}{BD} = \frac{5}{BD}$

$=> BD = 5\sqrt{3}$

$=> CD = BD - BC$

$= 5\sqrt{3} - \frac{5}{\sqrt{3}}$

$= 5 \times 1.73 - \frac{5}{1.73}$

$= 5.76$

$= \approx 6$

The correct answer is 6.

67. Let us drop a perpendicular from C on AB, as shown below:

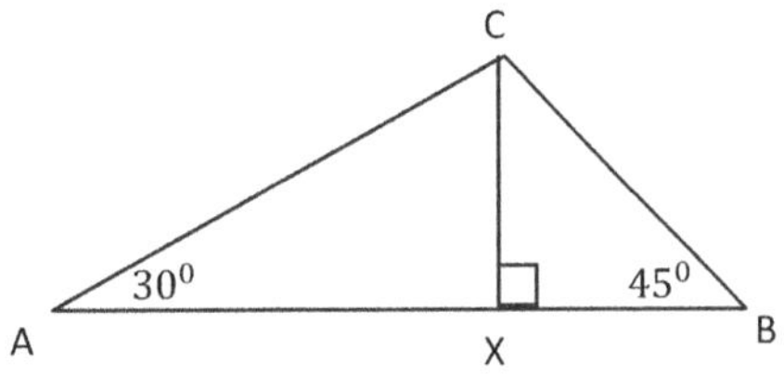

Let CX = x miles

From triangle BXC:

$\tan 45^\circ = \frac{CX}{BX} \Rightarrow 1 = \frac{CX}{BX}$

$\Rightarrow BX = CX = x$

From triangle AXC:

$\tan 30^\circ = \frac{CX}{AX} \Rightarrow \frac{1}{\sqrt{3}} = \frac{CX}{AX}$

$\Rightarrow AX = \sqrt{3} \times CX = x\sqrt{3} = 1.73x$

$\Rightarrow AB = AX + BX = 2.73x$

$\Rightarrow 2.73x = 273$

$\Rightarrow x = 100$ miles

Again, from triangle AXC:

$\text{Sin } 30^\circ = \frac{CX}{AC} \Rightarrow \frac{1}{2} = \frac{100}{AC}$

$\Rightarrow AC = 200$ miles

The correct answer is choice C.

68. Continuing from the above solution:

From triangle BXC:

$\sin 45^\circ = \frac{CX}{BC} \Rightarrow \frac{1}{\sqrt{2}} = \frac{100}{AC}$

$\Rightarrow AC = 100\sqrt{2} = 100 \times 1.41 = 141$ miles

The correct answer is choice D.

Made in the USA
Monee, IL
13 May 2026

50072880R00221